VOLUME ONE HUNDRED AND TWENTY TWO

# ADVANCES IN PROTEIN CHEMISTRY AND STRUCTURAL BIOLOGY

## Enzymes – Mechanisms, Dynamics and Inhibition

VOLUME ONE HUNDRED AND TWENTY TWO

# ADVANCES IN PROTEIN CHEMISTRY AND STRUCTURAL BIOLOGY

## Enzymes – Mechanisms, Dynamics and Inhibition

Edited by

**Tatyana Karabencheva-Christova**
*Associate Professor*
*Chemistry*
*Michigan Technological University*
*Houghton, MI, United States*

**Christo Christov**
*Associate Professor*
*Chemistry*
*Michigan Technological University*
*Houghton, MI, United States*

ELSEVIER

ACADEMIC PRESS
An imprint of Elsevier

Academic Press is an imprint of Elsevier
125 London Wall, London EC2Y 5AS, United Kingdom
525 B Street, Suite 1650, San Diego, CA 92101, United States
50 Hampshire Street, 5th Floor, Cambridge, MA 02139, United States
The Boulevard, Langford Lane, Kidlington, Oxford OX5 1GB, United Kingdom

First edition 2020



ISBN: 978-0-12-817762-4
ISSN: 1876-1623

For information on all Academic Press publications visit our website at https://www.elsevier.com/books-and-journals

*Publisher:* Zoe Kruze
*Acquisition Editor:* Ashlie Jackman
*Editorial Project Manager:* Leticia Lima
*Production Project Manager:* Abdulla Sait
*Cover Designer:* Miles Hitchen

Typeset by TNQ Technologies

# Contents

# Contributors

**Joaquín Ariño**
Institut de Biotecnologia i Biomedicina & Departament de Bioquímica i Biologia Molecular, Universitat Autònoma de Barcelona, Cerdanyola, del Vallès, Spain

**Carolina Barrientos-Salcedo**
Medicinal Chemistry Laboratory, School of Bioanalysis, University of Veracruz, Veracruz, Mexico

**Gary W. Black**
Hub for Biotechnology in the Built Environment, Faculty of Health and Life Sciences, Northumbria University, Newcastle upon Tyne, United Kingdom

**Antonio Casamayor**
Institut de Biotecnologia i Biomedicina & Departament de Bioquímica i Biologia Molecular, Universitat Autònoma de Barcelona, Cerdanyola, del Vallès, Spain

**Prashant Chandra Singh**
School of Chemical Sciences, Indian Association for the Cultivation of Science, Kolkata, West Bengal, India

**Simon J. Charnock**
Prozomix Limited, Haltwhistle, Northumberland, United Kingdom

**Darren J. Cook**
Prozomix Limited, Haltwhistle, Northumberland, United Kingdom

**Vasiliki E. Fadouloglou**
Department of Molecular Biology and Genetics, Democritus University of Thrace, Dragana University Campus, Alexandroupolis, Evros, Greece

**James D. Finnigan**
Prozomix Limited, Haltwhistle, Northumberland, United Kingdom

**Nicholas M. Glykos**
Department of Molecular Biology and Genetics, Democritus University of Thrace, Dragana University Campus, Alexandroupolis, Evros, Greece

**Alemayehu A. Gorfe**
McGovern Medical School, University of Texas Health Science Center at Houston, Department of Integrative Biology & Pharmacology, Houston, TX, United States

**Victor Guallar**
Barcelona Supercomputing Center (BSC), Barcelona, Spain; Institució Catalana de Recerca i Estudis Avançats (ICREA), Barcelona, Spain

**Cheng-Yang Huang**
School of Biomedical Sciences, Chung Shan Medical University, Taichung City, Taiwan; Department of Medical Research, Chung Shan Medical University Hospital, Taichung City, Taiwan

**Michael Kokkinidis**
Institute of Molecular Biology and Biotechnology, Heraklion, Crete, Greece; Department of Biology, University of Crete, Voutes University Campus, Heraklion, Crete, Greece

**Edgar López-López**
Department of Pharmacy, School of Chemistry, National Autonomous University of Mexico, Mexico City, Mexico; Department of Pharmacology, Center of Research and Advanced Studies of the National Polytechnic Institute (CINVESTAV), Mexico City, Mexico

**José L. Medina-Franco**
Department of Pharmacy, School of Chemistry, National Autonomous University of Mexico, Mexico City, Mexico

**Puspal Mukherjee**
School of Chemical Sciences, Indian Association for the Cultivation of Science, Kolkata, West Bengal, India

**Fernando D. Prieto-Martínez**
Department of Pharmacy, School of Chemistry, National Autonomous University of Mexico, Mexico City, Mexico

**Sergi Roda**
Barcelona Supercomputing Center (BSC), Barcelona, Spain

**Gerard Santiago**
Barcelona Supercomputing Center (BSC), Barcelona, Spain

**Abdallah Sayyed-Ahmad**
Department of Physics, Birzeit University, Birzeit, West Bank, Palestine

**Carl Young**
Prozomix Limited, Haltwhistle, Northumberland, United Kingdom

CHAPTER ONE

# Mapping enzyme-substrate interactions: its potential to study the mechanism of enzymes

**Sergi Roda[a], Gerard Santiago[a] and Victor Guallar[a,b,*]**
[a]Barcelona Supercomputing Center (BSC), Barcelona, Spain
[b]Institució Catalana de Recerca i Estudis Avançats (ICREA), Barcelona, Spain
*Corresponding author: E-mail: victor.guallar@bsc.es

## Contents



## Abstract

With the increase of the need to use more sustainable processes for the industry in our society, the modeling of enzymes has become crucial to fully comprehend their mechanism of action and use this knowledge to enhance and design their properties. A lot of methods to study enzymes computationally exist and they have been classified on sequence-based, structure-based, and the more new artificial intelligence-based ones. Albeit the abundance of methods to help predict the function of an enzyme, molecular modeling is crucial when trying to understand the enzyme mechanism, as they aim to correlate atomistic information with experimental data. Among them, methods that simulate the system dynamics at a molecular mechanics level of theory (classical force fields) have shown to offer a comprehensive study. In this book chapter, we will analyze these techniques, emphasizing the importance of precise modeling of enzyme-substrate interactions. In the end, a brief explanation of the transference of the information from research studies to the industry is given accompanied with two examples of family enzymes where their modeling has helped their exploitation.

*Advances in Protein Chemistry and Structural Biology*, Volume 122
ISSN 1876-1623
https://doi.org/10.1016/bs.apcsb.2020.06.001

## 1. Why modeling enzyme structure and dynamics

To face the current world challenges, human society has to adapt toward an eco-friendly based economy, where industrial production transformation is a milestone. In the field of chemistry, enzymes are becoming the alternative for inorganic catalysts, used in a wide variety of industrial applications (Chapman, Ismail, & Dinu, 2018; Schmid et al., 2001; Singh, Kumar, Mittal, & Mehta, 2016). In organisms, enzymes are the elements that make life kinetically possible. Otherwise, reactions needed for life would occur at rates in which life and the current atmosphere and ecosystems would not exist (Neet, 1998; Wolfenden, 2006; Wolfenden & Snider, 2001). For instance, ribulose-1,5-bisphosphate carboxylase/oxygenase (Rubisco), the most abundant enzyme on Earth (Dhingra, Portis, & Daniell, 2004), is the major catalyst involved in the carbon fixation of carbon dioxide.

Enzymes can be of other macromolecular nature besides the proteic one, including catalytic RNA and even designed catalytic DNA (Breaker, 1997; Guerrier-Takada, Gardiner, Marsh, Pace, & Altman, 1983; Kruger et al., 1982). However, protein-based catalysts are the most abundant in nature due to the fact that 21 amino acids offer incredibly a lot more combinations in comparison with just 4 nitrogenous bases (Böck et al., 1991; Doudna & Lorsch, 2005). As a result of this high variety of amino acids and of the presence of cofactors in a big part of enzymes, a large number of catalyzed reactions by enzymes exist (Fig. 1). This wide variety of protein-based enzymes brings in different types of activities along with many effects that can modulate them. In other words, enzyme activity is affected by pH, temperature, ionic concentration, presence or absence of the cofactor, and by the action of activator or inhibitor molecules (Robinson, 2015).

Notwithstanding, the mechanisms and factors that give enzymes the ability to catalyze reactions are still not fully comprehended (Knowles, 1991; Kraut, 1988). The widely known and most accepted idea to explain enzyme catalysis is the transition state theory (TST) (Eyring, 1935; Truhlar, 2015; Zinovjev & Tuñón, 2017). Enzymes can either decrease the Gibbs free energy of activation ($\Delta G^{\ddagger}$) or increase the transmission coefficient ($\kappa$), accelerating the rate of the reaction in comparison with an uncatalyzed reaction. Thus, enzymes either stabilize the transition state (TS) of the reaction or they enhance the productive cross of the TS barrier by the reactants (Agarwal, 2006; Truhlar, 2015; Zinovjev & Tuñón, 2017).

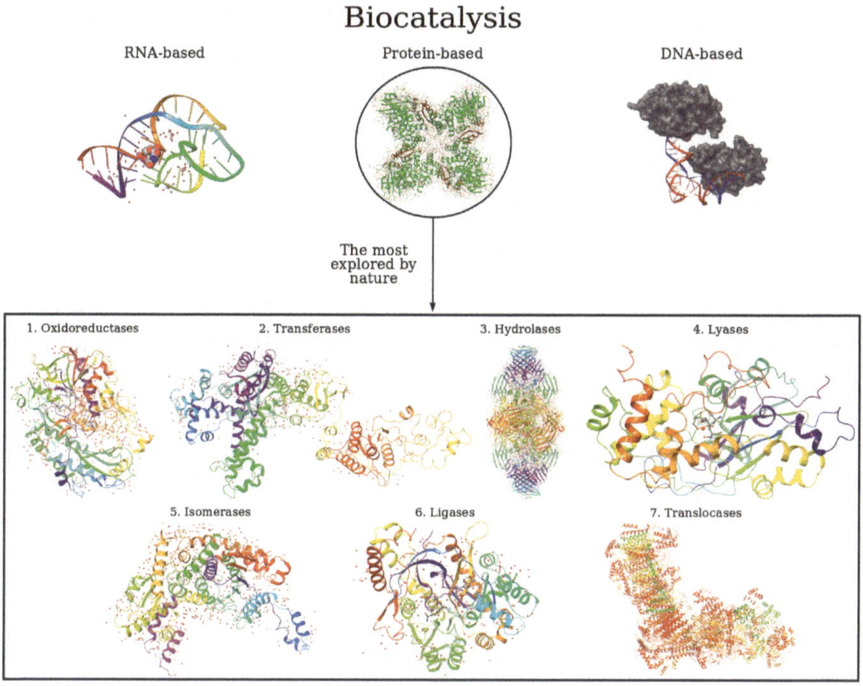


**Fig. 1** ***Macromolecular nature of biocatalysis.*** The figure represents the types of macromolecules where catalysis has been observed or designed, which are DNA, RNA, and proteins. Protein-based catalysis is the most explored by nature, due to the larger options that the number of residues and folding types give. At the bottom, the 7 types of global chemical reactions catalyzed by enzymes (classification of Enzyme Commission number) are represented with a 3D structure of each group. The rainbow labeling of the ribbon in all structures highlights from the N-terminal (or 5′ end in RNA) in red to the C-terminal (or 3′ end in RNA) in violet. The encircled enzyme is colored in green for the large subunit and in brown for the small one. The DNA of the DNAzyme has each strand stained with a different color. The PDB codes for all structures from the RNA-based to the Translocases are the following: 4OJI, 1GK8, 5XM9, 1CF3, 1TAQ, 6CVM, 1AHJ, 1IAT, 2HGS, and 6RFR.

Yet, the TST has some limitations, giving birth to other approaches and theories to study the enzyme catalysis of a reaction such as the Marcus theory (Marcus, 1956; Zinovjev & Tuñón, 2017). Moreover, the concrete way in which enzymes decrease $\Delta G^{\ddagger}$ or increase $\kappa$ of the TST remains a highly discussed topic. One of the important factors is the dynamics of proteins, which seems to be the missing piece to understand how enzymes catalyze reactions (Agarwal, 2006; Carvalho et al., 2014; Kamerlin & Warshel, 2010; Kohen, 2015; bib_Petrović_et_al_2018Petrović, Risso, Kamerlin, & Sanchez-Ruiz,

2018). Experimental studies enable to check the activity of an enzyme, its specificity or promiscuity, its optimal conditions, and even get the structure of the system. Still, a full mechanistic knowledge requires the study of the protein dynamics and of the molecular interaction with the substrate, a task that mandates the use of different computational methods (Carvalho et al., 2014) (Fig. 2).

The study of enzymes and their mechanisms has helped in comprehending evolution better, in designing systems with enzymes to diagnose, treat, and cure several diseases, and in synthesizing or refining a wide variety of materials and compounds, which are interesting for different industrial sectors (Chapman et al., 2018; Rajendra Singh et al., 2016; Robinson, 2015; Schmid et al., 2001). Some well-known examples include the use of glucose oxidase to measure and monitor the blood glucose level in patients with diabetes mellitus (Yoo & Lee, 2010), the production of lactose-free milk by using lactase for lactose-intolerant people (Soares, Tavora, Patera, & Baroni,

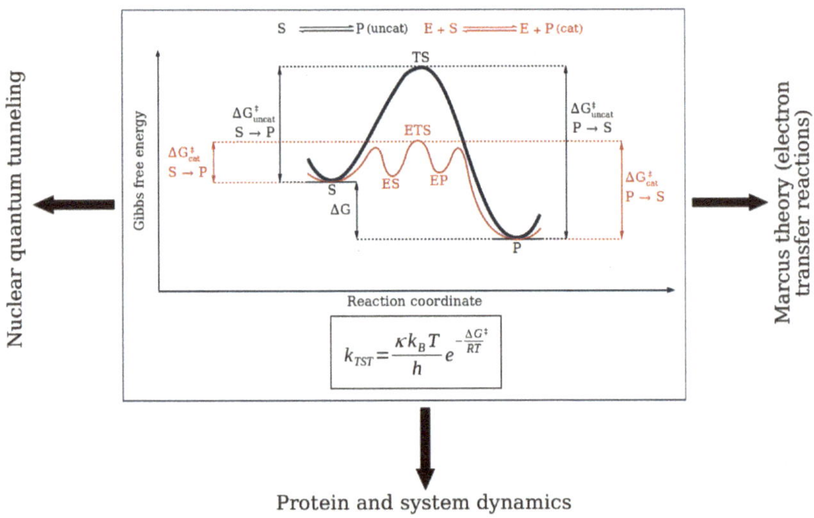


**Fig. 2** ***The state of the art and the current issues related to biocatalysis.*** The figure presents the typical free energy profile of a reaction uncatalyzed (in black) versus catalyzed (in red) followed by arrows pointing toward the current issues in the TST. Enzymes tend to decrease the value of $\Delta G^{\ddagger}$, increasing the reaction rate. A global mechanism is the formation of the enzyme-substrate (ES) complex, followed by the stabilization of the transition state (TS) of the reaction in the active site (ETS), ending with the releasing of the product (P) from the enzyme (EP). Enzymes can also increase the transmission coefficient (κ, the proportion of reactive trajectories that surpass the TS barrier and end as products). The reaction coordinate is a metric that represents the progress of the reaction according to some geometric parameters along the biochemical process.

2012), and the invention and success of the PCR technique (Mullis, 1990), which is crucial at any laboratory working with molecular biology, thanks to the discovery of the thermophile *Thermus aquaticus* DNA polymerase (Chien, Edgar, & Trela, 1976).

More examples are the synthesis of acrylamide, the precursor of the polymer famously used to perform Sodium Dodecyl Sulfate Polyacrylamide Gel Electrophoresis (SDS-PAGE) and water treatment (Laemmli, 1970; Smith & Oehme, 1991), is possible due to nitrile hydratase (Yamada & Kobayashi, 1996) and the common features observed in the ATP-grasp enzymes. These features are base residues to stabilize the $\alpha$ and $\beta$-phosphates, an acid residue that forms a hydrogen bond with the amino group in adenine, hydrophobic residues to stabilize adenine ring, three acid residues to coordinate $Mg^{2+}$ ions, and a shared catalytic mechanism (Galperin & Koonin, 1997, 2012; Thoden, Kappock, Stubbe, & Holden, 1999).

Moreover, for many systems such a gain of knowledge has been largely promoted by *in silico* studies. For instance, the design of DNA polymerases to accept unnatural nucleoside triphosphates would not have been able without enzyme modeling (F. Chen et al., 2010; Dunn, Otto, Fenton, & Chaput, 2016). Thus, addressing the enzymatic mechanisms of action has significantly been revamped through modeling their dynamics and enzyme-substrate interactions, leading to an increased understanding of the system and better design of their potential applications.

## 2. A method for everybody

As the title of the section entails, a large amount of methods used in science can be applied to the study of enzymes. Thus, enzymology requires physicists, mathematicians, and data scientists, besides the chemists and biologists, which are the most associated with the field. This chapter is focused on the computational tools used to study enzymes, many representative methods are summarized in Table 1.

Computational methods use either the sequence, the structure or experimental data, through analysis or simulations acting on it, to unveil the catalytic activity of enzymes; different levels might reveal key (different) features of the mechanism of action. Moreover, data extracted from these methods (along with experimental one) give the option to use artificial intelligence (AI), and more concretely machine learning (ML), to study them as well (Fig. 3).

Table 1 Recompilation of computational tools used to study enzyme mechanisms and dynamics.

| Method | Function | System size | Computational time | Group | Softwares/Examples |
|---|---|---|---|---|---|
| Sequence alignment | To align protein/DNA/RNA sequences. Phylogenetic analysis. | Any | Seconds | Sequence-based bioinformatics | CLUSTAL (Higgins & Sharp, 1988; Thompson et al., 1994), T-COFFEE (Notredame et al., 2000), Kalign (Lassmann & Sonnhammer, 2005), MAFFT (Katoh et al., 2002) MUSCLE (Edgar, 2004). |
| Database searching of sequences | To find homologous sequences to the user's target one. | Any | Seconds to hours | Sequence-based bioinformatics | BLAST (Altschul et al., 1990), SSEARCH (Pearson, 1991), USEARCH (Edgar, 2010), SANS (Koskinen & Holm, 2012). |
| Structural alignment | To align 3D structures. | Any | Seconds | Structure-based bioinformatics | DALI (Holm & Sander, 1996), combinatorial extension (Shindyalov & Bourne, 1998), sequential structure alignment program (Taylor & Orengo, 1989). |
| Homology modeling | To build the 3D structure from the target sequence and a template as a model. | Any | Minutes | Structure-based bioinformatics | MODELLER (Martí-Renom et al., 2000), Prime (Jacobson et al., 2004), SWISS-MODEL (Schwede et al., 2003). |

| | | | | | |
|---|---|---|---|---|---|
| Cavity & pocket detection | To find cavities and pockets in 3D structures. | Any | Minutes | Structure-based bioinformatics | Fpocket (Le Guilloux et al., 2009), GHECOM (Kawabata, 2010), KVFinder (Oliveira et al., 2014), GaussianFinder (Dias, Martins, Nguyen, & Gomes, 2017), SiteMap (Halgren, 2007). |
| MM minimization | To obtain the minimized/ relaxed 3D structure of the system based on classical physics. | $10^6$–$10^3$ atoms | Minutes | Molecular modeling | AMBER (Cornell et al., 1995), CHARMM (Brooks et al., 1983), OPLS (Jorgensen et al., 1983), GROMOS (van Gunsteren & Berendsen, 1987). |
| QM minimization | To obtain the minimized/ relaxed 3D structure of the system based on quantum physics. | $10^3$–$10^2$ atoms (semiempirical). $10^2$–10 atoms (DFT, HF). $\leq$ 10 atoms (Post-HF) | Minutes to hours (semiempirical). Hours to days (DFT, HF, Post-HF) | Molecular modeling | Gaussian (Binkley et al., 1978), ORCA (Neese, 2012), Qsite (Philipp & Friesner, 1999). |
| QM/MM minimization | To obtain the minimized/ relaxed 3D structure of the system based on classical physics for the vast majority of atoms and on quantum physics for the active site. | $10^6$–$10^3$ atoms (MM region). $10^3$–1 atoms (QM region) | Hours to days | Molecular modeling | MM: AMBER (Cornell et al., 1995), CHARMM (Brooks et al., 1983), OPLS (Jorgensen et al., 1983), GROMOS (van Gunsteren & Berendsen, 1987). QM: Gaussian (Binkley et al., 1978), ORCA (Neese, 2012), Qsite (Philipp & Friesner, 1999). |

*(Continued)*

Table 1 Recompilation of computational tools used to study enzyme mechanisms and dynamics.—cont'd

| Method | Function | System size | Computational time | Group | Softwares/Examples |
|---|---|---|---|---|---|
| MD | To simulate the dynamical behavior of the system in a particular environment over time. | MM or QM/MM size limits | Days to weeks | Molecular modeling | AMBER (Cornell et al., 1995), CHARMM (Brooks et al., 1983), GROMOS (van Gunsteren & Berendsen, 1987), Desmond (Bowers et al., 2006), PLUMED (Bonomi et al., 2009), Car–Parrinello MD (Car & Parrinello, 1985). |
| MC | To explore the conformational space of the system independently of time. | MM, QM, or QM/MM size limits | Hours to days | Molecular modeling | PELE (Borrelli, Vitalis, Alcantara, & Guallar, 2005; Car & Parrinello, 1985), Spartan (Lafayette & Wiebelt, 2017), Cassandra (Shah et al., 2017) |
| Molecular docking | To see the interaction mode of a molecule bound to another molecule (the system). | MM size limit | Seconds to minutes | Molecular modeling | AutoDock Vina (Trott & Olson, 2010), GOLD (Jones, Willett, Glen, Leach, & Taylor, 1997), Glide (Friesner et al., 2004), Haddock (Dominguez, Boelens, & Alexandre, 2003), SwissDock (bib_Grosdidier_et_al_2011Grosdidier, Zoete, & Michielin, 2011). |
| Protein folding prediction from sequence | To infer the 3D structure of a protein from its sequence of residues. | 50–300 residues | From minutes to days (Training). From seconds to days (Prediction) | AI-based approaches | AlphaFold (Senior et al., 2020) |

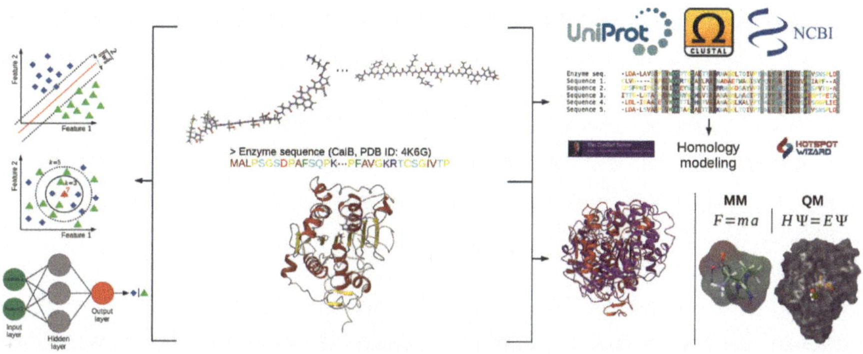


**Fig. 3** ***Computational approaches to study an enzyme.*** The figure summarizes the current methods used in computational enzymology to study catalytic mechanisms and more. The input of all methods consists in one of the levels of protein structure, from the primary level to the quaternary one. Thus, there are computational methods that work with the amino acid sequence of the enzyme to search for homologous sequences, perform a MSA, and use its information to infer properties of evolution/function (top right). Concerning the tertiary and quaternary structure of proteins, there are mainly two types of methods, structure-based bioinformatics and molecular modeling (bottom right). Lastly, the data obtained from computational and experimental studies related with enzymes and proteins can be used to train ML models to predict important features related with the mechanism of action of the enzyme (left). The sequence and structure can be found in the following PDB code: 4K6G. The aligned structure in the figure refers to this PDB code: 3GUU.

## 2.1 Sequence-based approaches

Even though enzymes tend to be studied using its 3D structure, a lot of information can be extracted from its sequence. Still, all the knowledge is obtained by comparison with one or other multiple sequences (Suplatov, Voevodin, & Švedas, 2015). The input of these techniques can be either the DNA coding for the mRNA of the enzyme or the protein primary structure.

Enzyme sequences can be found in several databases, but the main and best curated ones are NCBI and UniProtKB (National Center for Biotechnology Information, n.d.; UniProt Consortium, 2019). To search those homologous sequences to the query one in these databases, BLAST (Basic Local Alignment Search Tool) is commonly used (Altschul, Gish, Miller, Myers, & Lipman, 1990). Albeit the popularity of this algorithm of sequence comparison, many others exist including SSEARCH (Pearson, 1991), USEARCH (Edgar, 2010), and SANS (Koskinen & Holm, 2012).

Once you have the set of selected homologous sequences, they can be aligned with different multiple sequence alignment (MSA) tools like

CLUSTAL (Higgins & Sharp, 1988; Thompson, Higgins, & Gibson, 1994), T-COFFEE (Notredame, Higgins, & Heringa, 2000), and Kalign (Lassmann & Sonnhammer, 2005) in which the progressive alignment approach is used (Feng & Doolittle, 1987). Other approaches exist (Chowdhury & Garai, 2017; Daugelaite, O'Driscoll, & Sleator, 2013) and give room to other popular MSA softwares such as MAFFT (Katoh, Misawa, Kuma, & Miyata, 2002) or MUSCLE (Edgar, 2004).

MSA gives a lot of hidden structural information of the enzyme including correlated residues in the sequence that coevolve, the essential residues of the active site of a particular family of enzymes, the less conserved residues in the structure of the protein, and thus, more prone to be mutable. Thus, the comparison of homologous sequences with the studied one enables to predict substantial information of the enzyme's function. In fact, a lot of softwares used in enzyme characterization and engineering use MSA as the input. Some examples include: HotSpotWizard (Sumbalova, Stourac, Martinek, Bednar, & Damborsky, 2018), PoPMuSiC (Gilis & Rooman, 2000), or ConSurf server (Ashkenazy et al., 2016) to name a few.

## 2.2 Structure-based approaches

Enzymes structural features are key to decipher the catalytic mechanism. Inside the tools related with the structural rearrangement of the residues of the protein, two main groups have to be made. The first and most global one is molecular modeling, which refers to methods aiming at manipulating the structure of the molecules in a system at a particular level of theory of physics to infer its properties. The other subgroup encompasses other tools that work with structures at the level of analysis and data extraction, without performing large simulations, what is commonly known as structure-based bioinformatics. Still, the techniques of these two subgroups can be combined to study enzymes.

The use of structure-based bioinformatics to study enzymes and their mechanism of action is largely related with structural alignment to compare two similar proteins (Godzik, 1996; Ma & Wang, 2014), homology modeling to obtain a 3D structure from a target sequence and a template structure (Chothia & Lesk, 1986; Martí-Renom et al., 2000), detection of cavities and pockets (Le Guilloux, Schmidtke, & Tuffery, 2009), residue flexibility associated with the experimental B-factor (a parameter related with thermal motion and it is obtained in X-ray crystallography) (Debye, 1913), prediction of substrate promiscuity (Martínez-Martínez et al.,

2018), use of statistical potentials to score and rank protein-ligand interactions (Fan et al., 2011; Mooij & Verdonk, 2005), and more.

### 2.2.1 Molecular modeling

Molecular modeling is one of the most widespread techniques to study (bio) chemical and physical systems outside the wet laboratory. The methods can be classified either if they simulate the system with classical physics (referring to the Newton's laws), what it is called molecular mechanics (MM) or if they take quantum physics (referring to the Schrödinger's equation) into consideration, what it is known as quantum mechanics (QM). The treatment of the system in a more simple level of theory enables the computation to be less expensive compared with using the level of theory of quantum physics. Nonetheless, a higher level of theory returns more accurate calculi and better estimations of the user's parameters of interest (Jensen, 2007).

MM simplifies the system by treating atoms as the smallest and indivisible unit without taking into account subatomic particles. Atoms are connected by bonds, which are considered harmonic oscillators. This description of molecules with atoms and their bonds is known as the "ball-and-stick" model. Thus, the potential energy of the system is parametrized according to the atomic coordinates, the type of atoms, and the bonds made between them, what is commonly known as force fields (FF).

The numerical values assigned to the constants of the energy terms, the functional form used, and the number of parameters and terms taken into account is what defines and distinguishes a FF from another. Thus, the accuracy of the energy calculation of the user's system will depend in principle on the applied FF; at a practical level most FF give similar results on standard systems. FF are designed from experimental data or computational data obtained from a higher level of theory, such as QM; a handful of widely used FF exist, including AMBER (Cornell et al., 1995), CHARMM (Brooks et al., 1983), OPLS (Jorgensen, Chandrasekhar, Madura, Impey, & Klein, 1983), or GROMOS (van Gunsteren & Berendsen, 1987).

QM methods describe the system in the most inner part of particles, electrons and nuclei, and the Schrödinger equation must be solved. Different ways to solve the Schrödinger equation exist and they give room to different levels of accuracy and computational resources, depending on the approximation used to solve it. Semi-empirical methods, for example, use experimental data to reduce the computational cost of solving the Schrödinger equation; methods not using empirical data are called *ab initio* methods. These include Density Functional Theory (DFT) (Kohn & Sham, 1965) and Hartree Fock (HF) methods (Fock, 1930; Hartree, 1928). However, these methods usually only consider the average interaction between electrons. There are more precise methods that do take into account electron correlation, which are Configuration Interaction (Head-Gordon, Rico, Oumi, & Lee, 1994), Møller–Plesset perturbation theory (Møller & Plesset, 1934), or Coupled Cluster (Purvis & Bartlett, 1982).

Therefore, MM enables the study of large systems over a long period of time, while QM gives an accurate description of small systems at a stationary state or over (very) small periods of time. Computational enzymology also takes advantage of both ways, QM/MM techniques, to describe the system and to study its catalytic mechanism (Friesner & Guallar, 2005; Senn & Thiel, 2009; van der Kamp & Mulholland, 2013). The active site residues, (the cofactor), and the substrate/s are treated typically at the QM level and the rest of the system is described at the MM level.

Molecular modeling has led to two Nobel Prizes in Chemistry. One in 1998 for the development of DFT and the computational methods for QM by Walter Kohn and John A. Pople, respectively. The other in 2013 for the development of the QM/MM methods by Martin Karplus, Michael Levitt, and Arieh Warshel.

Typical problems in computational chemistry require the exploration of the potential energy of a system according to its spatial coordinates. The number of possible local minima, and thus, number of possible stable configurations of a molecule increase exponentially as the number of bonds increases (Howard & Kollman, 1988; Maranas & Floudas, 1994). This combinatorial explosion problem gives room to several approaches including molecular dynamics (MD), Monte Carlo (MC) methods, simulated annealing, genetic algorithms, diffusion methods, or distance geometry methods. From these, MD and MC are the best to sample locally, finding a good number of local minima. On the other hand, the remaining methods tend to be used to search for the global minimum of the system.

The concept of these approaches and their application to solve real problems in computational chemistry is a big topic. A lot of books and reviews exist to get further information (Hollingsworth & Dror, 2018; Jensen, 2007; Karplus & McCammon, 2002; Lei & Duan, 2007; Lonsdale, Harvey, & Mulholland, 2012; Maximova, Moffatt, Ma, Nussinov, & Shehu, 2016; Yang, Pan, Lightstone, & Merz, 2016).

Thus, the availability of the enzyme structure enables the proper modeling of a conformational sampling and/or dynamics related to the catalyzed reaction, gaining further insights of the key residues involved with the mechanism of action and substrate binding. Later on, this knowledge can be used to design a particular enzyme to enhance its activity against a specific family of substrates or to extend the conditions in which the enzyme can work (harsher pHs and temperatures), what it is known as enzyme engineering.

## 2.3 AI-based approaches

In spite of the fact that AI is still entering the field of enzymology, several number of applications to study protein properties in general have been developed (Larrañaga et al., 2006; Paladino, Marchetti, Rinaldi, & Colombo, 2017). One of the newest and most interesting ones refers to the AlphaFold built by Google to predict the 3D structure of a protein with its sequence as the only input (Senior et al., 2020). This method uses a convolutional neural network that predicts distance between residues and torsions of the backbone of the protein structure from its sequence. With these predictions, a protein-specific potential is created, and then, is further minimized by gradient descent, obtaining the final predicted structure. Another related method refers to the one published by AlQuraishi, which uses a recurrent neural network and parametrizes the local protein structure into torsional angles that take into account covalent chemistry. Subsequently, the predicted local protein structures are coupled into the overall protein structure with recurrent geometric units. Finally, the model uses a differentiable loss function to score the predicted structure over the experimental one and train the model. Thus, this model does not use co-evolution information and shares similar results with other protein structure prediction algorithms (AlQuraishi, 2019). The reach of predictions close to experimental structures would be a rather important aspect for the field of computational enzymology, since it would enable to perform molecular modeling studies of any enzyme, whether the experimental 3D structure is available or not.

The prediction of protein functions and the functional relevance of their amino acids in the 3D structure is crucial to understand and design enzymes. In that regard, DeeProtein software has been created, which also takes the sequence as input and infers the regions of the protein sequence that are important for the biological activity (Belzen et al., 2019). Another model regarding function has been created based on the available enzymes in the RCSB PDB to predict the Enzyme Commission number (Amidi et al., 2018).

The prediction of the optimal pH range has also been accomplished with the use of ML (Khan, Hayat, & Khan, 2015), which could be really useful in the field of enzyme engineering. Another model consists in the prediction of the solubility of proteins based on their structure (Hou, Kwasigroch, Rooman, & Pucci, 2020), which could be useful to avoid the trial of computationally promising variants but that would actually make the protein aggregate.

However, these techniques still need substantial improvement in order to really be used in the main projects of research centers and enterprises working with enzymes. These needs are often associated with the lack of enough high-quality datasets with no imbalances nor biases.

## 3. Enzymes mechanism: the potential of mapping enzyme-substrate interactions

As stated in the previous section, there are a plethora of different techniques to study and predict the relation between the sequence and/or structure with the function of enzymes. Moreover, these methods are often used in combination with amino acid mutations, in order to rationalize mechanistic data and, importantly, to design new variants with optimized (improved) properties. Still, an exhaustive analysis of the enzymatic mechanism demands typically the use of molecular modeling. Such study might be necessary when addressing substrate specificity, activity, etc.; sequence tools typically reduce to classification and conservation analysis.

A comprehensive molecular analysis should include three main aspects: enzyme dynamics, enzyme-substrate interactions, and electronic effects. The former aims at understanding the global dynamics of the enzyme, thus the flexibility associated with the active site, opening cavities, allosteric effects, etc. It is mainly studied through the use of MD and MC techniques (Cossins, Hosseini, & Guallar, 2012); (simpler) coarse grain techniques, such as ANM, might offer an initial quick view. The importance of an exhaustive dynamical study has been discussed in multiple studies and reviews (Kamerlin & Warshel, 2010; Kohen, 2015). Electronic effects are obviously addressed with QM techniques and are important when studying the chemical reactive step and if addressing possible (non-reactive) polarization aspects. However, in this section we want to underline the importance of an accurate analysis of enzyme-substrate interactions. Since we developed the Protein Energy Landscape Exploration (PELE) software, a specialized protein-ligand heuristic MC technique first introduced in 2005, we have applied it now to more than 100 systems, with more than half of them being enzymes (Gilabert, Lecina, Estrada, & Guallar, 2018). Our experience indicates that many properties can be studied with a robust analysis of how the substrate migrates and interacts with the enzyme. Obviously this includes aspects such as substrate specificity but also reactivity studies, typically studied with more expensive QM techniques. In most of the attempted studies, we obtain a quantitative correlation of the enzymatic efficiency constants from the (statistical) description of the active site enzyme-substrate induced fit. Examples include comparing different substrates against the same enzyme (and variations of it), as in a fungal aryl-alcohol oxidase (Serrano et al., 2019), laccases (Monza et al., 2015; Pardo et al., 2016) and unspecific peroxidases (Molina-Espeja et al., 2016)), or comparing the same substrate to different enzymes, as in unspecific peroxidases (Aranda et al., 2019) and peroxygenases (Lucas et al., 2016). These studies all seem to indicate the role of preorganization and near attack conformation (Hur & Bruice, 2003).

The sampling potential of PELE, capable of probing protein-ligand induced fit while mapping the entire protein surface, also allows for advanced mechanistic studies. When studying two different manganese peroxidases, for example, we could find that the ABTS active one binds the substrate in an unexpected (and fully solvated) surface site, using two histidines as electrostatic anchors. The inactive enzyme, however, did not show any significant predicted binding sites, presenting and asparagine and a glycine at those surface positions. Introducing these two solvent exposed mutations, we obtained *in vitro* ABTS activity in a similar fashion (similar $k_{cat}$ and $K_m$) to the active enzyme (Acebes et al., 2016). Global (full enzyme) substrate binding explorations have also allowed adding artificial active sites in enzymes: the design of PluriZymes. By identifying alternative substrate binding sites (complementary to the active site), and turning them into additional active centers, after insertion of catalytic triads, we could introduce a second hydrolase activity on several esterases (Alonso et al., 2020; Santiago et al., 2018) (Fig. 4).

Thus, it seems like obtaining an exhaustive structural sampling of the enzyme-substrate interaction, even at the MM level of theory, seems to be a key aspect when addressing enzymatic catalysis. Obviously, adding more sophisticated levels of theory, such as an electronic description through mixed QM/MM techniques, might turn into a more robust analysis. We have implemented such an approach when aiming at correlation between redox substrate potential with enzymatic activity (M. F. Lucas et al., 2017), or when increasing a laccase oxidation on a difficult to oxidize

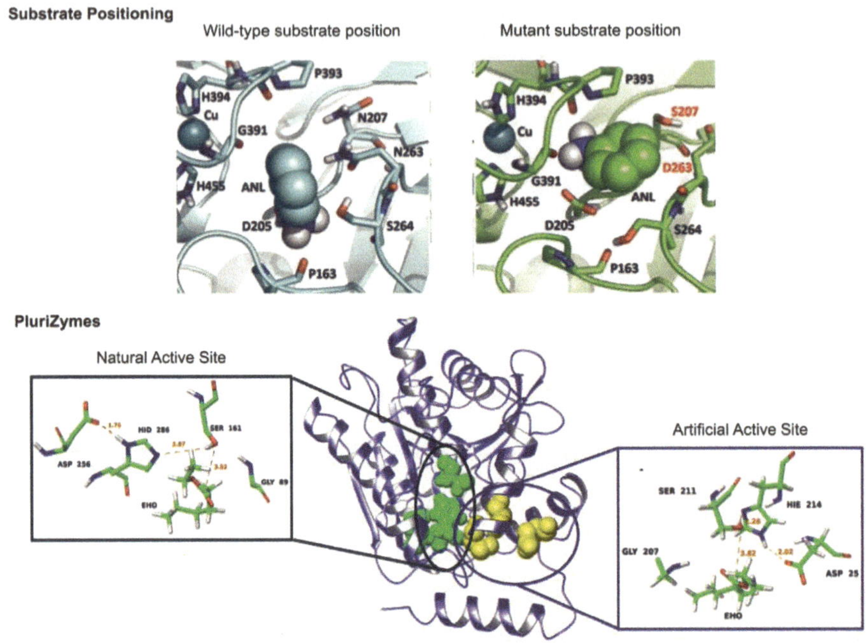


**Fig. 4** Examples of engineering success cases based on substrate positioning studies using classical force-fields. Introduced mutations are affecting the substrate position in the active site, creating a more favorable environment (substrate positioning) or to find a binding site where to create activity (PluriZymes). The substrate positioning panel has the substrate (aniline) displayed with the CPK model and the active site residues are displayed with the thick tube representation. The PluriZymes panel shows the protein scaffold and the zoomed areas highlight the natural and added active sites with the catalytic triad, the oxyanion hole, and the bound substrate predicted with PELE displayed with the thick tube representation. The PDB codes used are 5ANH (to perform homology modeling) and 5JD4, respectively.

substrate, the aniline cation (Santiago et al., 2016) (Fig. 4). Still, we are turning more and more to exclusively using classical FF approaches when addressing enzyme-substrate interactions for mechanistic and engineering studies.

## 4. From modeling to engineering

In the previous section, we already explored the potential of using a dynamical enzyme-substrate analysis to engineer enzymes. Here, we want to give a more broad perspective of how modeling can help and ease the enhancement of enzyme properties. In fact, different industrial sectors use the *in silico* modeling of enzymes combined with *in vitro* methods to optimize their biocatalysts to real bioprocessing (Bornscheuer, & Pohl, 2001; Jemli, Ayadi-Zouari, Hlima, & Bejar, 2016; Kirk, Borchert, & Fuglsang, 2002; Lutz, 2010). For instance, a wide list of engineered and newly characterized enzymes involved in the manufacturing of products of interest for the society exist, including different type of hydrolases for the detergent industry, hydrolases and some oxidoreductases for the food and beverage industry, and oxidoreductases, transferases, hydrolases, lyases, and isomerases for the pharmaceutical industry, the organic synthesis, and the waste management to name a few (Chapman et al., 2018; Rajendra Singh et al., 2016).

From the crystal structure (or homology model), key residues for substrate recognition can be identified. These identified residues are mutagenesis targets with beneficial effects toward enzyme activity, with some web tools already available for the *in silico* identification of these hotspots, like Swiss-Model (Schwede, Kopp, Guex, & Peitsch, 2003), Rosetta (Rohl, Strauss, Misura, & Baker, 2004), IntFOLD (McGuffin, Atkins, Salehe, Shuid, & Roche, 2015), Threading ASSEmblyRefinement (I-TASSER) (Roy, Kucukural, & Zhang, 2010), or Protein Homology/AnalogY Recognition Engine (Phyre) (Kelley & Sternberg, 2009). This approach is the first attempt to connect modeling with engineering.

The next step is to merge this structural information with the overall dynamic of the system. With this combination, modeling has become a powerful tool for enzyme engineering, as we will see in some examples focusing on cytochrome P450 (CYP) and transaminases (TA).

CYPs consist in a large superfamily of enzymes which can oxidize a broad range of substrates aiming at different physiological roles such as biosynthesis of sterols and other secondary metabolites or the clearance of toxic substances in the liver. Moreover, CYPs are enzymes that have been extensively studied with well-known catalytic mechanism and fold, including the heme cofactor (Denisov, Makris, Sligar, & Schlichting, 2005; Shaik, Kumar, de Visser, Altun, & Thiel, 2005). Thus, the knowledge of these biocatalysts and their capacity to catalyze oxidations of structurally complex molecules regioselectively and stereoselectively make them encouraging options for engineering of new chemical reactions using computational-based methods. Although directed evolution has had great success in the engineering of new chemistry in CYPs (Bajaj, Sreenilayam, Tyagi, & Fasan, 2016; Chen, Huang, Kan, Zhang, & Arnold, 2018; Farwell, Zhang, McIntosh, Hyster, & Arnold, 2015; Hernandez et al., 2016; Kan et al., 2016; Prier, Zhang, Buller, Brinkmann-Chen, & Arnold, 2017; Singh,

Kolev, Sutera, & Fasan, 2015; Zhang et al., 2019), several studies of rational design aided with *in silico* tools have been performed (Ba, Li, Zhang, Duan, & Lin, 2013; Dodani et al., 2016; Hayashi et al., 2008; Jung et al., 2018; Khatri et al., 2018; bib_Rühlmann_et_al_2017Rühlmann, Antovic, Müller, & Urlacher, 2017; Steck, Kolev, Ren, & Fasan, 2020; Syed, Porollo, Miller, & Yadav, 2013; Toporkova et al., 2013).

The complex nature of these enzymes has given researchers to modify and engineer different parts of the system including the substrate-binding active site, the substrate access channel, the residues involved in binding the heme cofactor, and more. Besides, these studies retail more insights of the structure-function relationships and have given substantial knowledge of the mechanism of action of CYPs. One remarkable example is the one where MD simulations revealed a key (unsolved crystallographically) loop that controlled its regioselectivity and by analyzing the different possible states of the loop, they obtained a mutant that switched the enzyme's regioselectivity (Dodani et al., 2016). A recent paper used a mechanism-guided approach to design CYPs to substantially enhance their C—H amination (the best nitrene transfer biocatalysts reported) by disrupting conserved structural elements related with the heme cofactor (Steck et al., 2020). Another study uses a particular CYP and the binding poses of progesterone predicted by molecular docking to engineer the regioselective hydroxylation of the substrate with the products having different medical applications (Khatri et al., 2018). Many other studies in CYPs use the information of the enzyme-substrate interactions obtained with molecular docking to select those residues to mutate in order to enhance the selectivity or activity toward a specific family of substrates (Ba et al., 2013; Hayashi et al., 2008; Jung et al., 2018; Syed et al., 2013).

Despite the many industrial applications of CYPs, the electron source constant requirement is a major limitation for full implementation. Constant amounts of expensive cofactors have to be introduced in the reaction, just to keep CYP's operating. But recently discovered unspecific peroxidases (UPOs), might overcome CYPs. UPOs can speed up similar chemical reactions as CYPs, but using hydrogen peroxide as an electron source, is cheaper and more accessible compared to NADP (Freakley et al., 2019; Hofrichter & Ullrich, 2014; Hrycay & Bandiera, 2012).

Like in CYPs, UPOs have been recently studied using *in silico* methods to enhance their reactivity, where their application toward fatty acids has been discussed. The study of substrate positing has proved to be of extreme importance for UPOs, with a crucial role defining the catalyzed reaction.

The global dynamic behavior of the enzyme and the enzyme-substrate complex formation will be also relevant for UPOs engineering (Aranda et al., 2019).

Other families have been the target of similar studies, like TAs. TAs are a superfamily of enzymes that catalyze the transfer of an amine group from a donor to an acceptor (usually a ketone or aldehyde) by using pyridoxal 5′-phosphate (PLP) to perform the oxidative deamination of the donor and the reductive amination of the acceptor (Ghislieri & Turner, 2014; Kelly et al., 2018; Malik, Park, & Shin, 2012). TAs are classified according to the relative position of the amine group to the carboxyl group of the substrate. αTAs require the presence of a carboxyl group next to the C atom bound to the amine group, while ωTAs does not have this requirement, making them more attractive for industrial bioprocessing due to the acceptance of a wider range of substrates. The active site geometry of ωTAs is fully comprehended with a two-site binding model with a large (L) and small (S) pocket (Shin & Kim, 2002). There is a concept of linking the S pocket with the place involved in substrate binding but not with the catalysis, meaning that this region could be redesigned to enhance the acceptance of bulkier residues by the TA (Park & Shin, 2011). Still, the L pocket can be modified too (Dourado et al., 2016; Savile et al., 2010).

The *in silico* rational design of ωTAs to enhance chiral amine synthesis has thrived with several examples. One of these successful cases refers to the engineering of a (R)-selective ωTA binding pocket to accept prositagliptin (the precursor of sitagliptin, an antidiabetic drug). The key mutation, G136F, changes the conformation of a loop near the active site, giving a broader binding site with bigger volume (it enlarges the L pocket) (Guan et al., 2015). Then, it was further evolved to gain activity against it. The result was a 10%–13% increase in overall yield, 53% increase in productivity, and 19% reduction in total waste, making the biocatalytic process replace the former rhodium-catalyzed asymmetric enamine hydrogenation, which needed toxic heavy metals and implied more expensive costs (Savile et al., 2010) (Fig. 5).

In the case of (S)-selective ωTAs, a study has also accomplished their rational design to accept bulky ketones with mutations aiming at enlarging the L pocket without compromising it, other mutations decreasing the charges near the active site (due to the nonpolar nature of the used ketone in the study), and a final mutation to improve the overall thermostability (Dourado et al., 2016). The design of the variants was based on molecular docking, MD simulations, and further structural analysis, highlighting the

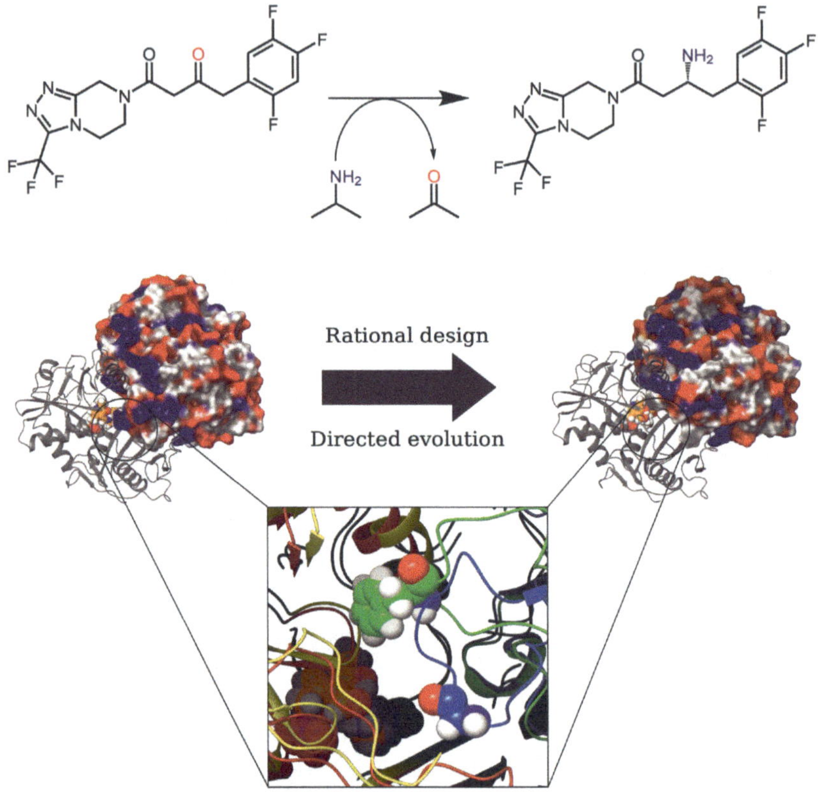


**Fig. 5** ***Engineering of a ωTA to enable the synthesis of sitagliptin with the ketone precursor.*** The top panel shows the asymmetric synthesis of sitagliptin with the use of isopropylamine as amine donor. The bottom panel showcases both the WT and mutant dimer structures with one chain (chain A) being represented with the surface labeled as residues charge (blue being positively charged and red being negatively charged) and the other chain (chain B) with ribbon representation, and the modified lysine and PLP are displayed with the CPK model and the C atoms stained in orange. The zoomed region is one of the binding pockets in both the WT and mutant enzymes; the critical residue (G136) that leads to the major change is displayed with the CPK model and with C atoms stained in blue on the WT and in green on the mutant. The color of the ribbon in chain A is red and blue in chain B in the WT enzyme, while in the mutant the color of the ribbon in chain A is yellow and green in chain B. The PDB codes for the WT and mutant enzymes are 3WWH and 3WWJ, respectively.

importance of modeling the interactions between the enzyme and a specific substrate.

Other studies of rational engineering in ωTAs exist (Midelfort et al., 2013; Nobili et al., 2015; Svedendahl, Branneby, Lindberg, & Berglund, 2010) and all they share the fact that the computational analyses are based on modeling the enzyme-substrate interactions with MM methods. Thus, the information these methods provide has been key to develop these applications of ωTAs in the pharmaceutical industry.

Finally, computational studies can also be applied when there is no enzyme as a starting point, but still imposing an enzyme-substrate organization (Welborn & Head-Gordon, 2019); through a *de novo* design. The transitions states are used to build the active site (or theozyme) to be completed with a protein scaffold. Next steps will be improving the binding mode of the active site and further characterization and optimization of the new enzyme (Malisi, Kohlbacher, & Höcker, 2009). One of the first attempts was a hydrolytic enzyme by Mayo in 2001 (Bolon & Mayo, 2001). More complex catalysis were engineered, consisting in, for example, a nucleophilic cysteine–histidine dyad with amides groups (backbones preferably) for oxyanion stabilization. The resulting enzymes were found to cleave activated ester substrates by a two-step acylation/deacylation reaction (Richter et al., 2012). In spite of these success cases, the main computational *de novo* design limitation still is the development of the required protein folding from scratch, typically resulting in low activity enzymes.

Overall, we have attempted to give a quick overview of different *in silico* options providing biocatalytic understanding and facilitating enzyme engineering, where we focused on underlying the potential of methods mapping the enzyme-substrate interaction. Still, these are mostly based on molecular modeling and bioinformatic techniques. In the (possibly very) near future, however, with the increase of experimental data and most likely with data augmentation from modeling, we expect to see the rise of machine learning techniques. In one form or the other, one thing appears clear to us, computational modeling will most likely guide most future engineering efforts.

## References

Acebes, S., Fernandez-Fueyo, E., Monza, E., Fatima Lucas, M., Almendral, D., Ruiz-Dueñas, F. J., et al. (2016). Rational enzyme engineering through biophysical and biochemical modeling. *ACS Catalysis, 6*(3), 1624–1629. http://dx.doi.org/10.1021/acscatal.6b00028.

Agarwal, P. K. (2006). Enzymes: An integrated view of structure, dynamics and function. *Microbial Cell Factories, 5*, 2.

Alonso, S., Santiago, G., Cea-Rama, I., Fernandez-Lopez, L., Coscolín, C., Modregger, J., et al. (2020). Genetically engineered proteins with two active sites for enhanced biocatalysis and synergistic chemo- and biocatalysis. *Nature Catalysis, 3*(3), 319–328. http://dx.doi.org/10.1038/s41929-019-0394-4.

AlQuraishi, M. (2019). End-to-End differentiable learning of protein structure. *Cell Systems, 8*(4), 292–301.e3.

Altschul, S. F., Gish, W., Miller, W., Myers, E. W., & Lipman, D. J. (1990). Basic local alignment search tool. *Journal of Molecular Biology, 215*(3), 403–410.

Amidi, A., Amidi, S., Vlachakis, D., Megalooikonomou, V., Paragios, N., & Zacharaki, E. I. (2018). EnzyNet: Enzyme classification using 3D convolutional neural networks on spatial representation. *PeerJ, 6*, e4750.

Aranda, C., Municoy, M., Guallar, V., Kiebist, J., Scheibner, K., Ullrich, R., et al. (2019). Selective synthesis of 4-hydroxyisophorone and 4-ketoisophorone by fungal peroxygenases. *Catalysis Science & Technology, 9*(6), 1398–1405. http://dx.doi.org/10.1039/c8cy02114g.

Ashkenazy, H., Abadi, S., Martz, E., Chay, O., Mayrose, I., Pupko, T., et al. (2016). ConSurf 2016: An improved methodology to estimate and visualize evolutionary conservation in macromolecules. *Nucleic Acids Research, 44*(W1), W344–W350.

Bajaj, P., Sreenilayam, G., Tyagi, V., & Fasan, R. (2016). Gram-scale synthesis of chiral cyclopropane-containing drugs and drug precursors with engineered myoglobin catalysts featuring complementary stereoselectivity. *Angewandte Chemie, 55*(52), 16110–16114.

Ba, L., Li, P., Zhang, H., Duan, Y., & Lin, Z. (2013). Semi-rational engineering of cytochrome P450sca-2 in a hybrid system for enhanced catalytic activity: Insights into the important role of electron transfer. *Biotechnology and Bioengineering, 110*(11), 2815–2825. http://dx.doi.org/10.1002/bit.24960.

Belzen, J. U. zu, Bürgel, T., Holderbach, S., Bubeck, F., Adam, L., Gandor, C., et al. (2019). Leveraging implicit knowledge in neural networks for functional dissection and engineering of proteins. *Nature Machine Intelligence, 1*(5), 225–235. http://dx.doi.org/10.1038/s42256-019-0049-9.

Binkley, J. S., Whiteside, R., Hariharan, P. C., Seeger, R., Hehre, W. J., Lathan, W. A., et al. (1978). *Gaussian 76: An ab initio molecular orbital program*. http://dx.doi.org/10.2172/6738000.

Böck, A., Forchhammer, K., Heider, J., Leinfelder, W., Sawers, G., Veprek, B., et al. (1991). Selenocysteine: The 21st amino acid. *Molecular Microbiology, 5*(3), 515–520. http://dx.doi.org/10.1111/j.1365-2958.1991.tb00722.x.

Bolon, D. N., & Mayo, S. L. (2001). Enzyme-like proteins by computational design. *Proceedings of the National Academy of Sciences of the United States of America, 98*(25), 14274–14279.

Bonomi, M., Branduardi, D., Bussi, G., Camilloni, C., Provasi, D., Raiteri, P., et al. (2009). Plumed: A portable plugin for free-energy calculations with molecular dynamics. *Computer Physics Communications, 180*(10), 1961–1972. http://dx.doi.org/10.1016/j.cpc.2009.05.011.

Bornscheuer, U. T., & Pohl, M. (2001). Improved biocatalysts by directed evolution and rational protein design. *Current Opinion in Chemical Biology, 5*(2), 137–143.

Borrelli, K. W., Vitalis, A., Alcantara, R., & Guallar, V. (2005). Pele: Protein energy Landscape exploration. A novel Monte Carlo based technique. *Journal of Chemical Theory and Computation, 1*(6), 1304–1311.

Bowers, K. J., Chow, D. E., Xu, H., Dror, R. O., Eastwood, M. P., Gregersen, B. A., et al. (2006). Scalable algorithms for molecular dynamics simulations on commodity clusters. In *ACM/IEEE SC 2006 conference (SC'06)*. http://dx.doi.org/10.1109/sc.2006.54.

Breaker, R. R. (1997). DNA enzymes. *Nature Biotechnology, 15*(5), 427–431.

Brooks, B. R., Bruccoleri, R. E., Olafson, B. D., States, D. J., Swaminathan, S., & Karplus, M. (1983). Charmm: A program for macromolecular energy, minimization, and dynamics calculations. *Journal of Computational Chemistry, 4*(2), 187–217. http://dx.doi.org/10.1002/jcc.540040211.

Car, R., & Parrinello, M. (1985). Unified approach for molecular dynamics and density-functional theory. *Physical Review Letters, 55*(22), 2471–2474.

Carvalho, A. T. P., Barrozo, A., Doron, D., Kilshtain, A. V., Major, D. T., & Kamerlin, S. C. L. (2014). Challenges in computational studies of enzyme structure, function and dynamics. *Journal of Molecular Graphics and Modelling, 54*, 62–79. http://dx.doi.org/10.1016/j.jmgm.2014.09.003.

Chapman, J., Ismail, A., & Dinu, C. (2018). Industrial applications of enzymes: Recent advances, techniques, and outlooks. *Catalysts, 8*(6), 238. http://dx.doi.org/10.3390/catal8060238.

Chen, F., Gaucher, E. A., Leal, N. A., Hutter, D., Havemann, S. A., Govindarajan, S., et al. (2010). Reconstructed evolutionary adaptive paths give polymerases accepting reversible terminators for sequencing and SNP detection. *Proceedings of the National Academy of Sciences of the United States of America, 107*(5), 1948–1953.

Chen, K., Huang, X., Kan, S. B. J., Zhang, R. K., & Arnold, F. H. (2018). Enzymatic construction of highly strained carbocycles. *Science, 360*(6384), 71–75.

Chien, A., Edgar, D. B., & Trela, J. M. (1976). Deoxyribonucleic acid polymerase from the extreme thermophile Thermus aquaticus. *Journal of Bacteriology, 127*(3), 1550–1557.

Chothia, C., & Lesk, A. M. (1986). The relation between the divergence of sequence and structure in proteins. *The EMBO Journal, 5*(4), 823–826. http://dx.doi.org/10.1002/j.1460-2075.1986.tb04288.x.

Chowdhury, B., & Garai, G. (2017). A review on multiple sequence alignment from the perspective of genetic algorithm. *Genomics, 109*(5–6), 419–431.

Cornell, W. D., Cieplak, P., Bayly, C. I., Gould, I. R., Merz, K. M., Ferguson, D. M., et al. (1995). A second generation force field for the simulation of proteins, nucleic acids, and organic molecules. *Journal of the American Chemical Society, 117*(19), 5179–5197. http://dx.doi.org/10.1021/ja00124a002.

Cossins, B. P., Hosseini, A., & Guallar, V. (2012). Exploration of protein conformational change with PELE and meta-dynamics. *Journal of Chemical Theory and Computation, 8*(3), 959–965.

Daugelaite, J., O' Driscoll, A., & Sleator, R. D. (2013). An overview of multiple sequence alignments and cloud computing in bioinformatics. *ISRN Biomathematics, 2013*, 1–14.

Debye, P. (1913). Interferenz von Röntgenstrahlen und Wärmebewegung. *Annalen der Physik, 348*(1), 49–92.

Denisov, I. G., Makris, T. M., Sligar, S. G., & Schlichting, I. (2005). Structure and chemistry of cytochrome P450. *Chemical Reviews, 105*(6), 2253–2277.

Dhingra, A., Portis, A. R., Jr., & Daniell, H. (2004). Enhanced translation of a chloroplast-expressed RbcS gene restores small subunit levels and photosynthesis in nuclear RbcS antisense plants. *Proceedings of the National Academy of Sciences of the United States of America, 101*(16), 6315–6320.

Dias, S. E. D., Martins, A. M., Nguyen, Q. T., & Gomes, A. J. P. (2017). GPU-based detection of protein cavities using Gaussian surfaces. *BMC Bioinformatics, 18*(1), 493.

Dodani, S. C., Kiss, G., Cahn, J. K. B., Su, Y., Pande, V. S., & Arnold, F. H. (2016). Discovery of a regioselectivity switch in nitrating P450s guided by molecular dynamics simulations and Markov models. *Nature Chemistry, 8*(5), 419–425.

Dominguez, C., Boelens, R., & Alexandre, M. J. (2003). Haddock: A Protein–Protein docking approach based on biochemical or biophysical information. *Journal of the American Chemical Society, 125*(7), 1731–1737. http://dx.doi.org/10.1021/ja026939x.

Doudna, J. A., & Lorsch, J. R. (2005). Ribozyme catalysis: Not different, just worse. *Nature Structural & Molecular Biology, 12*(5), 395–402. http://dx.doi.org/10.1038/nsmb932.

Dourado, D. F. A. R., Pohle, S., Carvalho, A. T. P., Dheeman, D. S., Caswell, J. M., Skvortsov, T., et al. (2016). *Rational Design of a (S)-Selective-Transaminase for asymmetric Synthesis of (1S)-1-(1,1′-biphenyl-2-yl)ethanamine [data set]*. http://dx.doi.org/10.1021/acscatal.6b02380.

Dunn, M. R., Otto, C., Fenton, K. E., & Chaput, J. C. (2016). Improving polymerase activity with unnatural substrates by sampling mutations in homologous protein architectures. *ACS Chemical Biology, 11*(5), 1210–1219.

Edgar, R. C. (2004). Muscle: Multiple sequence alignment with high accuracy and high throughput. *Nucleic Acids Research, 32*(5), 1792–1797.
Edgar, R. C. (2010). Search and clustering orders of magnitude faster than BLAST. *Bioinformatics, 26*(19), 2460–2461.
Eyring, H. (1935). The activated complex and the absolute rate of chemical reactions. *Chemical Reviews, 17*(1), 65–77. http://dx.doi.org/10.1021/cr60056a006.
Fan, H., Schneidman-Duhovny, D., Irwin, J. J., Dong, G., Shoichet, B. K., & Sali, A. (2011). Statistical potential for modeling and ranking of protein–ligand interactions. *Journal of Chemical Information and Modeling, 51*(12), 3078–3092. http://dx.doi.org/10.1021/ci200377u.
Farwell, C. C., Zhang, R. K., McIntosh, J. A., Hyster, T. K., & Arnold, F. H. (2015). Enantioselective enzyme-catalyzed aziridination enabled by active-site evolution of a cytochrome P450. *ACS Central Science, 1*(2), 89–93.
Feng, D.-F., & Doolittle, R. F. (1987). Progressive sequence alignment as a prerequisitetto correct phylogenetic trees. *Journal of Molecular Evolution, 25*(4), 351–360. http://dx.doi.org/10.1007/bf02603120.
Fock, V. (1930). Näherungsmethode zur Lösung des quantenmechanischen Mehrkörperproblems. *Zeitschrift für Physik, 61*(1–2), 126–148. http://dx.doi.org/10.1007/bf01340294.
Freakley, S. J., Kochius, S., van Marwijk, J., Fenner, C., Lewis, R. J., Baldenius, K., et al. (2019). A chemo-enzymatic oxidation cascade to activate C–H bonds with in situ generated H2O2. *Nature Communications, 10*(1). http://dx.doi.org/10.1038/s41467-019-12120-w.
Friesner, R. A., Banks, J. L., Murphy, R. B., Halgren, T. A., Klicic, J. J., Mainz, D. T., et al. (2004). Glide: A new approach for rapid, accurate docking and scoring. 1. Method and assessment of docking accuracy. *Journal of Medicinal Chemistry, 47*(7), 1739–1749.
Friesner, R. A., & Guallar, V. (2005). Ab initio quantum chemical and mixed quantum mechanics/molecular mechanics (QM/MM) methods for studying enzymatic catalysis. *Annual Review of Physical Chemistry, 56*, 389–427.
Galperin, M. Y., & Koonin, E. V. (1997). A diverse superfamily of enzymes with ATP-dependent carboxylate-amine/thiol ligase activity. *Protein Science: A Publication of the Protein Society, 6*(12), 2639–2643.
Galperin, M. Y., & Koonin, E. V. (2012). Divergence and convergence in enzyme evolution. *Journal of Biological Chemistry, 287*(1), 21–28.
Ghislieri, D., & Turner, N. J. (2014). Biocatalytic approaches to the synthesis of enantiomerically pure chiral amines. *Topics in Catalysis, 57*(5), 284–300. http://dx.doi.org/10.1007/s11244-013-0184-1.
Gilabert, J. F., Lecina, D., Estrada, J., & Guallar, V. (2018). Monte Carlo techniques for drug design: The success case of PELE. *Biomolecular Simulations in Structure-Based Drug Discovery*, 87–103. http://dx.doi.org/10.1002/9783527806836.ch5.
Gilis, D., & Rooman, M. (2000). PoPMuSiC, an algorithm for predicting protein mutant stability changes: Application to prion proteins. *Protein Engineering, 13*(12), 849–856.
Godzik, A. (1996). The structural alignment between two proteins: Is there a unique answer? *Protein Science, 5*(7), 1325–1338. http://dx.doi.org/10.1002/pro.5560050711.
Grosdidier, A., Zoete, V., & Michielin, O. (2011). SwissDock, a protein-small molecule docking web service based on EADock DSS. *Nucleic Acids Research, 39*(Web Server issue), W270–W277.
Guan, L.-J., Ohtsuka, J., Okai, M., Miyakawa, T., Mase, T., Zhi, Y., et al. (2015). A new target region for changing the substrate specificity of amine transaminases. *Scientific Reports, 5*, 10753.

Guerrier-Takada, C., Gardiner, K., Marsh, T., Pace, N., & Altman, S. (1983). The RNA moiety of ribonuclease P is the catalytic subunit of the enzyme. *Cell, 35*(3 Pt 2), 849–857.

van Gunsteren, W. F., & Berendsen, H. J. (1987). Groningen molecular simulation (GROMOS) library manual. *Biomos, Groningen, 24*(682704), 13.

Halgren, T. (2007). New method for fast and accurate binding-site identification and analysis. *Chemical Biology & Drug Design, 69*(2), 146–148.

Hartree, D. R. (1928). The wave mechanics of an atom with a non-coulomb central field. Part I. Theory and methods. *Mathematical Proceedings of the Cambridge Philosophical Society, 24*(1), 89–110. http://dx.doi.org/10.1017/s0305004100011919.

Hayashi, K., Sugimoto, H., Shinkyo, R., Yamada, M., Ikeda, S., Ikushiro, S., et al. (2008). Structure-based design of a highly active vitamin D hydroxylase fromStreptomyces griseolusCYP105A1†‡. *Biochemistry, 47*(46), 11964–11972. http://dx.doi.org/10.1021/bi801222d.

Head-Gordon, M., Rico, R. J., Oumi, M., & Lee, T. J. (1994). A doubles correction to electronic excited states from configuration interaction in the space of single substitutions. *Chemical Physics Letters, 219*(1–2), 21–29. http://dx.doi.org/10.1016/0009-2614(94)00070-0.

Hernandez, K. E., Renata, H., Lewis, R. D., Jennifer Kan, S. B., Zhang, C., Forte, J., et al. (2016). Highly stereoselective biocatalytic synthesis of key cyclopropane intermediate to ticagrelor. *ACS Catalysis, 6*(11), 7810–7813.

Higgins, D. G., & Sharp, P. M. (1988). Clustal: A package for performing multiple sequence alignment on a microcomputer. *Gene, 73*(1), 237–244.

Hofrichter, M., & Ullrich, R. (2014). Oxidations catalyzed by fungal peroxygenases. *Current Opinion in Chemical Biology, 19*, 116–125.

Hollingsworth, S. A., & Dror, R. O. (2018). Molecular dynamics simulation for all. *Neuron, 99*(6), 1129–1143.

Holm, L., & Sander, C. (1996). Mapping the protein universe. *Science, 273*(5275), 595–602. http://dx.doi.org/10.1126/science.273.5275.595.

Hou, Q., Kwasigroch, J. M., Rooman, M., & Pucci, F. (2020). SOLart: A structure-based method to predict protein solubility and aggregation. *Bioinformatics, 36*(5), 1445–1452.

Howard, A. E., & Kollman, P. A. (1988). An analysis of current methodologies for conformational searching of complex molecules. *Journal of Medicinal Chemistry, 31*(9), 1669–1675.

Hrycay, E. G., & Bandiera, S. M. (2012). The monooxygenase, peroxidase, and peroxygenase properties of cytochrome P450. *Archives of Biochemistry and Biophysics, 522*(2), 71–89.

Hur, S., & Bruice, T. C. (2003). The near attack conformation approach to the study of the chorismate to prephenate reaction. *Proceedings of the National Academy of Sciences, 100*(21), 12015–12020. http://dx.doi.org/10.1073/pnas.1534873100.

Jacobson, M. P., Pincus, D. L., Rapp, C. S., Day, T. J. F., Honig, B., Shaw, D. E., et al. (2004). A hierarchical approach to all-atom protein loop prediction. *Proteins, 55*(2), 351–367.

Jemli, S., Ayadi-Zouari, D., Hlima, H. B., & Bejar, S. (2016). Biocatalysts: Application and engineering for industrial purposes. *Critical Reviews in Biotechnology, 36*(2), 246–258.

Jensen, F. (2007). *Introduction to computational chemistry*. John Wiley & Sons.

Jones, G., Willett, P., Glen, R. C., Leach, A. R., & Taylor, R. (1997). Development and validation of a genetic algorithm for flexible docking. *Journal of Molecular Biology, 267*(3), 727–748.

Jorgensen, W. L., Chandrasekhar, J., Madura, J. D., Impey, R. W., & Klein, M. L. (1983). Comparison of simple potential functions for simulating liquid water. *The Journal of Chemical Physics, 79*(2), 926–935. http://dx.doi.org/10.1063/1.445869.

Jung, E., Park, B. G., Yoo, H.-W., Kim, J., Choi, K.-Y., & Kim, B.-G. (2018). Semi-rational engineering of CYP153A35 to enhance ω-hydroxylation activity toward palmitic acid. *Applied Microbiology and Biotechnology, 102*(1), 269–277.

Kamerlin, S. C. L., & Warshel, A. (2010). At the dawn of the 21st century: Is dynamics the missing link for understanding enzyme catalysis? *Proteins, 78*(6), 1339–1375.

van der Kamp, M. W., & Mulholland, A. J. (2013). Combined quantum mechanics/molecular mechanics (QM/MM) methods in computational enzymology. *Biochemistry, 52*(16), 2708–2728.

Kan, S. B. J., Lewis, R. D., Chen, K., & Arnold, F. H. (2016). Directed evolution of cytochrome c for carbon-silicon bond formation: Bringing silicon to life. *Science, 354*(6315), 1048–1051.

Karplus, M., & McCammon, J. A. (2002). Molecular dynamics simulations of biomolecules. *Nature Structural Biology, 9*(9), 646–652.

Katoh, K., Misawa, K., Kuma, K.-I., & Miyata, T. (2002). Mafft: A novel method for rapid multiple sequence alignment based on fast fourier transform. *Nucleic Acids Research, 30*(14), 3059–3066.

Kawabata, T. (2010). Detection of multiscale pockets on protein surfaces using mathematical morphology. *Proteins, 78*(5), 1195–1211.

Kelley, L. A., & Sternberg, M. J. E. (2009). Protein structure prediction on the web: A case study using the phyre server. *Nature Protocols, 4*(3), 363–371. http://dx.doi.org/10.1038/nprot.2009.2.

Kelly, S. A., Pohle, S., Wharry, S., Mix, S., Allen, C. C. R., Moody, T. S., et al. (2018). Application of ω-transaminases in the pharmaceutical industry. *Chemical Reviews, 118*(1), 349–367.

Khan, Z. U., Hayat, M., & Khan, M. A. (2015). Discrimination of acidic and alkaline enzyme using Chou's pseudo amino acid composition in conjunction with probabilistic neural network model. *Journal of Theoretical Biology, 365*, 197–203. http://dx.doi.org/10.1016/j.jtbi.2014.10.014.

Khatri, Y., Jóźwik, I. K., Ringle, M., Ionescu, I. A., Litzenburger, M., Hutter, M. C., et al. (2018). Structure-based engineering of steroidogenic CYP260A1 for stereo- and regioselective hydroxylation of progesterone. *ACS Chemical Biology, 13*(4), 1021–1028. http://dx.doi.org/10.1021/acschembio.8b00026.

Kirk, O., Borchert, T. V., & Fuglsang, C. C. (2002). Industrial enzyme applications. *Current Opinion in Biotechnology, 13*(4), 345–351.

Knowles, J. R. (1991). Enzyme catalysis: Not different, just better. *Nature, 350*(6314), 121–124. http://dx.doi.org/10.1038/350121a0.

Kohen, A. (2015). Role of dynamics in enzyme catalysis: Substantial versus semantic controversies. *Accounts of Chemical Research, 48*(2), 466–473.

Kohn, W., & Sham, L. J. (1965). Self-consistent equations including exchange and correlation effects. *Physical Review, 140*(4A), A1133–A1138. http://dx.doi.org/10.1103/physrev.140.a1133.

Koskinen, J. P., & Holm, L. (2012). Sans: High-throughput retrieval of protein sequences allowing 50% mismatches. *Bioinformatics, 28*(18), i438–i443.

Kraut, J. (1988). How do enzymes work? *Science, 242*(4878), 533–540. http://dx.doi.org/10.1126/science.3051385.

Kruger, K., Grabowski, P. J., Zaug, A. J., Sands, J., Gottschling, D. E., & Cech, T. R. (1982). Self-splicing RNA: Autoexcision and autocyclization of the ribosomal RNA intervening sequence of Tetrahymena. *Cell, 31*(1), 147–157.

Laemmli, U. K. (1970). Cleavage of structural proteins during the assembly of the head of bacteriophage T4. *Nature, 227*(5259), 680–685.

Lafayette, L., & Wiebelt, B. (2017). Spartan and NEMO: Two HPC-cloud hybrid implementations. In 2017 *IEEE 13th International Conference on e-Science (e-Science)*. http://dx.doi.org/10.1109/escience.2017.70.

Larrañaga, P., Calvo, B., Santana, R., Bielza, C., Galdiano, J., Inza, I., et al. (2006). Machine learning in bioinformatics. *Briefings in Bioinformatics*, 7(1), 86–112.

Lassmann, T., & Sonnhammer, E. L. L. (2005). Kalign–an accurate and fast multiple sequence alignment algorithm. *BMC Bioinformatics*, *6*, 298.

Le Guilloux, V., Schmidtke, P., & Tuffery, P. (2009). Fpocket: An open source platform for ligand pocket detection. *BMC Bioinformatics*, *10*, 168.

Lei, H., & Duan, Y. (2007). Improved sampling methods for molecular simulation. *Current Opinion in Structural Biology*, *17*(2), 187–191. http://dx.doi.org/10.1016/j.sbi.2007.03.003.

Lonsdale, R., Harvey, J. N., & Mulholland, A. J. (2012). A practical guide to modelling enzyme-catalysed reactions. *Chemical Society Reviews*, *41*(8), 3025–3038.

Lucas, F., Babot, E. D., Cañellas, M., del Río, J. C., Kalum, L., Ullrich, R., et al. (2016). Molecular determinants for selective C25-hydroxylation of vitamins D2and D3by fungal peroxygenases. *Catalysis Science & Technology*, *6*(1), 288–295. http://dx.doi.org/10.1039/c5cy00427f.

Lucas, M. F., Monza, E., Jørgensen, L. J., Ernst, H. A., Piontek, K., Bjerrum, M. J., et al. (2017). Simulating substrate recognition and oxidation in laccases: From description to design. *Journal of Chemical Theory and Computation*, *13*(3), 1462–1467.

Lutz, S. (2010). Beyond directed evolution–semi-rational protein engineering and design. *Current Opinion in Biotechnology*, *21*(6), 734–743.

Malik, M. S., Park, E.-S., & Shin, J.-S. (2012). Features and technical applications of ω-transaminases. *Applied Microbiology and Biotechnology*, *94*(5), 1163–1171.

Malisi, C., Kohlbacher, O., & Höcker, B. (2009). Automated scaffold selection for enzyme design. *Proteins: Structure, Function, and Bioinformatics*, 77(1), 74–83. http://dx.doi.org/10.1002/prot.22418.

Maranas, C. D., & Floudas, C. A. (1994). A deterministic global optimization approach for molecular structure determination. *The Journal of Chemical Physics*, *100*(2), 1247–1261. http://dx.doi.org/10.1063/1.467236.

Marcus, R. A. (1956). On the theory of oxidation-reduction reactions involving electron transfer. I. *The Journal of Chemical Physics*, *24*(5), 966–978. http://dx.doi.org/10.1063/1.1742723.

Martí-Renom, M. A., Stuart, A. C., Fiser, A., Sánchez, R., Melo, F., & Šali, A. (2000). Comparative protein structure modeling of genes and genomes. *Annual Review of Biophysics and Biomolecular Structure*, *29*(1), 291–325. http://dx.doi.org/10.1146/annurev.biophys.29.1.291.

Martínez-Martínez, M., Coscolín, C., Santiago, G., Chow, J., Stogios, P. J., Bargiela, R., et al. (2018). Determinants and prediction of esterase substrate promiscuity patterns. *ACS Chemical Biology*, *13*(1), 225–234.

Ma, J., & Wang, S. (2014). Algorithms, applications, and challenges of protein structure alignment. *Advances in Protein Chemistry and Structural Biology*, 121–175. http://dx.doi.org/10.1016/b978-0-12-800168-4.00005-6.

Maximova, T., Moffatt, R., Ma, B., Nussinov, R., & Shehu, A. (2016). Principles and overview of sampling methods for modeling macromolecular structure and dynamics. *PLoS Computational Biology*, *12*(4), e1004619. http://dx.doi.org/10.1371/journal.pcbi.1004619.

McGuffin, L. J., Atkins, J. D., Salehe, B. R., Shuid, A. N., & Roche, D. B. (2015). IntFOLD: An integrated server for modelling protein structures and functions from amino acid sequences. *Nucleic Acids Research*, *43*(W1), W169–W173.

Midelfort, K. S., Kumar, R., Han, S., Karmilowicz, M. J., McConnell, K., Gehlhaar, D. K., et al. (2013). Redesigning and characterizing the substrate specificity and activity of Vibrio fluvialis aminotransferase for the synthesis of imagabalin. *Protein Engineering Design and Selection: PEDS, 26*(1), 25–33.

Molina-Espeja, P., Cañellas, M., Plou, F. J., Hofrichter, M., Lucas, F., Guallar, V., et al. (2016). Synthesis of 1-naphthol by a natural peroxygenase engineered by directed evolution. *ChemBioChem: A European Journal of Chemical Biology, 17*(4), 341–349.

Møller, C., & Plesset, M. S. (1934). Note on an approximation treatment for many-electron systems. *Physical Review, 46*(7), 618–622. http://dx.doi.org/10.1103/physrev.46.618.

Monza, E., Fatima Lucas, M., Camarero, S., Alejaldre, L. C., Martínez, A. T., & Guallar, V. (2015). Insights into laccase engineering from molecular simulations: Toward a binding-focused strategy. *Journal of Physical Chemistry Letters, 6*(8), 1447–1453. http://dx.doi.org/10.1021/acs.jpclett.5b00225.

Mooij, W. T. M., & Verdonk, M. L. (2005). General and targeted statistical potentials for protein-ligand interactions. *Proteins, 61*(2), 272–287.

Mullis, K. B. (1990). The unusual origin of the polymerase chain reaction. *Scientific American, 262*(4), 56–65. http://dx.doi.org/10.1038/scientificamerican0490-56.

National Center for Biotechnology Information. (n.d.). Retrieved February 24, 2020, from https://www.ncbi.nlm.nih.gov/.

Neese, F. (2012). The ORCA program system. *WIREs Computational Molecular Science, 2*(1), 73–78.

Neet, K. E. (1998). Enzyme catalytic power minireview series. *Journal of Biological Chemistry, 273*(40), 25527–25528.

Nobili, A., Steffen-Munsberg, F., Kohls, H., Trentin, I., Schulzke, C., Höhne, M., et al. (2015). Engineering the active site of the amine transaminase fromVibrio fluvialisfor the asymmetric synthesis of aryl-alkyl amines and amino alcohols. *ChemCatChem,* 7(5), 757–760. http://dx.doi.org/10.1002/cctc.201403010.

Notredame, C., Higgins, D. G., & Heringa, J. (2000). T-coffee: A novel method for fast and accurate multiple sequence alignment. *Journal of Molecular Biology, 302*(1), 205–217.

Oliveira, S. H. P., Ferraz, F. A. N., Honorato, R. V., Xavier-Neto, J., Sobreira, T. J. P., & de Oliveira, P. S. L. (2014). KVFinder: Steered identification of protein cavities as a PyMOL plugin. *BMC Bioinformatics, 15*, 197.

Paladino, A., Marchetti, F., Rinaldi, S., & Colombo, G. (2017). Protein design: >From computer models to artificial intelligence. *Wiley Interdisciplinary Reviews: Computational Molecular Science,* 7(5), e1318. http://dx.doi.org/10.1002/wcms.1318.

Pardo, I., Santiago, G., Gentili, P., Lucas, F., Monza, E., Medrano, F. J., et al. (2016). Redesigning the substrate binding pocket of laccase for enhanced oxidation of sinapic acid. *Catalysis Science & Technology, 6*(11), 3900–3910. http://dx.doi.org/10.1039/c5cy01725d.

Park, E.-S., & Shin, J.-S. (2011). Free energy analysis of ω-transaminase reactions to dissect how the enzyme controls the substrate selectivity. *Enzyme and Microbial Technology, 49*(4), 380–387.

Pearson, W. R. (1991). Searching protein sequence libraries: Comparison of the sensitivity and selectivity of the smith-waterman and FASTA algorithms. *Genomics, 11*(3), 635–650.

Petrović, D., Risso, V. A., Kamerlin, S. C. L., & Sanchez-Ruiz, J. M. (2018). Conformational dynamics and enzyme evolution. *Journal of The Royal Society Interface, 15*(144), 20180330. http://dx.doi.org/10.1098/rsif.2018.0330.

Philipp, D. M., & Friesner, R. A. (1999). Mixedab initio QM/MM modeling using frozen orbitals and tests with alanine dipeptide and tetrapeptide. *Journal of Computational Chemistry, 20*(14), 1468–1494. http://dx.doi.org/10.1002/(sici)1096-987x(19991115)20:14<1468::aid-jcc2>3.0.co;2-0.

Prier, C. K., Zhang, R. K., Buller, A. R., Brinkmann-Chen, S., & Arnold, F. H. (2017). Enantioselective, intermolecular benzylic C–H amination catalysed by an engineered iron-haem enzyme. *Nature Chemistry, 9*(7), 629–634. http://dx.doi.org/10.1038/nchem.2783.

Purvis, G. D., & Bartlett, R. J. (1982). A full coupled-cluster singles and doubles model: The inclusion of disconnected triples. *The Journal of Chemical Physics, 76*(4), 1910–1918. http://dx.doi.org/10.1063/1.443164.

Richter, F., Blomberg, R., Khare, S. D., Kiss, G., Kuzin, A. P., Smith, A. J. T., et al. (2012). Computational design of catalytic dyads and oxyanion holes for ester hydrolysis. *Journal of the American Chemical Society, 134*(39), 16197–16206.

Robinson, P. K. (2015). Enzymes: Principles and biotechnological applications. *Essays in Biochemistry, 59*, 1–41.

Rohl, C. A., Strauss, C. E. M., Misura, K. M. S., & Baker, D. (2004). Protein structure prediction using Rosetta. In *Numerical computer methods, Part D* (Vol. 383, pp. 66–93). Elsevier.

Roy, A., Kucukural, A., & Zhang, Y. (2010). I-TASSER: A unified platform for automated protein structure and function prediction. *Nature Protocols, 5*(4), 725–738.

Rühlmann, A., Antovic, D., Müller, T. J. J., & Urlacher, V. B. (2017). Regioselective hydroxylation of stilbenes by engineered cytochrome P450 fromThermobifida fuscaYX. *Advanced Synthesis & Catalysis, 359*(6), 984–994. http://dx.doi.org/10.1002/adsc.201601168.

Santiago, G., de Salas, F., Fátima Lucas, M., Monza, E., Acebes, S., Martinez, Á. T., et al. (2016). Computer-aided laccase engineering: Toward biological oxidation of arylamines. *ACS Catalysis, 6*(8), 5415–5423. http://dx.doi.org/10.1021/acscatal.6b01460.

Santiago, G., Martínez-Martínez, M., Alonso, S., Bargiela, R., Coscolín, C., Golyshin, P. N., et al. (2018). Rational engineering of multiple active sites in an ester hydrolase. *Biochemistry, 57*(15), 2245–2255. http://dx.doi.org/10.1021/acs.biochem.8b00274.

Savile, C. K., Janey, J. M., Mundorff, E. C., Moore, J. C., Tam, S., Jarvis, W. R., et al. (2010). Biocatalytic asymmetric synthesis of chiral amines from ketones applied to sitagliptin manufacture. *Science, 329*(5989), 305–309.

Schmid, A., Dordick, J. S., Hauer, B., Kiener, A., Wubbolts, M., & Witholt, B. (2001). Industrial biocatalysis today and tomorrow. *Nature, 409*(6817), 258–268.

Schwede, T., Kopp, J., Guex, N., & Peitsch, M. C. (2003). SWISS-MODEL: An automated protein homology-modeling server. *Nucleic Acids Research, 31*(13), 3381–3385.

Senior, A. W., Evans, R., Jumper, J., Kirkpatrick, J., Sifre, L., Green, T., et al. (2020). Improved protein structure prediction using potentials from deep learning. *Nature, 577*(7792), 706–710.

Senn, H. M., & Thiel, W. (2009). QM/MM methods for biomolecular systems. *Angewandte Chemie, 48*(7), 1198–1229.

Serrano, A., Sancho, F., Viña-González, J., Carro, J., Alcalde, M., Guallar, V., et al. (2019). Switching the substrate preference of fungal aryl-alcohol oxidase: Towards stereoselective oxidation of secondary benzyl alcohols. *Catalysis Science & Technology, 9*(3), 833–841. http://dx.doi.org/10.1039/c8cy02447b.

Shah, J. K., Marin-Rimoldi, E., Mullen, R. G., Keene, B. P., Khan, S., Paluch, A. S., et al. (2017). Cassandra: An open source Monte Carlo package for molecular simulation. *Journal of Computational Chemistry, 38*(19), 1727–1739.

Shaik, S., Kumar, D., de Visser, S. P., Altun, A., & Thiel, W. (2005). Theoretical perspective on the structure and mechanism of cytochrome P450 enzymes. *Chemical Reviews, 105*(6), 2279–2328.

Shindyalov, I. N., & Bourne, P. E. (1998). Protein structure alignment by incremental combinatorial extension (CE) of the optimal path. *Protein Engineering, 11*(9), 739–747.

Shin, J.-S., & Kim, B.-G. (2002). Exploring the active site of amine:pyruvate aminotransferase on the basis of the substrate structure-reactivity relationship: How the enzyme controls substrate specificity and stereoselectivity. *Journal of Organic Chemistry, 67*(9), 2848–2853.

Singh, R., Kolev, J. N., Sutera, P. A., & Fasan, R. (2015). Enzymatic C(sp)-H amination: P450-Catalyzed conversion of carbonazidates into oxazolidinones. *ACS Catalysis, 5*(3), 1685–1691.

Singh, R., Kumar, M., Mittal, A., & Mehta, P. K. (2016). Microbial enzymes: Industrial progress in 21st century. *3 Biotech, 6*(2), 174.

Smith, E. A., & Oehme, F. W. (1991). Acrylamide and polyacrylamide: A review of production, use, environmental fate and neurotoxicity. *Reviews on Environmental Health, 9*(4), 215–228.

Soares, I., Tavora, Z., Patera, R., & Baroni, S. (2012). Microorganism-produced enzymes in the food industry. *Scientific, Health and Social Aspects of the Food Industry*, 83–94. http://dx.doi.org/10.5772/31256.

Steck, V., Kolev, J. N., Ren, X., & Fasan, R. (2020). Mechanism-guided design and discovery of efficient P450-derived C—H amination biocatalysts. *Journal of the American Chemical Society*. http://dx.doi.org/10.1021/jacs.9b12859.

Sumbalova, L., Stourac, J., Martinek, T., Bednar, D., & Damborsky, J. (2018). HotSpot wizard 3.0: Web server for automated design of mutations and smart libraries based on sequence input information. *Nucleic Acids Research, 46*(W1), W356–W362.

Suplatov, D., Voevodin, V., & Švedas, V. (2015). Robust enzyme design: Bioinformatic tools for improved protein stability. *Biotechnology Journal, 10*(3), 344–355.

Svedendahl, M., Branneby, C., Lindberg, L., & Berglund, P. (2010). Reversed enantiopreference of an ω-transaminase by a single-point mutation. *ChemCatChem, 2*(8), 976–980. http://dx.doi.org/10.1002/cctc.201000107.

Syed, K., Porollo, A., Miller, D., & Yadav, J. S. (2013). Rational engineering of the fungal P450 monooxygenase CYP5136A3 to improve its oxidizing activity toward polycyclic aromatic hydrocarbons. *Protein Engineering Design and Selection: PEDS, 26*(9), 553–557.

Taylor, W. R., & Orengo, C. A. (1989). Protein structure alignment. *Journal of Molecular Biology, 208*(1), 1–22.

Thoden, J. B., Kappock, T. J., Stubbe, J., & Holden, H. M. (1999). Three-dimensional structure of N5-carboxyaminoimidazole ribonucleotide synthetase: A member of the ATP grasp protein superfamily. *Biochemistry, 38*(47), 15480–15492.

Thompson, J. D., Higgins, D. G., & Gibson, T. J. (1994). Clustal W: Improving the sensitivity of progressive multiple sequence alignment through sequence weighting, position-specific gap penalties and weight matrix choice. *Nucleic Acids Research, 22*(22), 4673–4680.

Toporkova, Y. Y., Ermilova, V. S., Gorina, S. S., Mukhtarova, L. S., Osipova, E. V., Gogolev, Y. V., et al. (2013). Structure-function relationship in the CYP74 family: Conversion of divinyl ether synthases into allene oxide synthases by site-directed mutagenesis. *FEBS Letters, 587*(16), 2552–2558.

Trott, O., & Olson, A. J. (2010). AutoDock vina: Improving the speed and accuracy of docking with a new scoring function, efficient optimization, and multithreading. *Journal of Computational Chemistry, 31*(2), 455–461.

Truhlar, D. G. (2015). Transition state theory for enzyme kinetics. *Archives of Biochemistry and Biophysics, 582*, 10–17. http://dx.doi.org/10.1016/j.abb.2015.05.004.

UniProt Consortium. (2019). UniProt: A worldwide hub of protein knowledge. *Nucleic Acids Research, 47*(D1), D506–D515.

Welborn, V. V., & Head-Gordon, T. (2019). Computational design of synthetic enzymes. *Chemical Reviews, 119*(11), 6613–6630. http://dx.doi.org/10.1021/acs.chemrev.8b00399.

Wolfenden, R. (2006). Degrees of difficulty of water-consuming reactions in the absence of enzymes. *Chemical Reviews, 106*(8), 3379–3396.
Wolfenden, R., & Snider, M. J. (2001). The depth of chemical time and the power of enzymes as catalysts. *Accounts of Chemical Research, 34*(12), 938–945.
Yamada, H., & Kobayashi, M. (1996). Nitrile hydratase and its application to industrial production of acrylamide. *Bioscience Biotechnology and Biochemistry, 60*(9), 1391–1400.
Yang, Y., Pan, L., Lightstone, F. C., & Merz, K. M., Jr. (2016). The role of molecular dynamics potential of mean force calculations in the investigation of enzyme catalysis. *Methods in Enzymology, 577*, 1–29.
Yoo, E.-H., & Lee, S.-Y. (2010). Glucose biosensors: An overview of use in clinical practice. *Sensors, 10*(5), 4558–4576.
Zhang, R. K., Chen, K., Huang, X., Wohlschlager, L., Renata, H., & Arnold, F. H. (2019). Enzymatic assembly of carbon–carbon bonds via iron-catalysed sp3 C–H functionalization. *Nature, 565*(7737), 67–72. http://dx.doi.org/10.1038/s41586-018-0808-5.
Zinovjev, K., & Tuñón, I. (2017). Quantifying the limits of transition state theory in enzymatic catalysis. *Proceedings of the National Academy of Sciences of the United States of America, 114*(47), 12390–12395.

CHAPTER TWO

# Experimental insight into enzyme catalysis and dynamics: A review on applications of state of art spectroscopic methods

**Puspal Mukherjee and Prashant Chandra Singh***
School of Chemical Sciences, Indian Association for the Cultivation of Science, Kolkata, West Bengal, India
*Corresponding author: E-mail: sppcs@iacs.res.in

## Contents



## Abstract

Enzymes are dynamic in nature and understanding their activity depends on exploring their overall structural fluctuation as well as transformation at the active site in free state as well as turnover conditions. In this chapter, the application of several different spectroscopy techniques *viz.* single molecule spectroscopy, ultrafast spectroscopy and Raman spectroscopy in the context of enzyme dynamics and catalysis are discussed. The importance of such studies are significant in the understanding of new discoveries of drugs, cure for some lethal diseases, gene modification as well as in industrial applications.

## 1. Introduction

Enzymes, which are typically proteins, act as biocatalysts and as same as any other catalyst, they speed up biological reactions (Robinson, 2015). The term was first coined by German scientist Wilhelm Kühne in 1878 and it

*Advances in Protein Chemistry and Structural Biology*, Volume 122
ISSN 1876-1623
https://doi.org/10.1016/bs.apcsb.2020.06.003

literally means "within yeast" (Robinson, 2015). Enzymes are classified based on the reactions they catalyze and by principle, enzymes are named in singular manner based upon to the reaction (Classification and Nomenclature of Enzymes, 1992). However, they are grouped according to broad category of reactions. The main classes of enzymes are oxidoreductases (catalyze oxidation-reduction reactions), transferases (transfers one or more functional groups), hydrolases (catalysehydrolysis), lyases (takes part in elimination reactions), isomerases (catalyze isomerization reactions) and ligases (catalyzes addition/joining reactions) (Classification and Nomenclature of Enzymes, 1992). Enzymes in each of this category varies in their specific functionalities; however, while discussing enzymes, sometimes it is easier to refer them as per their broad classes. Enzymes forms a vital part of life itself but apart from the biological functionalities of enzymes, they have been commercialized for industrial reactions in food, beverages, pharmaceutical fields because of the mild reaction conditions, reactant and product specific functions, low toxicity etc.(Chapman, Ismail, & Dinu, 2018). The biological and commercial aspect of enzymes point to the fact that the understanding the process of catalysis and subsequent structural changes in enzymes is extremely important. A crucial question in this regard is that the structural change happens when the substrate binds with the active site of the enzyme or the electrostatic interactions induces structural changes which leads to substrate binding (Gärtner, 2009)?Macroscopically, the catalytic power of enzymes is described by turnover number or turnover rate. It is defined as the number of substrate molecule converted to product by one molecule of enzyme per second (Roskoski, 2015). Turnover number have a huge variation from values in thousands to one for different kinds of enzyme (Roskoski, 2015) which points to the fact that the mechanism of enzyme kinetics is not a straightforward phenomenon. As we know, one of the earliest models of enzyme catalysis is the lock and key model which was postulated by Emil Fisher in 1894 (van Gelderen, Mayer, & Testa, 1994). In this model, the specificity of enzyme reactions originates from the perfect fit of substrate in the active site of the enzyme (van Gelderen et al., 1994). It may seem that a particular substrate can attach to an enzyme anytime but application of transition state theory to enzyme kinetics implies that the substrate-enzyme interaction happens in the transition state and not in the ground state and thus the transition state is stabilized by the enzyme (Benkovic & Hammes-Schiffer, 2003; van Gelderen et al., 1994; Wolfenden, 2003). This idea led to development of transition state inhibitors (van Gelderen et al., 1994). Then there is induced fit model which describes that the substrates induce structural changes in the enzyme which let it bind and undergo subsequent reactions (van Gelderen et al., 1994). These models were based on the thermodynamics and kinetic exploration of enzyme catalysis reactions. Now enzymes like all other proteins undergo dynamic spontaneous motions (Mulholland, Roitberg, & Tuñón, 2012). Whether dynamics is important to catalytic rate or turnover is debated between researchers for long with arguments in both sides (Olsson, Parson, & Warshel, 2006; Pineda, Antoniou, & Schwartz, 2010). First, we should emphasize on the term catalysis in case of enzymes. Although, the trivial definition of "catalysis" meaning "acceleration of reaction" does apply in this case but the question is acceleration compared to which uncatalyzed reaction (Lipscomb, 1983). Many biological reactions simply do not occur without enzymes. Even if they occur is water solutions, the cell environment is quite different compared to bulk solvent environment. Moreover, enzyme may need cofactors, or specific transport to operate and they work like a complex structure with the substrate (Mulholland et al., 2012). So, when we say enzyme catalysis, we generally compare the reaction in water at certain conditions (pH, pressure, temperature etc.) (Mulholland et al., 2012). The term dynamic also means two things. It can mean the vibrations of chemical bond before chemical reaction but in case of enzyme kinetics it means the equilibrium and non-equilibrium conformational fluctuations which may or may not lead to a chemical reaction (Greaves et al., 2008). So, while comparing the enzyme catalysis reactions with uncatalyzed reactions it means comparing the reaction happening in solvent cage and in enzyme cage (Olsson et al., 2006). The achievement of the transition state can occur due to the thermal fluctuations of the solvents, specific bond vibrations and enzyme specific vibrational coupling between frequencies. Now induced fit model may not hold true for many cases but it gives an overall idea about the process. Several other propositions have been made such as electrostatic preorganization which means that before the binding of the ligand with enzyme, an electrostatic organization occurs which facilitates this process (Rao, Singh, Bash, & Kollman, 1987). Studies on lysozyme also showed that enzyme catalysis can originate from steric strains in the ground state but in later years this idea was forfeited due to very small amount of energy changes in this interaction (Olsson et al., 2006). There is also propositions based on entropic considerations of the ground state (Kollman et al., 2001).

Enzymes are essentially proteins and conformational fluctuations in proteins can occur from femtosecond to microsecond (or more) time scale

(Cannon & Benkovic, 1998; Agarwal, 2006). Biochemical reactions have multiple vibrational frequencies associated with them and analysis them in association with enzymes is a difficult job. But modern computational chemistry especially quantum mechanical/molecular mechanical (QM/MM) methods are showing great promises in this field. For example, recent computations have revealed that the electric perturbation from the enzyme-solvent complex is the trigger for the catalysis of horseradish peroxidase (Zazza et al., 2010). Computational studies elucidated the mechanism of enzyme catalyzed cycloaddition reactions and this particular set of enzymes are termed as Diels-Alderase (Gordeev & Ananikov, 2015). One of the most commonly studied enzyme is Cytochrome P450 which are heme based enzymes and used for phase I metabolism (Dutkiewicz & Mikstacka, 2018). Computation al chemistry has been applied for drug design for inhibitors, exploration of barrier crossing in the potential energy surface etc. (Dutkiewicz & Mikstacka, 2018). One interesting observation is of β-lactamase in which the experimental studies have shown that structural changes can affect the turnover number profoundly but computational studies predicted that there is little correlation between the structures and the turnover rate (Jackson et al., 2009; Tokuriki & Jackson, 2014). This long-standing debate is refueled by NMR studies of cis—transisomerase cyclophilin A (Eisenmesser et al., 2005). NMR spectroscopy revealed that the different conformations necessary for this particular enzyme to catalyze the reaction not only in a substrate bound state but also in free states. (Eisenmesser et al., 2005). This idea is also supported by other computational studies which suggested that the ground state structure of enzyme is not a single one but an ensemble of various sub structures (Osuna, Jiménez-Osés, Noey, & Houk, 2015). Thus enzymes can display a broad range of structures which corresponds to a broad free energy landscape comprised of several global and local minima (Maria-Solano, Serrano-Hervás, Romero-Rivera, Iglesias-Fernández, & Osuna, 2018). This conformational energy landscape is essential in understanding how substrates attach to enzymes and reaction trajectories.

Computational chemistry can provide us with many information but it has some underlying assumptions. The real systems can differ significantly from the computed environments. So, it is highly desirable to develop experimental techniques to observe an enzyme. Spectroscopic and microscopic methods were found to be well suited for this kind of study. Development of state of art spectroscopy techniques such as single molecule, confocal microscopy, ultrafast and Raman spectroscopy has opened new doors in enzymology. All these experimental methods are shedding new light on the enzyme structure, time fluctuations and enzyme-substrate equilibria.

## 2. Single molecule detection

The structural dynamics of enzymes has been investigated by attaching fluorophore and combined with single molecule fluorescence techniques. Let's say that the fluorescence of fluorophore is quenched when the enzyme binds with the substrate. Now, if we take a steady state emission spectra of the enzyme —substrate solution, we will see a time and ensemble average picture where it is impossible to determine the fraction of enzyme molecules that are free and the fraction which is bound to the substrates (Lakowicz, 2006). We assume that the medium properties surrounding all the molecules are exactly same and there is no heterogeneity in the solution. However, observing molecules at single molecule level can yield completely different results since each of the molecule has a local environment. Single molecule enzyme catalysis was initially studied by Rotman who attached 6-hydroxy-fluoran (6-HF) with galactose (Rotman, 1961). 6-HF is highly fluorescent in its bound state. On reaction with β-D-galactosidase, 6-HF undergoes hydrolyzes which releases the 6-HF and it became non-fluorescent (Rotman, 1961). In the following sections, we will discuss some of the important aspects of single molecule spectroscopy in relation to enzymology.

### 2.1 Fluorescence correlation spectroscopy

The inception of fluorescence correlation spectroscopy (FCS) was realized at the same time by Douglas Magde, Elliot Eison, and W. W. Webb at Cornell University, USA, and by Ehrenberg and Rigler in Karolinska Institute, Sweden' (Ehrenberg & Rigler, 1974; Elson & Magde, 1974; Magde, Elson, & Webb, 1972, 1974; Magde, Webb, & Elson, 1978). In 1990, with the incorporation of confocal microscopy and two-photon fluorescence, a better experimental setup for FCS was designed and after that FCS became a familiar and extremely popular tool for the molecular level understanding of the dynamics of enzyme.

FCS in a non-invasive technique applied to fluorescence molecule labeled systems in dilute solutions. When a system is in thermal equilibrium, the average thermodynamic properties of the system, as well as overall concentrations, does not change. However, if we observe a tiny portion of the system i.e., a local volume, we will see that the properties are spontaneously varying in space and time due to the Brownian motion of the molecules diffusing through the small volume. For a chemical reaction, the probability of change in concentration in the "local volume" is described by Poisson distribution (Elson, 2011; Lakowicz, 2006). Thus, FCS can be described to observe a steady-state within a small volume for an overall molecular ensemble is in thermal and/or chemical equilibrium. The immediate implications are achievement of small observational volume, very low fluorescence intensity detection over noise from scattered light, background emission etc. A schematic experimental set-up of FCS has been shown in Fig. 1.

A laser beam is directed toward the objective lens using a dichroic filter. The objective lens focuses the excitation beam just above the cover slip in a very small volume where the sample resides. The emission is also collected by the objective lens. The dichroic mirror passes the emission toward an emission filter and then a lens is used to focus the emission light onto the detector through a pin-hole. The use of the pin-hole is extremely important

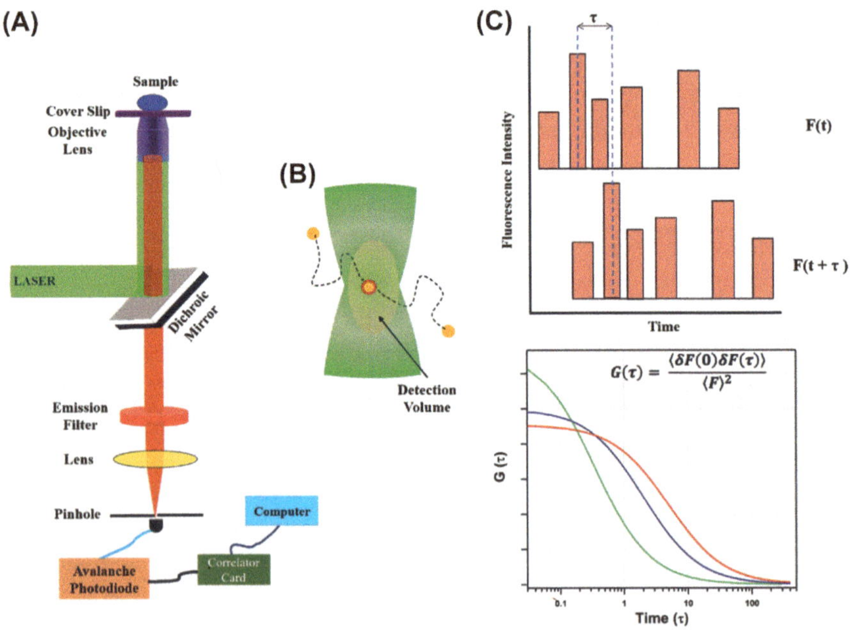


**Fig. 1** (A) Schematic Diagram of setup of FCS with all its essential components; (B) Focal volume and detection volume depicting the motion of fluorophore; (C) A schematic of data and autocorrelation curve.

since it selects the light coming from the detection volume only and rejects all the other light. The detector used generally is an avalanche photodiode (APD) which passes the photocurrent data to the correlator card and then to the computer. When fluorescent molecules diffuse through the observation volume, emitted light is collected as photon bursts for a period of time in the APD. This varying fluorescence intensity is the primary data of FCS. The data is analyzed in terms of autocorrelation which means the similarity of the signal with itself over time. The autocorrelation function is given by (Lakowicz, 2006)

$$G(\tau) = \frac{\delta F(0)\delta F(\tau)}{F^2} \tag{1}$$

Here, $\delta F$ is the fluctuation from the mean value of intensity. The autocorrelation data has to be fitted in order to extract information and it has contribution from translational diffusion and triplet state (Günther, Börsch, & Fischer, 2018).

$$G(\tau) = G(\tau)_{\text{Diff}} G(\tau)_{\text{Triplet}} \tag{2}$$

FCS is primarily used to measure translational diffusion and the contribution from the triplet state is neglected and thus (Günther et al., 2018; Lakowicz, 2006)

$$G(\tau) \approx G(\tau)_{\text{Diff}} = \frac{1}{N}\left(\frac{1}{1+\frac{\tau}{\tau_D}}\right)\left(\frac{1}{1+\left(\frac{\omega_0}{z_0}\right)^2\frac{\tau}{\tau_D}}\right)^{\frac{1}{2}} \tag{3}$$

Here, $N$ is the average number of fluorescent molecules in the observation volume, $\tau_D$ is the translation diffusion time $\omega_0$ is the radius in the radial direction and $z_0$ is the radius in the axial direction (Günther et al., 2018). The translation diffusion coefficient of the fluorophore is defined as

$$D = \frac{\omega_0^2}{4\tau_D} \tag{4}$$

As can be seen from the above equations, FCS data analysis requires knowledge about $\omega_0$ and $z_0$ which is not readily available. That's why, FCS parameters such as diffusion coefficients are measured relative to a reference. The $\tau_D$ of a dye whose diffusion coefficient in that particular solvent is known, is measured and the radius of the focal spot is calculated.

To understand the applicability of the technique we decided to go through few examples with one particular problem related to structural evolution during enzyme catalysis described in more details. For example, one of the earliest studies in this field was by Boërsch et al. who studied binding of $H^+$-ATPase to nucleotide (Börsch et al., 1998). They labeled the H + -ATPase obtained from *Escherichia coli* with tetramethylrhodamine and inspected the conformational dynamics by observing the diffusion time in FCS (Börsch et al., 1998). There was a distinct increase in the diffusion constant which was attributed to overall size reduction of the enzyme (Börsch et al., 1998). In another FCS study of catalytic reaction of F1-ATPase enzyme (labeled with Alexa dye 488) with substrate adenosine triphosphate (MgATP), it has been showed that the autocorrelation curves are showing faster decay with the progress of reaction (Shah, Hutcheon, Haarer, & Duncan, 2013). This phenomenon was predicted to be caused by dissociation of subunits of enzyme ATPase at lower experimental concentrations (Shah et al., 2013). Previously, to establish the reactivity of adenylate kinase, a phenomenon known as "cracking" was predicted theoretically which sort of describes the unfolding of enzyme during reaction and may be a cause of enhanced diffusion (Whitford, Miyashita, Levy, & Onuchic, 2007). Ultimately, this observation was rationalized through FCS studies in the following manner.

In an enzyme catalyzed reaction, when enzyme and substrate bind to each other heat is released specifically from the active site of the enzyme (Riedel et al., 2015). It was long assumed that this heat does not affect the chemical reaction nor the enzyme in any way (Riedel et al., 2015). However, studies including FCS in 2010 and 2013 by researchers from Pennsylvania state university suggested that when an enzyme reacts with a substrate, its diffusivity increases (Muddana, Sengupta, Mallouk, Sen, & Butler, 2010; Sengupta et al., 2013). The enzymes used in these experiments were urease and catalase (Riedel et al., 2015). In 2015, Riedel et al. carried out FCS studies with four enzymes viz. catalase, urease, alkaline phosphatase and triose phosphate isomerase labeled with fluorescent dye Atto 647—maleimide and Atto 647-NHS (Riedel et al., 2015). The substrates used were hydrogen peroxide, urea, p-nitrophenylphosphate and D-glyceraldehyde 3-phosphate for catalase, urease, alkaline phosphatase and triose phosphate isomerase respectively (Riedel et al., 2015). The FCS plots has been reproduced in Fig. 2. The zoomed portion of the data as shown by authors confirms the inflection points.

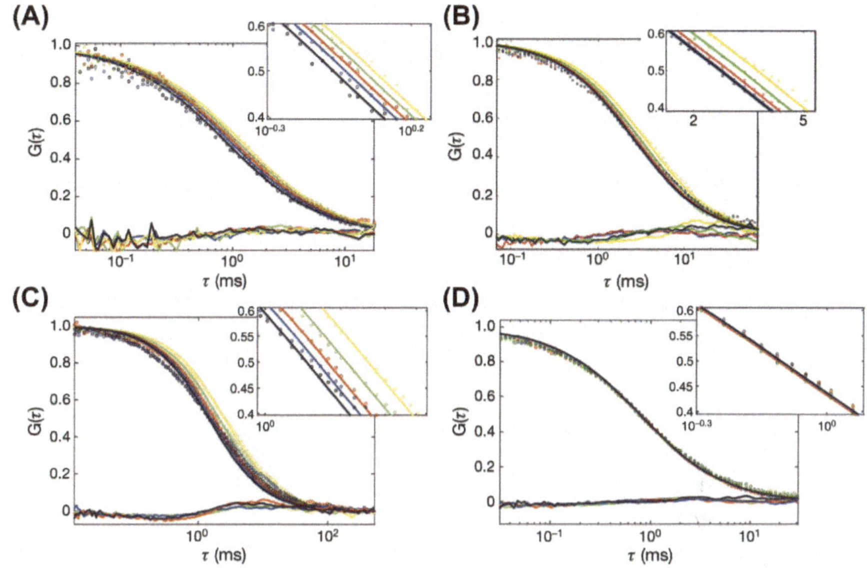


**Fig. 2** FCS autocorrelation plot along with fitting using single diffusion showing gradual addition of (A) Catalase. Hydrogen peroxideconcentration: yellow, 0 mM; green, 3 mM; red, 6 mM; blue, 12 mM; black, 25 mM (B), Urease. Urea concentration: yellow, 0 mM; green, 1 mM; red, 10 mM; blue, 100 mM; black, 1 M. (C), Alkaline phosphatase. p-Nitrophenylphosphateconcentration: yellow, 0 mM; green, 0.44 mM; red, 0.88 mM; blue, 1.74 mM; black, 2.64 mM (D), TIM. D-Glyceraldehyde 3-phosphateconcentration: yellow, 0 mM; green, 0.15 mM; red, 0.31 mM; blue, 0.62 mM; black, 1.3 mM. *This has been reused with permission from Riedel, C., Gabizon, R., Wilson, C. A. M., Hamadani, K., Tsekouras, K., Marqusee, S., et al. (2015). The heat released during catalytic turnover enhances the diffusion of an enzyme.* Nature, 517*(7533), 227–230. https://doi.org/10.1038/nature14043.*

The FCS data (Fig. 2) shows variation with increase in substrate concentration and clearly there is an increase of diffusion coefficient when the substrate concentration is increased. This signifies that the heat released during the chemical reaction taking place at the active site of the enzyme is accelerating its center of mass. This is not a mere result of enzyme substrate binding/unbinding because the fcs experiments with non-competitive reversible inhibitor showed no increase in diffusion (Riedel et al., 2015). This means the heat released during the turnover event may unfold local structure of the enzyme which can influence the catalytic activity as discussed before (Riedel et al., 2015). The enhanced diffusion upon substrate binding effect was later confirmed in hexokinase and aldolase enzyme by monitoring local concentration fluctuations in a microfluidic channel using FCS (Zhao et al., 2018). Stimulated emission-depletion fluorescence correlation spectroscopy (STED-FCS) experiments on urease and acetylcholinesterase showed that it can migrate to lower substrate concentration region (Jee, Dutta, Cho, Tlusty, & Granick, 2018). The enhancement of diffusion in urease in two stages. When substrate concentration is increased upto the Michaelis constant, the rate of diffusion increases almost proportionally and after that it reaches a plateau (Jee, Cho, Granick, & Tlusty, 2018). When the substrate concentration is increased further then there is a second enhancement in diffusion which is much faster compared to the former (Jee, Cho, et al., 2018).

To understand the enzyme-substrate reaction process in the biological environment, it is required to examine the contribution of dissociation of oligomeric structures during the process before assigning the diffusivity to the chemical reaction itself. Urease has been a choice of enzyme for these studies and urea has been used as a substrate and a denaturing agent (Jee, Chen, Tlusty, Zhao, & Granick, 2019). The autocorrelation curves for addition of 1 mM and 100 mM urea is same as in buffer (Fig. 3), however, the actual fluorescence intensity recorded over time diminished with real measurement time in presence of 1 M urea (Fig. 3, second panel) (Jee et al., 2019). The data shows a Gaussian type of distribution upto 10 mM urea and bimodal distribution for more than 100 mM urea (Jee et al., 2019). The substrate urea concentration within the Michaelis constant did not promoted any dissociation, a fact supported by DLS measurements (Jee et al., 2019). Thus, it was concluded that the enhancement of diffusion must have originated from the enzyme-substrate complex formation and chemical reaction.

Although FCS is extremely useful in studying enzymes, although it has certain limitations. Enzymes like aldolase contain oligomeric subunits which tend to dissociate at the concentration at which FCS is performed (Liu et al., 2018). Doubts has been raised over the artifacts in common FCS measurements originating from triplet conversion and other photo-physical processes and contaminate true nature of diffusion information (Bai & Wolynes, 2015; Günther et al., 2018). Moreover, most of the FCS experiments involve tagging the biomolecule with fluorophore. In this regard, it is extremely important to maintain the purity of the solution otherwise quenching of fluorescence can take place.

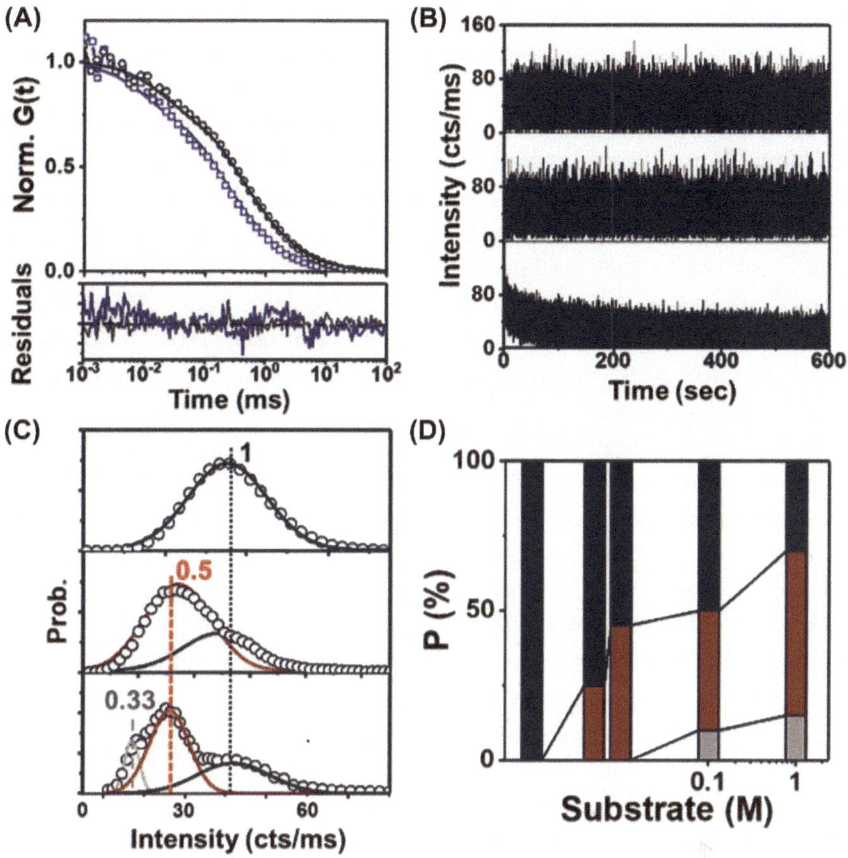


**Fig. 3** FCS experiments with urease. (A) Normalized autocorrelation function *G*(*t*) of Ur1f in 1 mM urea (black) and 100 mM urea (blue) in PBS buffer. The bottom panel shows fitted residuals. (B) Fluorescence intensity time trace of urease with no substrate present; 1 mM urea; and 1 M urea, from top panel to bottom. (C) Fluorescence intensity distribution of urease sample Ur1u at different urea concentration regime. (D) The dissociation from hexameric urease to trimer and dimer according to substrate concentration obtained from area fraction of each distribution. Black, red, and gray shows hexamer, trimer, and dimer, respectively.

## 2.2 Single molecule Förster Resonance Energy Transfer

A more sensitive and informative single molecule technique which became very famous in the last decade is single molecule Förster Resonance Energy Transfer (smFRET). FRET is a well-known method in fluorescence spectroscopy and it is widely applied as a "spectroscopic ruler" (Lakowicz, 2006). In this method, a donor molecule is excited and in the excited state it transfers its energy to an acceptor molecule which in turn goes to an excited electronic state (Lakowicz, 2006). The efficiency of the energy transfer depends on the distance between the donor and acceptor molecules. If the donor and acceptor molecules are separated by a certain distance then the FRET efficiency will provide the distance information at molecular level and typically it is in the order of nanometers (Lakowicz, 2006). There can be more factors like orientation of the donor and acceptor molecules which can also affect the FRET. All the process which induces a change in the biological molecules such as ligand binding, denaturation, ion-channel dynamics, inhibition etc. are all can be tracked at single molecule level (Sasmal, Pulido, Kasal, & Huang, 2016; Turunen, Rowan, & Blank, 2014). Unfortunately, the major problem with the technique is same as of its advantage. We need to label one single biomolecule such as a protein with two fluorescence markers and we need to some prior knowledge about their location (Sasmal et al., 2016; Turunen et al., 2014). Nonetheless, the potential for this method is immense and it was first recognized by Ha et al.(Ha et al., 1999). They studied the conformational dynamics and cleavage mechanism of enzyme staphylococcal nuclease. The enzyme was tagged with tetramethyl rhodamine as donor and Cy5 as acceptor molecule (Ha et al., 1999). They first obtained the fluctuation time and then to study the DNA cleavage reaction by this enzyme, about 200 events were recorded to measure the kinetics of the reaction (Ha et al., 1999).

The earliest papers on smFRET described observing a freely diffusing macromolecule tagged by two fluophores. Some of the early works consists of enzyme cleavage of DNA by tagging with TMR malemide and Cy5; study of enzyme functionality of enzyme staphylococcal nuclease; colocalization processes of different ligands on receptor etc. (Deniz et al., 1999; Ha et al., 1999; Schütz, Trabesinger, & Schmidt, 1998). Human immunodeficiency virus -1 protease is an enzyme that catalyzes HIV maturation (Yoo, Louis, & Chung, 2019). It is thus very important to find out its forms of activity. However, the enzyme has nanomolar dissociation constant which makes it difficult to follow its unfolding dynamics with bulk methods (Yoo et al., 2019). In order to understand its mechanism, smFRET was employed by tagging immunodeficiency virus -1 protease with Alexa Fluor dyes and studies it at picomolar concentration to avoid any dimerization etc. (Yoo et al., 2019). The author found wide ranges of FRET signals which is indicative of slow conformational dynamics and their analyses showed a rugged energy landscape (Yoo et al., 2019). Human immunodeficiency virus -1 protease is a natural protein, however, unlike to the most of protein it has rugged energy landscape. Nevertheless, the rugged energy landscape of virus -1 protease is not unusual as it helps the enzyme to enhance the adaptability of the virus (Yoo et al., 2019). In a similar way, Bavishi et al. have studied cytochrome P450 oxidoreductase enzyme by tagging it with Cy3 and Cy5 malemide dyes (Bavishi et al., 2018). They studied the effect of ionic strength on the unfolding dynamics of the enzyme too (Bavishi et al., 2018). smFRET allowed them to directly observe and quantify the number of conformational states in the enzyme which turned out to be three (Bavishi et al., 2018).

Enzymes may undergo open and close form in free states and out of these many have been found to remain in the open state for longer periods of time (Aviram et al., 2018). However, the timescale of these processes changes dramatically when a substrate binds to the enzyme and the reaction occurs (Aviram et al., 2018). This is known as the turnover condition. Single molecule studies allow to track these conformational changes in real time as reaction progresses. One of such study is the dynamics of protein adenylate kinase from *E. coli* (Aviram et al., 2018). It was found by smFRET studies that upon substrate (ADP and ATP) binding to the enzyme reduces the opening and closing time of the domain (Aviram et al., 2018). Substrate activity can also be site specific in enzymes. For example, in DNA polymerases the enzymatic process substrate binding occurs at different terminals (Lamichhane, Berezhna, Gill, Van der Schans, & Millar, 2013). Lamichhane et al. recorded the smFRET trajectories during the catalysis and were able to monitor the movement of the DNA between the polymerase and exonuclease sites of the enzyme (Lamichhane et al., 2013).

In all these examples, the enzyme has been tagged with two fluorescence markers at different positions. Generally, a single marker does not work but the experiments can be modified in certain way to attach the enzymes to a surface and substrate can be freely diffusing (Lamichhane et al., 2013). In this type of experiments, the enzyme is tagged using a single fluorescence marker and the substrate has one fluorophore (Lamichhane et al., 2013). The scanning is done by moving the stage and the fluorescence of enzyme is observed during the scanning process. The freely diffusing substrate molecule comes into contact of the enzyme and thus energy transfer can take place which leads to emission from the substrate acceptor (Lamichhane et al., 2013). This type of experiments has been performed on Cytochrome P450 Oxidoreductase by Laursen et al. (Laursen et al., 2014). Rothwell et al. studied HIV-1 reverse transcriptase enzyme and its interaction with DNA by attaching an Alexa-C5 malemide dye with a cystine residue of the enzyme and a red colored acceptor dye Cy5-N-hydroxysuccinimide ester with the DNA primer (Rothwell et al., 2003). By observing several FRET parameters, they were able to distinguish three complexes between the enzyme and the DNA out of which they found two are productive in strand transfer (Rothwell et al., 2003).

## 3. Ultrafast spectroscopy

Ultrafast spectroscopy measures events happening extremely quickly in a system and it is achieved by using lasers having pulse width in femtosecond time domain. Transient absorption spectroscopy is one of the main examples of pump probe spectroscopy. In this method, a short femtosecond "pump" pulse is used an excitation light and then a frequency domain broad pulse is used to detect the changes in the system (Fig. 4). Generally, the change in the intensity of the probe pulse is monitored before and after the application of the pump pulse in the sample. The probe is delayed mechanically/optically compared to the pump and by adjusting the delay, we can observe time evolution of a particular spectral component. Depending on the nature of the pump and the probe pulse several types of transitions can be observed in the system. So, this spectroscopy provides two datasets simultaneously, one is the evolution of the spectra with time in wavelength domain and the kinetics of particular wavelength in time domain. The basic schematic has been shown in Fig. 4. As depicted in the figure, there are four different processes viz. ground state bleaching, stimulated emission, excited state absorption and product absorption which can be observed. Ground state bleaching is the signal of the depleted ground state population by the action of the pump and stimulated emission originates from the excited state population created by pump. These are the two negative signals in the spectra. On the other hand, the newly formed excited state can absorb the probe light and give a positive signal known as excited state absorption. Similarly, if a product state is formed by means of excited state reactions then it can also absorb the probe light and give a positive signal known as product absorption. Detailed molecular level descriptions and applications of these processes in exploring the excited state dynamics in several systems can be found in the previous literature (Berera, van Grondelle, & Kennis, 2009).

Application of transient absorption spectroscopy in enzymology is two folds. One type of study focuses on the active site of enzymes only by mimicking it as performed by Caplins, Lomont, Nguyen and Harris

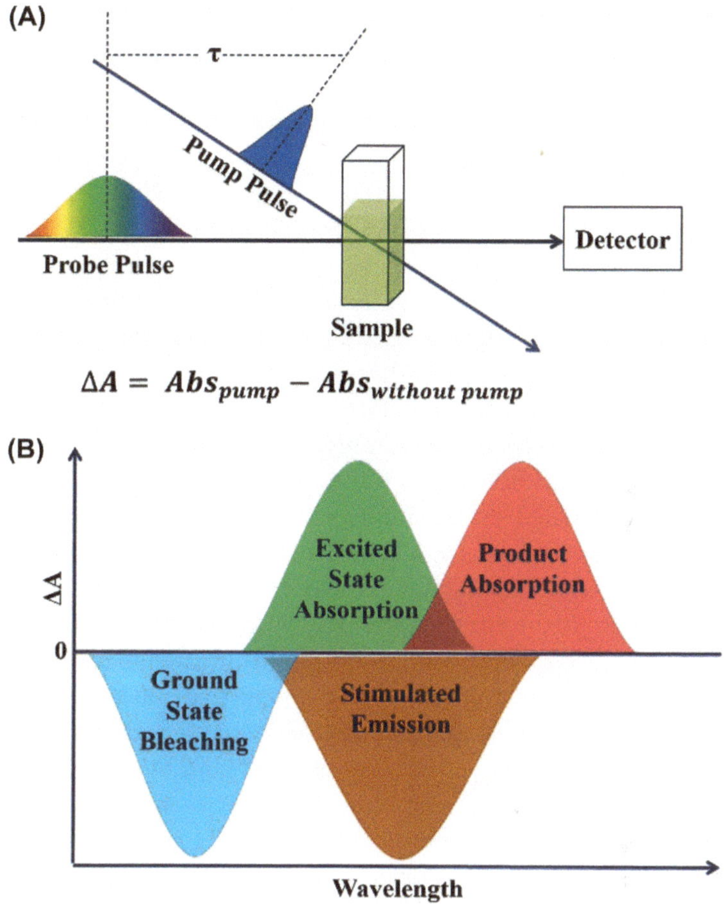


**Fig. 4** Schematic Diagram of (A) the basic pump-probe setup and (B) processes observed in transient absorption spectroscopy.

(2014). The importance of the model active site compounds is that they are becoming potential agents for photocatalytic hydrogen production. In this case, they have studied a mimic compound of active site of the enzyme [FeFe]-hydrogenase using 400 nm pump light and IR probe light (Caplins et al., 2014). They observed that vibrational cooling in the compound happens in picosecond time scale and the corresponding carbonyl stretching frequency is in agreement with solvent dependent longer lifetime component (Caplins et al., 2014). The second type of experiments is to apply pump-probe spectroscopy directly to enzyme. For example, the role of light driven enzyme protochlorophyllide oxidoreductase has been studied by Heyes, Hardman, Mansell, Gardiner and Scrutton (2012). This particular enzyme catalyze the reduction of protochlorophyllide, the chlorophyll precursor, to chlorophyllide (Heyes et al., 2012). They observed this process using femtosecond pump probe having 475 nm pump light and visible range probe light (Heyes et al., 2012). They revealed that previous reports on the short-lived excited state species is essentially a dimer of the substrate and product and it is not vital for the activity of the enzyme (Heyes et al., 2012). Transient absorption spectroscopy with 600 nm pump and broad visible probe has been used to study the enzyme laccase from Pleurotusostreatus (Delfino, Viola, Cerullo, & Lepore, 2015). Laccase is part of blue multicopper oxidase type enzymes and they help in reduction of oxygen to water (Delfino et al., 2015). Using a 15 fs pulse, the authors were able to observe the vibrational coherence and the initial charge transfer process which lead to the conclusion that light induced charge transfer process originates at the T1 site copper center and the reduction happens thereafter (Delfino et al., 2015). One of the recent and detailed examination of kinetics of enzyme using pump-probe spectroscopy is done by Sanchez et al. (2019). They have used time resolved transient visible and IR spectroscopy along with laser induced potential jump to study the enzyme [FeFe] hydrogenase from Chlamydomonas reinhardtii (Sanchez et al., 2019) using a photo-initiated mediator 2,2′-propylbipyridinium. All the intermediates of the oxidation-reduction cycle have been tracked using the ground state bleach and the excited state absorption signals (Sanchez et al., 2019). For example, the biexponential ground state recovery at 1941 $cm^{-1}$ signifies the first electron reduction of the enzyme, the excited state absorption signal at 1934 and 1891 $cm^{-1}$ corresponds to reduced state and the proton clustered state of the enzyme respectively (Sanchez et al., 2019). Thus the kinetics of each of the kinetic components were tracked and the time constants are within the turnover time of the enzyme. This led to a detailed kinetic scheme of the enzyme reaction pathway which the author believed is a pH dependent one (Sanchez et al., 2019). Hydrogenase enzymes along with many others are particularly detectable in IR frequency range since the active sites of enzymes contain carbonyl and cyanide moiety (Horch et al., 2019). As an example, Fig. 5 depicts several vibrational transitions of the CO and $CN^-$ group residing in the active site of [NiFe] hydrogenase from Ralstoniaeutropha (ReRH). It can be seen, several optical transitions can be tracked for this enzyme via time resolved IR spectroscopy. In Fig. 6, the actual spectra obtained of the hydrogenase from ReRH has been shown (Horch et al., 2019). Clearly, the negative features in the CO and $CN^-$ spectra are due to $v = 1$ to $v = 1$ transition and the stimulated emission part. The excited state absorption

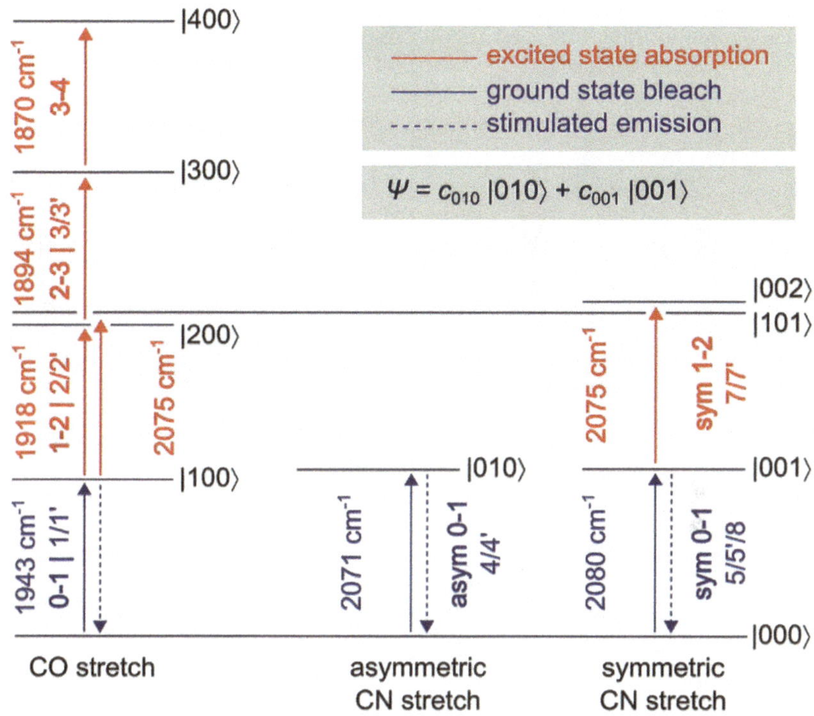


**Fig. 5** Vibrational frequencies and eigen states for CO/CN− in [NiFe] hydrogenase from Ralstoniaeutropha (ReRH). *Reproduced with permission from Horch, M., Schoknecht, J., Wrathall, S. L. D., Greetham, G. M., Lenz, O., & Hunt, N. T. (2019).* Chemical Science, 10*(39), 8981–8989.*

signals are due to transitions in higher vibrational states. The kinetics at different wavelengths will provide the lifetime of the states. During the reduction, the active site binds with $H^+$ which changes the vibrational frequency and/or the lifetime. Thus, these spectra elucidate the mechanism of the reaction (Horch et al., 2019). The interesting observation is the coherent oscillation in the CN− signal which originates due to superposition (anharmonic coupling) of excited states of two CN− vibrations (Horch et al., 2019). The authors have concluded the excess vibrational energy in CO bond is dissipated in the protein matrix and the damping oscillations in CN− designates the two redox sites in the enzyme (Horch et al., 2019).

2D IR spectroscopy is an optical analog of two-dimensional NMR spectroscopy (Le Sueur, Horness, & Thielges, 2015; Noda, 1989). In 2D IR measurements, the time dependent response of the normal modes of the systems to an external perturbation is measured. (Le Sueur et al., 2015; Noda,

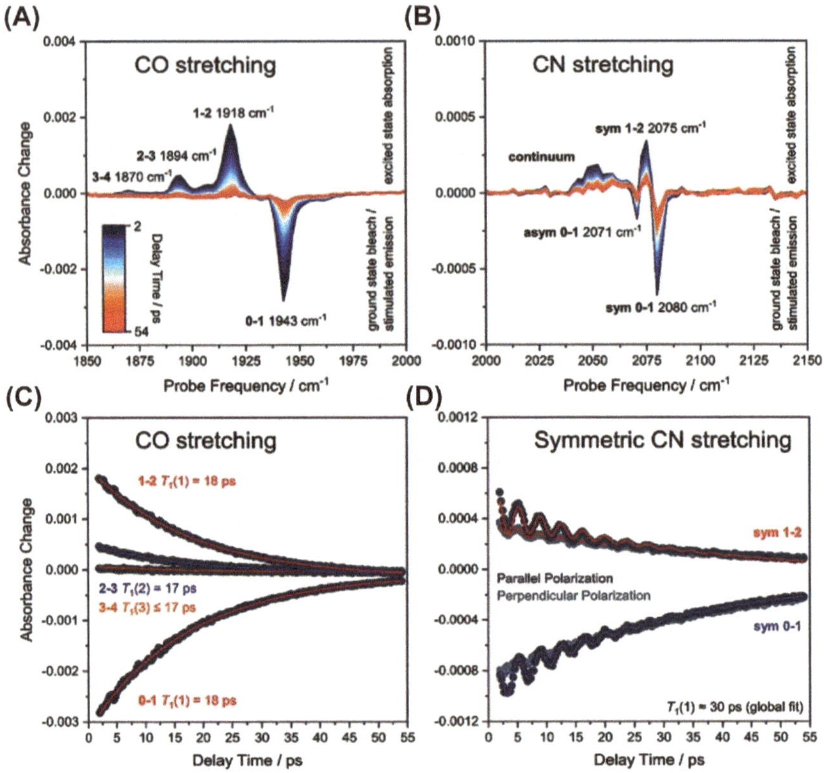


**Fig. 6** Broadband pump–probe spectra (A and B) and time evolution of selected signals (C and D). Data referring to CO and CN stretch modes are plotted separately on the left (A and C) and right (B and D), respectively. *Reproduced with permission from Horch, M., Schoknecht, J., Wrathall, S. L. D., Greetham, G. M., Lenz, O., & Hunt, N. T. (2019).* Chemical Science, 10*(39), 8981–8989.*

1989). 2D IR spectroscopy employs a four pulse sequence, they detect & follow the vibrations of the systems and employ external perturbation (Le Sueur et al., 2015; Noda, 1989). The response of the system is recorded in the form of heterodyne detection by combining with a fourth pulse or echo and the two-dimensional data is plotted as final recorded frequencies versus initial input frequencies (Le Sueur et al., 2015; Noda, 1989). This technique essentially finds the correlation between the different vibrations and frequencies that exists in the system (Le Sueur et al., 2015; Noda, 1989). The peaks observed in 2D-IR spectra are correlation intensities. There are two types of peaks and two types of times involved in 2D IR experiments. Diagonal peaks or auto peaks are the ones which appear around the diagonal axis of the two-dimensional data and generally signify number of vibrations and their temporal evolution. Cross peaks appear when two auto peaks are coupled with each other (Le Sueur et al., 2015; Noda, 1989). This coupling can signify chemical coupling, energy dissipation or in-phase vibration of two normal modes (Le Sueur et al., 2015; Noda, 1989). So, the appearance of cross-peaks gives us a way to understand the inter and intra molecular interactions. The intensity of this band depends on the strength of interaction and therefore are really important for understanding the 2D IR data which is otherwise not straightforward. The time difference between the first and the second pulse is called coherence time and same between the second and the third pulse is known as the waiting time (Le Sueur et al., 2015; Noda, 1989). 2D IR spectra are accumulated as a snapshot at different waiting times and scanning along coherence time (Le Sueur et al., 2015; Noda, 1989). Also, the time delay between the third pulse and the echo pulse is termed as detection time.

One of the most recent examples of employment of 2D IR spectroscopy for understanding enzyme activity has been reported by Osawa et al. (Osawa, Kossowska, Park, Kwak, & Cho, 2019). The enzyme studied in this report is cytochrome *c* (cyt *c*) which is very commonly known for vital role in the synthesis of adenosine tri-phosphate in cell (Osawa et al., 2019). The structure of cyt *c* is pH dependent and the alkaline form shows peroxide activity (Osawa et al., 2019). However, the binding of carbon monoxide in this catalytic cycle to cyt *c* brings about conformational change which is not well understood. Moreover, this enzyme is attracting attention as a potential bio-catalyst and it has been found that deep eutectic solvents can increase its activity, the reason for which is not so clear (Osawa et al., 2019). The authors recognized three distinguished peaks corresponding to three forms i.e. C1, C2 and C3 of the enzyme through steady state IR spectra (Osawa et al., 2019). Deep eutectic solvents affect the kinetics of the enzymes by changing the local structure of the enzyme (Osawa et al., 2019). To explore this scenario, the authors has performed 2D IR spectroscopy and here we have reproduced one of their data sets in Fig. 7. This data is for the carbon monoxide bound cyt *c* (carboxycytochrome c) in urea/buffer and deep eutectic solvent/buffer system (Osawa et al., 2019). The positive red peaks are due to ground state bleaching and stimulated emission for the vibrational transition of $0 \rightarrow 1$ level. The negative blue peak represents the simultaneous excited state absorption process for vibrational transition of $1 \rightarrow 2$ level (Osawa et al., 2019). The elongation of peaks is due to inhomogeneous broadening (Osawa et al., 2019). The authors have observed that the peak corresponding to C3 state is more prominent compared to C2 state (Osawa et al., 2019).

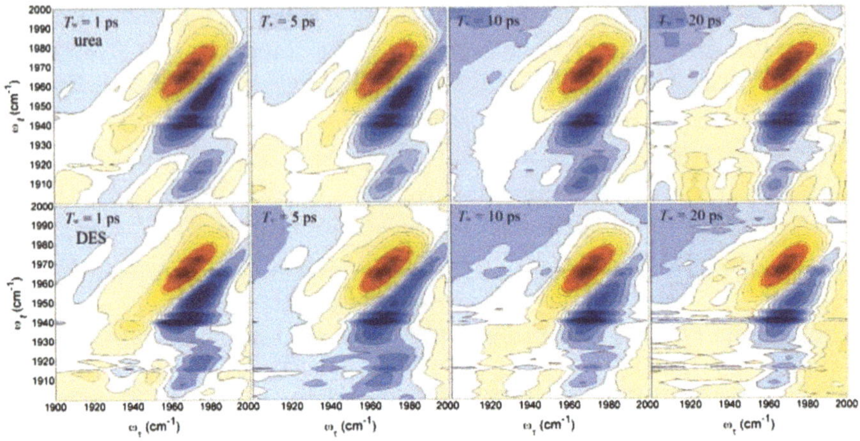


**Fig. 7** 2D IR spectra of cytochrome *c* peroxidase in urea (top panel) and DES/buffer (bottom panel) solutions. *Reproduced with permission from Osawa, K., Kossowska, D., Park, K., Kwak, K., & Cho, M. (2019). Two-dimensional infrared spectroscopic study of cytochrome c peroxidase activity in deep eutectic solvent.* Structural Dynamics, *6(6), 064703. https://doi.org/10.1063/1.5130940.*

Ruling out the viscosity effect through more analysis, it was concluded that the C3 state of the enzyme is the cause of enhanced peroxidase activity (Osawa et al., 2019).

It has been said before that the model active site compound of the enzymes is studied to explore its properties. A similar work has been done for the active site of hydrogenase enzymes using 2D IR spectroscopy. The actual compounds studied are μ-pdt-[Fe(CO)3]2 and μ-edt-[Fe(CO)3]2 where pdt and edt are propanedithiolate, and ethanedithiolate respectively (Eckert & Kubarych, 2017). The authors observed the dynamics fluctuations of the ligands in these two compounds which is responsible for the activity of the enzyme (Eckert & Kubarych, 2017). Another crucial problem where 2D IR has been applied is binding of antibiotic and enzyme in the perspective of resistance to antibiotic medicine (Shaw et al., 2017). Shaw et al. studied the use of pro-drug isoniazid resistance in case of tuberculosis enzymeInhA (Shaw et al., 2017). Interestingly, they found that there is no overall structural change of enzyme upon ligand binding but the vibrational peak at 1620 cm$^{-1}$ of the amide band increases in intensity (Shaw et al., 2017). They move on to track the vibrational modes of amide I of the enzyme-ligand complex using 2D IR spectroscopy and found that there are multiple components associated with the band corresponding to the secondary structures of the protein as shown in Fig. 8 (Shaw et al., 2017). The off-diagonal

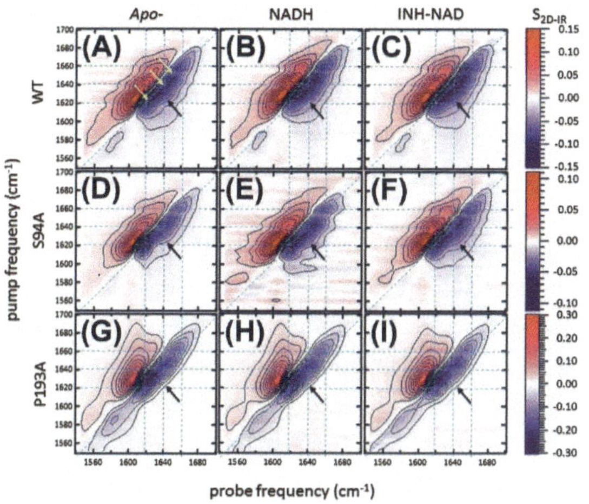


**Fig. 8** 2D-IR spectra of tuberculosis enzyme (A–C) wild-type InhA (D–F) S94A–InhA and (G–I) P193A–InhA. NADH means cofactor NADH present and INH-NAD means inhibitor present. *Reproduced with permission from Shaw, D. J., Hill, R. E., Simpson, N., Husseini, F. S., Robb, K., Greetham, G. M., et al. (2017). Examining the role of protein structural dynamics in drug resistance in Mycobacterium tuberculosis.* Chemical Science, *8(12), 8384–8399. https://doi.org/10.1039/C7SC03336B.*

peak corresponding to the 1620 $cm^{-1}$ frequency showed broadening when co-factor binds with enzyme and further broadening due to inhibition (Shaw et al., 2017). Further analysis and molecular modeling helped the authors to conclude that the inhibitor locks the structure of the enzyme in a particular fashion which decreases in the flexibility of enzyme. (Shaw et al., 2017). 2D IR spectroscopy has also been applied in the backbone structural studies, ion-channel transport and ligand binding of enzyme (Petti, Lomont, Maj, & Zanni, 2018).

## 4. Raman and Resonance Raman spectroscopy

Raman spectroscopy is a very common technique in many fields of science like crystal engineering, biological and medical fields, nanomaterials and many more (Hellerer et al., 2007; Malard, Pimenta, Dresselhaus, & Dresselhaus, 2009; Venezuela, Lazzeri, & Mauri, 2011). The advantage of Raman spectroscopy over fluorescence-based techniques is that it does not require a tagged internal probe molecule. The fundamental principle of Raman spectroscopy is based on the idea that when light-matter interaction takes place, the charge distributions of the matter can exchange energy and momentum with the photon and thus change the frequency. It is not required that the interaction should lead to absorption or emission. Basically, any light can be scattered by molecule but mostly the collisions are elastic which means the frequency of the fundamental light does not change. This is known as Rayleigh scattering (Jones, Hooper, Zhang, Wolverson, & Valev, 2019). However, a very small fraction of the collisions are inelastic in nature and leads to frequency shifted Raman scattering (Jones et al., 2019). The inelastically scattered photon can have energy greater or lower than the elastically scattered photons and the Raman signals originating because of that is termed as ani-Stokes' and Stokes' signal (Jones et al., 2019). The interaction of the incoming photon with the molecule is imagined as promoting the molecule to a virtual state and then it comes back to a higher or lower energy rotational or vibrational energy level. In case of spontaneous Raman scattering, this intermediate state does not correspond to any real energy state of the system. However, in case of Resonance Raman spectroscopy, it is actually a real electronic energy state of the system (Jones et al., 2019). Although, it seems that any kind of scattering will lead to a Raman spectra but there are symmetry considerations in place and the thermal distribution can control the intensity of the transitions (Jones et al., 2019). Spontaneous Raman is an incoherent process and the signals are very weak. So monochromatic, high intensity lasers are used to record the Raman spectra of the molecules (Jones et al., 2019). Raman spectra can give us information and identification about the vibrational and rotational transitions present within a molecule and thus lead to study the molecules in a material or other environment. The intensity of the signal in Resonance Raman spectroscopy is almost $10^6$ times stronger than that of spontaneous Raman spectroscopy which means much lower concentration of sample can be studied using the former technique. This is particularly advantageous while studying biomolecules.

These techniques have been applied for the understanding of enzymology from a long time. The early works in the field consists of identifying the vibrational and electronic bands associated with the enzymes. For example, resonance Raman spectroscopy of Achromobactercy cloclastes nitrite reductase by exciting the enzyme at the two main absorption bands at 458 nm and 585 nm revealed that the transitions are due to the cystine sulfur to Cu (II) ligand to metal charge transfer in the active site of the enzyme (Dooley, Moog, Liu, Payne, & LeGall, 1988). Another example is the characterization of cytochrome *c* oxidase where visible and near IR transitions were assigned to the Cu-cystine and Cu-histidine ligand to metal charge transfer by observing the stretching modes of Cu—S and Cu—N (amide) vibrations (Takahashi, Ogura, Kitagawa, Itoh-Shinzawa, & Yoshikawa, 1991). Resonance Raman has also been used to examine the oxidation-reduction property and structural change in certain enzymes such as biotin sulfoxide reductase (Garton et al., 2000). The diversity of molybdenum containing enzymes has been studied with Raman spectroscopy. The active site of human sulfite oxidase enzyme contains Mo site coordinated by two oxo groups out of which only one is exchanged with water during the catalytic mechanism (Garton, Garrett, Rajagopalan, & Johnson, 1997). This particular information was obtained through identification of vibrational shift due to isotopic exchanges in the active site by resonance Raman spectroscopy at the cystine to Mo charge transfer band excitation (Garton et al., 1997). Westley et al. followed the conversion of nitrile to amide using nitrile hydratase (Westley et al., 2017). The substrate benzonitriles UV pumped resonance Raman spectra were monitored while enzyme was introduced to initialize the reaction. As the reaction progresses, the shift in the bands of the spectra were monitored to know the kinetics of the reaction (Westley et al., 2017). The resonance Raman spectra of the free enzyme via excitation at Soret band of the heme showed four major vibrational peaks (Martins et al., 2010). Upon the spin transition of the six coordinated low spin Fe-heme to five coordinated high spin complex, one of the Raman peak shows a downfield shift (Martins et al., 2010). It was proposed that the substrate binds to the free axial position of the high spin Fe(II)-heme active site and on the reaction of enzyme, the nitrite frequency shift was found to be 40% (Martins et al., 2010). This clearly proved that spin transition takes place when the enzyme forms complex with substrate (Martins et al., 2010).

The Lytic polysaccharide monooxygenases (LPMOs) is a single copper ion containing enzyme responsible for oxidative cleavage of polysaccharides by dioxygen or hydrogen peroxide (Paradisi et al., 2019). The complicacy of studying the reactivity of the enzyme is that it exists in 6 active forms which are slightly different from each other which lead to the belief that the secondary coordination sphere of the Cu ion can be an important factor for the reactivity of the ligands bound to the active site (Paradisi et al., 2019). X-ray absorption studies revealed the oxidation state of Cu ion to be II and UV visible revealed six different absorption peak of the complex (Paradisi et al., 2019). Next resonance Raman spectroscopy (Fig. 9) performed under isotopic substitution of water and hydrogen peroxide to

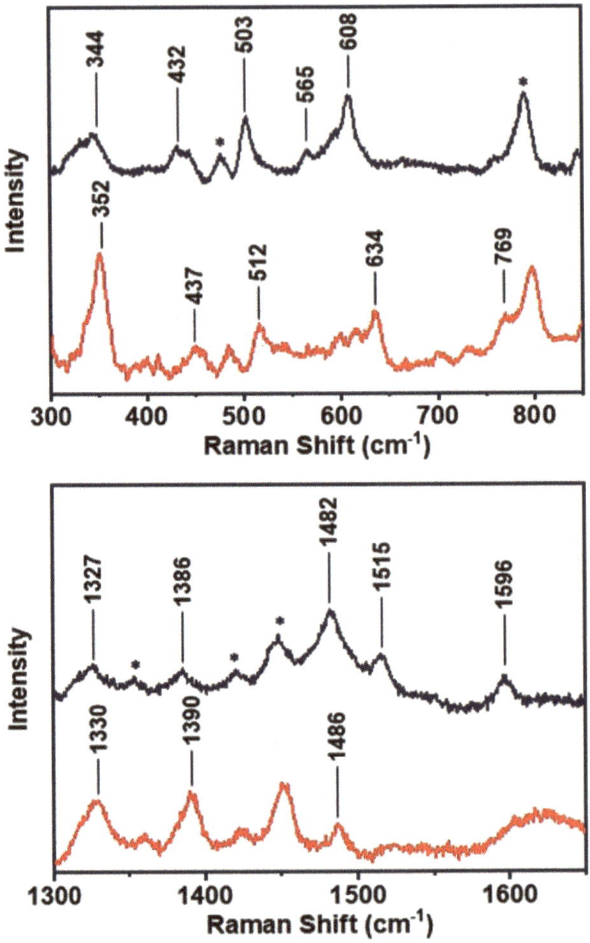


**Fig. 9** Resonance Raman spectra of the LsAA9 purple species obtained with 532 (black) and 785 nm (red) excitation, at 293 K, 50 mM CAPS pH 10.0. Asterisk (*) denotes vibrations due to CAPS buffer (Paradisi et al., 2019).

find that the peaks were not sensitive suggesting that there is no Cu-peroxo or Cu-superoxo type units in the complex (Paradisi et al., 2019). The 785 nm excitation gives rise to a prominent 352 $cm^{-1}$ peak which is very weak in 532 nm excitation (Paradisi et al., 2019). Comparison of the spectra with the same for oxidized form of galactose oxidase tells that the 1596, 1482, and 1386 $cm^{-1}$ peaks are due to Cu(II)–tyrosyl radical formation (Paradisi et al., 2019).

## 5. Conclusion

There are several aspects of enzyme ranging from structure determination, functions, biological activity, and inhibition to industrial applications. Each of the topics generates a field of study. Experimental and theoretical chemistry has been employed from long time to understand the properties of enzymes. Some parts of enzymology are well described whereas some parts are still debatable. Herein we have described the employment of spectroscopic techniques which reveal the temporal evolution of enzymes in free form and in binding with substrate. Obviously, this review does not cover the full spectrum of spectroscopy techniques such as impedance spectroscopy which has been used to understand the oxidation reduction and charge transfer associated with the enzymes. One more important point is only experiment or computation alone cannot properly capture the wide variety of enzymes and their activity. It is better to combine the two methods which is currently undertaken by many researchers and promises to shed new light in understanding the dynamics and structural change of enzymes.

## References

Agarwal, P. K. (2006). Enzymes: An integrated view of structure, dynamics and function. *Microbial Cell Factories, 5*(2). http://dx.doi.org/10.1186/1475-2859-5-2.

Aviram, H. Y., Pirchi, M., Mazal, H., Barak, Y., Riven, I., & Haran, G. (2018). Direct observation of ultrafast large-scale dynamics of an enzyme under turnover conditions. *Proceedings of the National Academy of Sciences of the United States of America, 115*(13), 3243–3248. http://dx.doi.org/10.1073/pnas.1720448115.

Bai, X., & Wolynes, P. G. (2015). On the hydrodynamics of swimming enzymes. *The Journal of Chemical Physics, 143*(16), 165101. http://dx.doi.org/10.1063/1.4933424.

Bavishi, K., Li, D., Eiersholt, S., Hooley, E. N., Petersen, T. C., Møller, B. L., et al. (2018). Direct observation of multiple conformational states in Cytochrome P450 oxidoreductase and their modulation by membrane environment and ionic strength. *Scientific Reports, 8*(1), 6817. http://dx.doi.org/10.1038/s41598-018-24922-x.

Benkovic, S. J., & Hammes-Schiffer, S. (2003). A perspective on enzyme catalysis. *Science, 301*(5637), 1196–1202. http://dx.doi.org/10.1126/science.1085515.

Berera, R., van Grondelle, R., & Kennis, J. T. M. (2009). Ultrafast transient absorption spectroscopy: Principles and application to photosynthetic systems. *Photosynthesis Research, 101*(2-3), 105–118. http://dx.doi.org/10.1007/s11120-009-9454-y.

Börsch, M., Turina, P., Eggeling, C., Fries, J. R., Seidel, C. A., Labahn, A., et al. (1998). Conformational changes of the H + -ATPase from *Escherichia coli* upon nucleotide binding detected by single molecule fluorescence. *FEBS Letters, 437*(3), 251–254. http://dx.doi.org/10.1016/S0014-5793(98)01247-2.

Cannon, W. R., & Benkovic, S. J. (1998). Solvation, reorganization energy, and biological catalysis. *Journal of Biological Chemistry, 273*(41), 26257–26260. http://dx.doi.org/10.1074/jbc.273.41.26257.

Caplins, B. W., Lomont, J. P., Nguyen, S. C., & Harris, C. B. (2014). Vibrational cooling dynamics of a [FeFe]-hydrogenase mimic probed by time-resolved infrared spectroscopy. *Journal of Physical Chemistry, 118*(49), 11529–11540. http://dx.doi.org/10.1021/jp510517z.

Chapman, J., Ismail, A. E., & Dinu, C. Z. (2018). Industrial applications of enzymes: Recent advances, techniques, and outlooks. *Catalysts, 8*(6), 238. http://dx.doi.org/10.3390/catal8060238.

Classification and Nomenclature of Enzymes. (1992). *Enzyme nomenclature* (pp. 5–22). http://dx.doi.org/10.1016/b978-0-12-227164-9.50005-4.

Delfino, I., Viola, D., Cerullo, G., & Lepore, M. (2015). Ultrafast excited-state charge-transfer dynamics in laccase type I copper site. *Biophysical Chemistry, 200–201*, 41–47. http://dx.doi.org/10.1016/j.bpc.2015.03.003.

Deniz, A. A., Dahan, M., Grunwell, J. R., Ha, T., Faulhaber, A. E., Chemla, D. S., et al. (1999). Single-pair fluorescence resonance energy transfer on freely diffusing molecules: Observation of Förster distance dependence and subpopulations. *Proceedings of the National Academy of Sciences of the United States of America, 96*(7), 3670–3675. http://dx.doi.org/10.1073/pnas.96.7.3670.

Dooley, D. M., Moog, R. S., Liu, M. Y., Payne, W. J., & LeGall, J. (1988). Resonance Raman spectra of the copper-sulfur chromophores in Achromobactercycloclastes nitrite reductase. *Journal of Biological Chemistry, 263*(29), 14625–14628.

Dutkiewicz, Z., & Mikstacka, R. (2018). Structure-based drug design for cytochrome P450 family 1 inhibitors. *Bioinorganic Chemistry and Applications, 2018*, 1–21. http://dx.doi.org/10.1155/2018/3924608.

Eckert, P. A., & Kubarych, K. J. (2017). Dynamic flexibility of Hydrogenase active site models studied with 2D-IR spectroscopy. *Journal of Physical Chemistry, 121*(3), 608–615. http://dx.doi.org/10.1021/acs.jpca.6b11962.

Ehrenberg, M., & Rigler, R. (1974). Rotational brownian motion and fluorescence intensify fluctuations. *Chemical Physics, 4*(3), 390–401. http://dx.doi.org/10.1016/0301-0104(74)85005-6.

Eisenmesser, E. Z., Millet, O., Labeikovsky, W., Korzhnev, D. M., Wolf-Watz, M., Bosco, D. A., et al. (2005). Intrinsic dynamics of an enzyme underlies catalysis. *Nature, 438*(7064), 117–121. http://dx.doi.org/10.1038/nature04105.

Elson, E. L. (2011). Fluorescence correlation spectroscopy: Past, present, future. *Biophysical Journal, 101*(12), 2855–2870. http://dx.doi.org/10.1016/j.bpj.2011.11.012.

Elson, E. L., & Magde, D. (1974). Fluorescence correlation spectroscopy. I. Conceptual basis and theory. *Biopolymers, 13*(1), 1–27. http://dx.doi.org/10.1002/bip.1974.360130102.

Gärtner, W. (2009). Enzyme catalysis "reilluminated.". *Angewandte Chemie International Edition, 48*(25), 4484–4485. http://dx.doi.org/10.1002/anie.200900840.

Garton, S. D., Garrett, R. M., Rajagopalan, K. V., & Johnson, M. K. (1997). Resonance Raman characterization of the molybdenum center in sulfite oxidase: Identification of Mo=O stretching modes. *Journal of the American Chemical Society, 119*(10), 2590–2591. http://dx.doi.org/10.1021/ja963931c.

Garton, S. D., Temple, C. A., Dhawan, I. K., Barber, M. J., Rajagopalan, K. V., & Johnson, M. K. (2000). Resonance Raman characterization of biotin sulfoxide reductase. Comparing oxomolybdenum enzymes in the Me2SO reductase family. *Journal of Biological Chemistry, 275*(10), 6798–6805. http://dx.doi.org/10.1074/jbc.275.10.6798.

van Gelderen, H., Mayer, J. M., & Testa, B. (1994). Thermodynamic studies of enzyme catalysis. What can we learn about specificity and mechanism? *Biochemical Pharmacology, 47*(5), 753–756. http://dx.doi.org/10.1016/0006-2952(94)90473-1.

Gordeev, E. G., & Ananikov, V. P. (2015). Computational study of a model system of enzyme-mediated [4+2] cycloaddition reaction. *PLoS One, 10*(4), e0119984. http://dx.doi.org/10.1371/journal.pone.0119984.

Greaves, S. J., Wrede, E., Goldberg, N. T., Zhang, J., Miller, D. J., & Zare, R. N. (2008). Vibrational excitation through tug-of-war inelastic collisions. *Nature, 454*(7200), 88–91. http://dx.doi.org/10.1038/nature07079.

Günther, J.-P., Börsch, M., & Fischer, P. (2018). Diffusion measurements of swimming enzymes with fluorescence correlation spectroscopy. *Accounts of Chemical Research, 51*(9), 1911–1920. http://dx.doi.org/10.1021/acs.accounts.8b00276.

Ha, T., Ting, A. Y., Liang, J., Caldwell, W. B., Deniz, A. A., Chemla, D. S., et al. (1999). Single-molecule fluorescence spectroscopy of enzyme conformational dynamics and cleavage mechanism. *Proceedings of the National Academy of Sciences, 96*(3), 893–898. http://dx.doi.org/10.1073/pnas.96.3.893.

Hellerer, T., Axäng, C., Brackmann, C., Hillertz, P., Pilon, M., & Enejder, A. (2007). Monitoring of lipid storage in *Caenorhabditis elegans* using coherent anti-Stokes Raman scattering (CARS) microscopy. *Proceedings of the National Academy of Sciences of the United States of America, 104*(37), 14658–14663. http://dx.doi.org/10.1073/pnas.0703594104.

Heyes, D. J., Hardman, S. J. O., Mansell, D., Gardiner, J. M., & Scrutton, N. S. (2012). Mechanistic reappraisal of early stage photochemistry in the light-driven enzyme protochlorophyllide oxidoreductase. *PLoS One, 7*(9), e45642. http://dx.doi.org/10.1371/journal.pone.0045642.

Horch, M., Schoknecht, J., Wrathall, S. L. D., Greetham, G. M., Lenz, O., & Hunt, N. T. (2019). Understanding the structure and dynamics of hydrogenases by ultrafast and two-dimensional infrared spectroscopy. *Chemical Science, 10*(39), 8981–8989. http://dx.doi.org/10.1039/C9SC02851J.

Jackson, C. J., Foo, J.-L., Tokuriki, N., Afriat, L., Carr, P. D., Kim, H.-K., et al. (2009). Conformational sampling, catalysis, and evolution of the bacterial phosphotriesterase. *Proceedings of the National Academy of Sciences, 106*(51), 21631–21636. http://dx.doi.org/10.1073/pnas.0907548106.

Jee, A. Y., Chen, K., Tlusty, T., Zhao, J., & Granick, S. (2019). Enhanced diffusion and oligomeric enzyme dissociation. *Journal of the American Chemical Society, 141*(51), 20062–20068. http://dx.doi.org/10.1021/jacs.9b06949.

Jee, A. Y., Cho, Y. K., Granick, S., & Tlusty, T. (2018). Catalytic enzymes are active matter. *Proceedings of the National Academy of Sciences of the United States of America, 115*(46), E10812–E10821. http://dx.doi.org/10.1073/pnas.1814180115.

Jee, A. Y., Dutta, S., Cho, Y. K., Tlusty, T., & Granick, S. (2018). Enzyme leaps fuel antichemotaxis. *Proceedings of the National Academy of Sciences of the United States of America, 115*(1), 14–18. http://dx.doi.org/10.1073/pnas.1717844115.

Jones, R. R., Hooper, D. C., Zhang, L., Wolverson, D., & Valev, V. K. (2019). Raman techniques: Fundamentals and frontiers. *Nanoscale Research Letters, 14*(1), 231. http://dx.doi.org/10.1186/s11671-019-3039-2.

Kollman, P. A., Kuhn, B., Donini, O., Perakyla, M., Stanton, R., & Bakowies, D. (2001). Elucidating the nature of enzyme catalysis utilizing a new twist on an old methodology: Quantum Mechanical–Free energy calculations on chemical reactions in enzymes and in aqueous solution. *Accounts of Chemical Research, 34*(1), 72–79. http://dx.doi.org/10.1021/ar000032r.

Lakowicz, J. R. (2006). Principles of fluorescence spectroscopy. In J. R. Lakowicz (Ed.), *Principles of fluorescence spectroscopy*. http://dx.doi.org/10.1007/978-0-387-46312-4.

Lamichhane, R., Berezhna, S. Y., Gill, J. P., Van der Schans, E., & Millar, D. P. (2013). Dynamics of site switching in DNA polymerase. *Journal of the American Chemical Society, 135*(12), 4735–4742. http://dx.doi.org/10.1021/ja311641b.

Laursen, T., Singha, A., Rantzau, N., Tutkus, M., Borch, J., Hedegård, P., et al. (2014). Single molecule activity measurements of cytochrome P450 oxidoreductase reveal the existence of two discrete functional states. *ACS Chemical Biology, 9*(3), 630–634. http://dx.doi.org/10.1021/cb400708v.

Le Sueur, A. L., Horness, R. E., & Thielges, M. C. (2015). Applications of two-dimensional infrared spectroscopy. *The Analyst, 140*(13), 4336–4349. http://dx.doi.org/10.1039/c5an00558b.

Lipscomb, W. N. (1983). Structure and catalysis of enzymes. *Annual Review of Biochemistry, 52*(1), 17–34. http://dx.doi.org/10.1146/annurev.bi.52.070183.000313.
Liu, Y., Zhang, Y., Ortega, S., Ibáñez, M., Lim, K. H., Grau-Carbonell, A., et al. (2018). Crystallographically textured nanomaterials produced from the liquid phase sintering of Bi x Sb 2– x Te 3 nanocrystal building blocks. *Nano Letters, 18*(4), 2557–2563. http://dx.doi.org/10.1021/acs.nanolett.8b00263.
Magde, D., Elson, E., & Webb, W. W. (1972). Thermodynamic fluctuations in a reacting system measurement by fluorescence correlation spectroscopy. *Physical Review Letters, 29*(11), 705–708. http://dx.doi.org/10.1103/PhysRevLett.29.705.
Magde, D., Elson, E. L., & Webb, W. W. (1974). Fluorescence correlation spectroscopy. II. An experimental realization. *Biopolymers, 13*(1), 29–61. http://dx.doi.org/10.1002/bip.1974.360130103.
Magde, D., Webb, W. W., & Elson, E. L. (1978). Fluorescence correlation spectroscopy. III. Uniform translation and laminar flow. *Biopolymers, 17*(2), 361–376. http://dx.doi.org/10.1002/bip.1978.360170208.
Malard, L. M., Pimenta, M. A., Dresselhaus, G., & Dresselhaus, M. S. (2009). Raman spectroscopy in graphene. *Physics Reports, 473*(5–6), 51–87. http://dx.doi.org/10.1016/j.physrep.2009.02.003.
Maria-Solano, M. A., Serrano-Hervás, E., Romero-Rivera, A., Iglesias-Fernández, J., & Osuna, S. (2018). Role of conformational dynamics in the evolution of novel enzyme function. *Chemical Communications, 54*(50), 6622–6634. http://dx.doi.org/10.1039/C8CC02426J.
Martins, G., Rodrigues, L., Cunha, F. M., Matos, D., Hildebrandt, P., Murgida, D. H., et al. (2010). Substrate binding to a nitrite reductase induces a spin transition. *Journal of Physical Chemistry B, 114*(16), 5563–5566. http://dx.doi.org/10.1021/jp9118502.
Muddana, H. S., Sengupta, S., Mallouk, T. E., Sen, A., & Butler, P. J. (2010). Substrate catalysis enhances single-enzyme diffusion. *Journal of the American Chemical Society, 132*(7), 2110–2111. http://dx.doi.org/10.1021/ja908773a.
Mulholland, A. J., Roitberg, A. E., & Tuñón, I. (2012). Enzyme dynamics and catalysis in the mechanism of DNA polymerase. *Theoretical Chemistry Accounts, 131*(12), 1286. http://dx.doi.org/10.1007/s00214-012-1286-8.
Noda, I. (1989). Two-dimensional infrared spectroscopy. *Journal of the American Chemical Society, 111*(21), 8116–8118. http://dx.doi.org/10.1021/ja00203a008.
Olsson, M. H. M., Parson, W. W., & Warshel, A. (2006). Dynamical contributions to enzyme catalysis: Critical tests of a popular hypothesis. *Chemical Reviews, 106*(5), 1737–1756. http://dx.doi.org/10.1021/cr040427e.
Osawa, K., Kossowska, D., Park, K., Kwak, K., & Cho, M. (2019). Two-dimensional infrared spectroscopic study of cytochrome c peroxidase activity in deep eutectic solvent. *Structural Dynamics, 6*(6), 064703. http://dx.doi.org/10.1063/1.5130940.
Osuna, S., Jiménez-Osés, G., Noey, E. L., & Houk, K. N. (2015). Molecular dynamics explorations of active site structure in designed and evolved enzymes. *Accounts of Chemical Research, 48*(4), 1080–1089. http://dx.doi.org/10.1021/ar500452q.
Paradisi, A., Johnston, E. M., Tovborg, M., Nicoll, C. R., Ciano, L., Dowle, A., et al. (2019). Formation of a copper(II)-Tyrosyl complex at the active site of lytic polysaccharide monooxygenases following oxidation by H2O2. *Journal of the American Chemical Society, 141*(46), 18585–18599. http://dx.doi.org/10.1021/jacs.9b09833.
Petti, M. K., Lomont, J. P., Maj, M., & Zanni, M. T. (2018). Two-dimensional spectroscopy is being used to address core scientific questions in biology and materials science. *Journal of Physical Chemistry B, 122*(6), 1771–1780. http://dx.doi.org/10.1021/acs.jpcb.7b11370.
Pineda, J. R. E. T., Antoniou, D., & Schwartz, S. D. (2010). Slow conformational motions that favor sub-picosecond motions important for catalysis. *The Journal of Physical Chemistry B, 114*(48), 15985–15990. http://dx.doi.org/10.1021/jp1071296.

Rao, S. N., Singh, U. C., Bash, P. A., & Kollman, P. A. (1987). Free energy perturbation calculations on binding and catalysis after mutating Asn 155 in subtilisin. *Nature, 328*(6130), 551–554. http://dx.doi.org/10.1038/328551a0.

Riedel, C., Gabizon, R., Wilson, C. A. M., Hamadani, K., Tsekouras, K., Marqusee, S., et al. (2015). The heat released during catalytic turnover enhances the diffusion of an enzyme. *Nature, 517*(7533), 227–230. http://dx.doi.org/10.1038/nature14043.

Robinson, P. K. (2015). Enzymes: Principles and biotechnological applications. *Essays in Biochemistry, 59*, 1–41. http://dx.doi.org/10.1042/BSE0590001.

Roskoski, R. (2015). Michaelis-menten kinetics. In *Reference module in biomedical sciences*. http://dx.doi.org/10.1016/b978-0-12-801238-3.05143-6.

Rothwell, P. J., Berger, S., Kensch, O., Felekyan, S., Antonik, M., Wöhrl, B. M., et al. (2003). Multiparameter single-molecule fluorescence spectroscopy reveals heterogeneity of HIV-1 reverse transcriptase:primer/template complexes. *Proceedings of the National Academy of Sciences of the United States of America, 100*(4), 1655–1660. http://dx.doi.org/10.1073/pnas.0434003100.

Rotman, B. (1961). Measurement of activity of single molecules of beta-D-galactosidase. *Proceedings of the National Academy of Sciences of the United States Of, 47*(12), 1981–1991. http://dx.doi.org/10.1073/pnas.47.12.1981.

Sanchez, M. L. K., Sommer, C., Reijerse, E., Birrell, J. A., Lubitz, W., & Dyer, R. B. (2019). Investigating the kinetic competency of CrHydA1 [FeFe] hydrogenase intermediate states via time-resolved infrared spectroscopy. *Journal of the American Chemical Society, 141*(40), 16064–16070. http://dx.doi.org/10.1021/jacs.9b08348.

Sasmal, D. K., Pulido, L. E., Kasal, S., & Huang, J. (2016). Single-molecule fluorescence resonance energy transfer in molecular biology. *Nanoscale, 8*(48), 19928–19944. http://dx.doi.org/10.1039/c6nr06794h.

Schütz, G. J., Trabesinger, W., & Schmidt, T. (1998). Direct observation of ligand colocalization on individual receptor molecules. *Biophysical Journal, 74*(5), 2223–2226. http://dx.doi.org/10.1016/S0006-3495(98)77931-7.

Sengupta, S., Dey, K. K., Muddana, H. S., Tabouillot, T., Ibele, M. E., Butler, P. J., et al. (2013). Enzyme molecules as nanomotors. *Journal of the American Chemical Society, 135*(4), 1406–1414. http://dx.doi.org/10.1021/ja3091615.

Shah, N. B., Hutcheon, M. L., Haarer, B. K., & Duncan, T. M. (2013). F1-ATPase of *Escherichia coli*: The ε-inhibited state forms after ATP hydrolysis, is distinct from the ADP-inhibited state, and responds dynamically to catalytic site ligands. *Journal of Biological Chemistry, 288*(13), 9383–9395. http://dx.doi.org/10.1074/jbc.M113.451583.

Shaw, D. J., Hill, R. E., Simpson, N., Husseini, F. S., Robb, K., Greetham, G. M., et al. (2017). Examining the role of protein structural dynamics in drug resistance in *Mycobacterium tuberculosis*. *Chemical Science, 8*(12), 8384–8399. http://dx.doi.org/10.1039/C7SC03336B.

Takahashi, S., Ogura, T., Kitagawa, T., Itoh-Shinzawa, K., & Yoshikawa, S. (1991). Observation of the CuA-ligand stretching resonance Raman band for cytochrome c oxidase. *Journal of the American Chemical Society, 113*(24), 9400–9401. http://dx.doi.org/10.1021/ja00024a073.

Tokuriki, N., & Jackson, C. J. (2014). Enzyme dynamics and engineering: One step at a time. *Chemistry & Biology, 21*(10), 1259–1260. http://dx.doi.org/10.1016/j.chembiol.2014.10.003.

Turunen, P., Rowan, A. E., & Blank, K. (2014). Single-enzyme kinetics with fluorogenic substrates: Lessons learnt and future directions. *FEBS Letters, 588*(19), 3553–3563. http://dx.doi.org/10.1016/j.febslet.2014.06.021.

Venezuela, P., Lazzeri, M., & Mauri, F. (2011). Theory of double-resonant Raman spectra in graphene: Intensity and line shape of defect-induced and two-phonon bands. *Physical*

*Review B: Condensed Matter and Materials Physics, 84*(3), 035433. http://dx.doi.org/10.1103/PhysRevB.84.035433.

Westley, C., Fisk, H., Xu, Y., Hollywood, K. A., Carnell, A. J., Micklefield, J., et al. (2017). Real-time monitoring of enzyme-catalysed reactions using deep UV resonance Raman spectroscopy. *Chemistry - A European Journal, 23*(29), 6983–6987. http://dx.doi.org/10.1002/chem.201701388.

Whitford, P. C., Miyashita, O., Levy, Y., & Onuchic, J. N. (2007). Conformational transitions of adenylate kinase: Switching by cracking. *Journal of Molecular Biology, 366*(5), 1661–1671. http://dx.doi.org/10.1016/j.jmb.2006.11.085.

Wolfenden, R. (2003). Thermodynamic and extrathermodynamic requirements of enzyme catalysis. *Biophysical Chemistry, 105*(2–3), 559–572. http://dx.doi.org/10.1016/S0301-4622(03)00066-8.

Yoo, J., Louis, J. M., & Chung, H. S. (2019). Diverse folding pathways of HIV-1 protease monomer on a rugged energy landscape. *Biophysical Journal, 117*(8), 1456–1466. http://dx.doi.org/10.1016/j.bpj.2019.09.015.

Zazza, C., Palma, A., Amadei, A., Sanna, N., Tatoli, S., & Aschi, M. (2010). On the catalytic role of structural fluctuations in enzyme reactions: Computational evidence on the formation of compound 0 in horseradish peroxidase. *Faraday Discussions, 145*, 107–119. http://dx.doi.org/10.1039/B906614D.

Zhao, X., Palacci, H., Yadav, V., Spiering, M. M., Gilson, M. K., Butler, P. J., et al. (2018). Substrate-driven chemotactic assembly in an enzyme cascade. *Nature Chemistry, 10*(3), 311–317. http://dx.doi.org/10.1038/NCHEM.2905.

CHAPTER THREE

# Structure, catalytic mechanism, posttranslational lysine carbamylation, and inhibition of dihydropyrimidinases

**Cheng-Yang Huang**[a, b, *]
[a]School of Biomedical Sciences, Chung Shan Medical University, Taichung City, Taiwan
[b]Department of Medical Research, Chung Shan Medical University Hospital, Taichung City, Taiwan
*Corresponding author: E-mail: cyhuang@csmu.edu.tw

## Contents



## Abstract

Dihydropyrimidinase catalyzes the reversible hydrolytic ring opening of dihydrouracil and dihydrothymine to *N*-carbamoyl-β-alanine and *N*-carbamyl-β-aminoisobutyrate, respectively. Dihydropyrimidinase from microorganisms is normally known as hydantoinase because of its role as a biocatalyst in the synthesis of D- and L-amino acids for the industrial production of antibiotic precursors and its broad substrate specificity. Dihydropyrimidinase belongs to the cyclic amidohydrolase family, which also includes imidase, allantoinase, and dihydroorotase. Although these metal-dependent enzymes share low levels of amino acid sequence homology, they possess similar active site architectures and may use a similar mechanism for catalysis. By contrast, the five human dihydropyrimidinase-related proteins possess high amino acid sequence identity and are structurally homologous to dihydropyrimidinase, but they are neuronal proteins with no dihydropyrimidinase activity. In this chapter, we summarize and discuss current knowledge and the recent advances on the structure, catalytic mechanism, and inhibition of dihydropyrimidinase.

*Advances in Protein Chemistry and Structural Biology*, Volume 122
ISSN 1876-1623
https://doi.org/10.1016/bs.apcsb.2020.05.002

## 1. Introduction

Pyrimidines are the key components in nucleic acids (Fig. 1) and are essential for the replication of genetic information in all biological systems (Del Cano-Ochoa, Moreno-Morcillo, & Ramon-Maiques, 2019; Evans & Guy, 2004; O'Donovan & Neuhard, 1970; Pardee & Yates, 1956). Pyrimidine is catabolized via oxidative and reductive pathways (Hayaishi & Kornberg, 1952; Vogels & Van der Drift, 1976). The reductive catabolism of pyrimidines proceeds in three sequential enzymatic steps via the enzymes dihydropyrimidine dehydrogenase (EC 1.3.1.2), dihydropyrimidinase

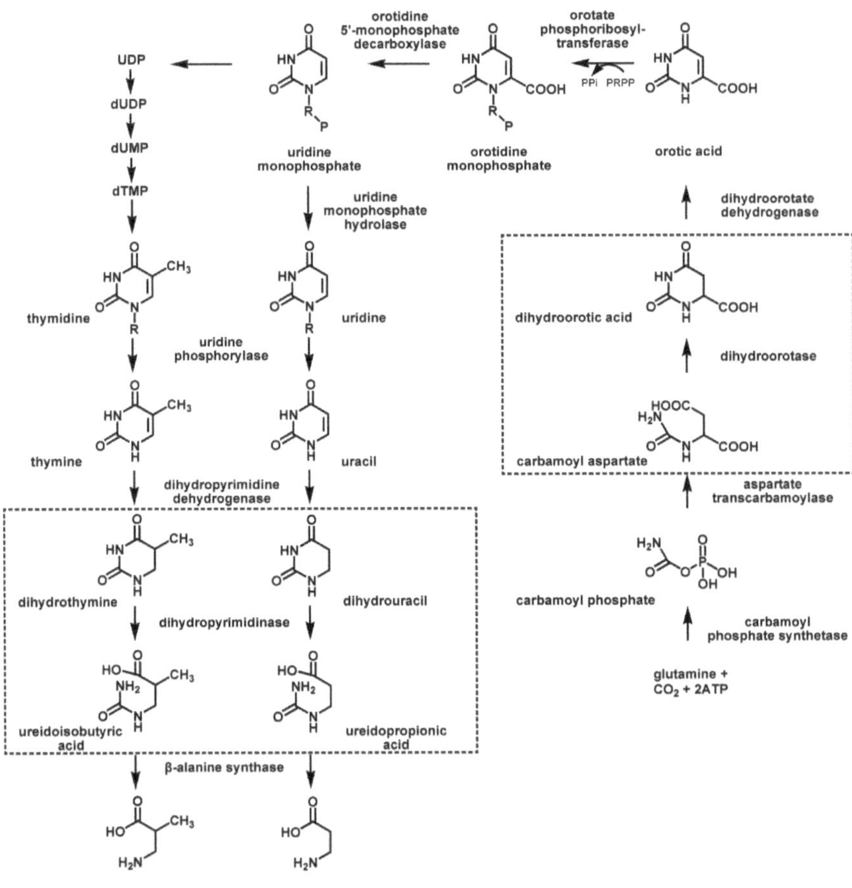


**Fig. 1** Pathways of pyrimidine nucleotide synthesis and degradation. Chemical reactions catalyzed by dihydropyrimidinase and dihydroorotase are highlighted.

(EC 3.5.2.2), and β-alanine synthase (EC 3.5.1.6) (Schnackerz & Dobritzsch, 2008). In the second step catalyzed by dihydropyrimidinase, dihydrouracil and dihydrothymine are converted to *N*-carbamoyl-β-alanine and *N*-carbamyl-β-aminoisobutyrate, respectively. This reaction is reversible, so dihydropyrimidinase also catalyzes the cyclization reaction in acidic environments.

Owing to its broad substrate specificity (Fig. 2), dihydropyrimidinase is also known as dihydropyrimidine hydrase (Wallach & Grisolia, 1957),

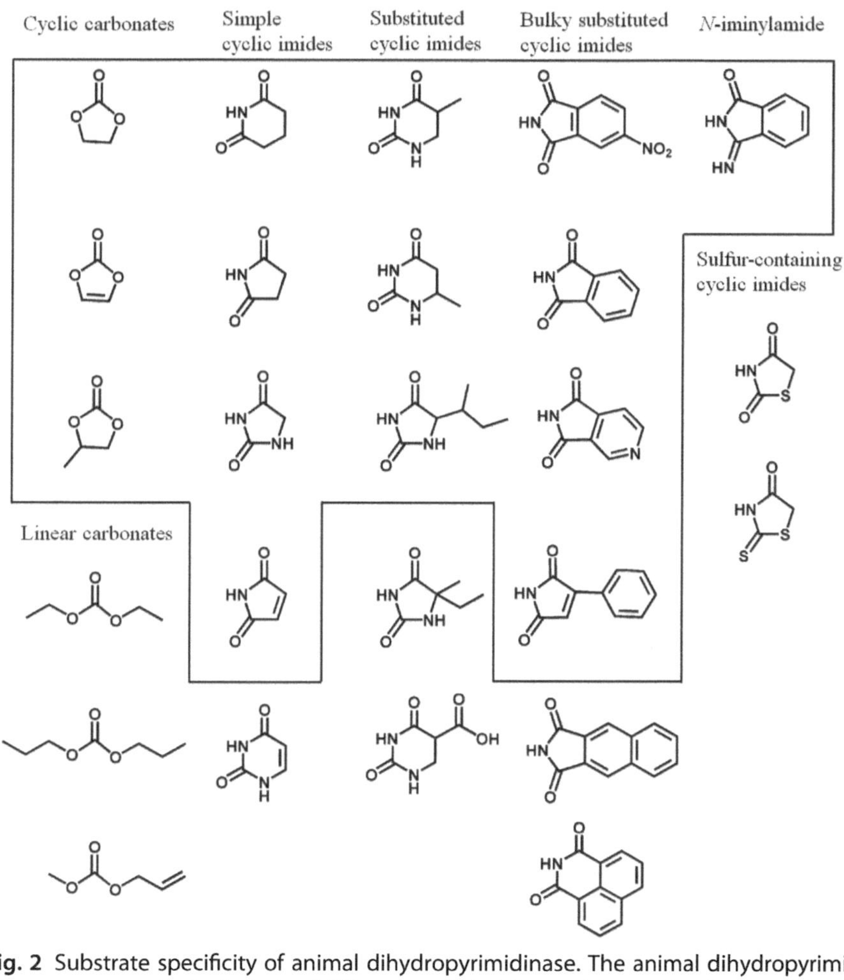


**Fig. 2** Substrate specificity of animal dihydropyrimidinase. The animal dihydropyrimidinases catalyze the hydrolysis of different cyclic imides, ureides, and carbonates. Some prokaryotic dihydropyrimidinases (hydantoinases) do not prefer simple cyclic imides as substrates. Some sulfur-containing cyclic imides can be hydrolyzed by imdase from *Blastobacter* sp. strain A17p-4.

5,6-dihydropyrimidine amidohydrolase (Kautz & Schnackerz, 1989), hydantoinase (Eadie, Bernheim, & Bernheim, 1949), and imidase (Yang, Ramaswamy, & Jakoby, 1993). In addition to the physiological substrates dihydrouracil and dihydrothymine, dihydropyrimidinase catalyzes the hydrolytic cleavage of the linear and the heterocyclic imides (Yang et al., 1993). Phthalimide, 3,4-pyridine dicarboximide, 2,3-pyridine dicarboximide, 6-methyl dihydrouracil, 5-methyl dihydrouracil, succinimide, 2-phenylsuccinimide, glutarimide, adipimide, 1-methyl hydantoin, and 5-monosubstituted hydantoins are substrates of animal dihydropyrimidinase (Yang et al., 1993). The substrate spectrum of animal dihydropyrimidinase also includes cyclic carbonate (Yang, Ramaswamy, & Jakoby, 1998), maleimide (Huang & Yang, 2003), and *N*-iminylamide (Huang & Yang, 2005). Some sulfur-containing cyclic imides can be hydrolyzed by imidase from *Blastobacter* sp. strain A17p-4 (Ogawa, Soong, Honda, & Shimizu, 1997).

Dihydropyrimidinase from microorganisms is normally known as hydantoinase because of its role as a biocatalyst (Fig. 3) in the synthesis of D- and L-amino acids for the industrial production of precursors for the semisynthesis of antibiotics, active peptides, hormones, and pesticides (Altenbuchner, Siemann-Herzberg, & Syldatk, 2001; Clemente-Jimenez, Martinez-Rodriguez, Rodriguez-Vico, & Heras-Vazquez, 2008; Schoemaker, Mink, & Wubbolts, 2003). Optically pure D-amino acids can be produced by the hydrolysis of D,L-5-monosubstituted hydantoins through the "hydantoinase process." For example, D-*p*-hydroxyphenylglycine, a widely used intermediate for the synthesis of semi-synthetic antibiotics, can be produced from D,L-5-*p*-hydroxyphenylhydantoin by using D-hydantoinase in combination with D-carbamoylase. Alternatively, hydantoin racemase is used in this process for actively catalyzing the racemization of the unconverted L-hydantoin compound.

Almost all dihydropyrimidinases are tetramers (Fig. 4A). The global architecture of dihydropyrimidinase monomer consists of two domains,

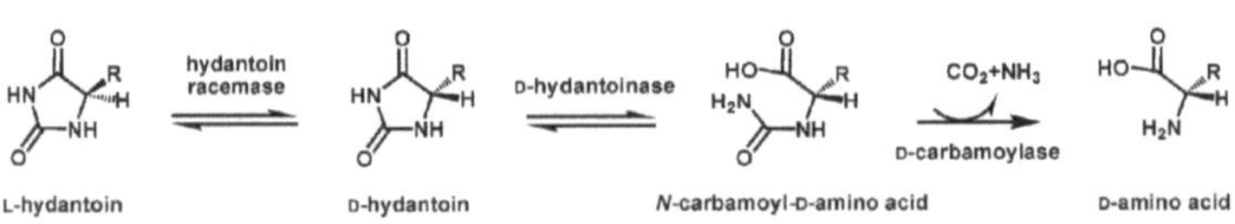


**Fig. 3** Production of optically pure D-amino acids through the hydantoinase process. Optically pure D-amino acids can be produced by the hydrolysis of D,L-5-monosubstituted hydantoins through two sequential enzymatic reactions catalyzed by D-hydantoinase and D-carbamoylase. Alternatively, hydantoin racemase is also used in this process for actively catalyzing the racemization of the unconverted L-hydantoin compound.

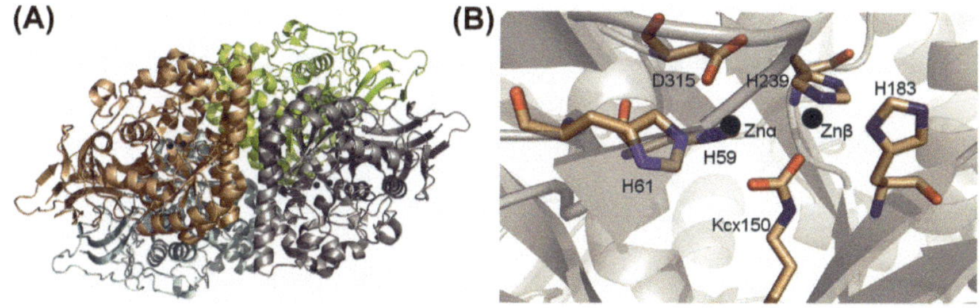


**Fig. 4** Crystal structure of *Thermus* sp. dihydropyrimidinase. (A) Ribbon diagram of a dihydropyrimidinase tetramer (PDB entry 1GKQ). Each dihydropyrimidinase monomer is color-coded. Two zinc ions in the active site are presented as black spheres. (B) The binuclear metal center of dihydropyrimidinase. The active site contains four His, one Asp, and one post-carbamylated Lys (Kcx) residues.

namely, a large domain with a classic $(\beta/\alpha)_8$-barrel structure core embedding the catalytic dimetal center (Fig. 4B) and a small β-sandwich domain. The binuclear metal center in dihydropyrimidinase consists of four His, one Asp, and one post-translational carbamylated Lys residues (Abendroth, Niefind, & Schomburg, 2002). Dihydropyrimidinase is a member of the cyclic amidohydrolase family (Gerlt & Babbitt, 2001; Kim & Kim, 1998b), which also includes dihydroorotase (Gojkovic et al., 2003; Grande-Garcia, Lallous, Diaz-Tejada, & Ramon-Maiques, 2014; Lee et al., 2007; Thoden, Phillips, Neal, Raushel, & Holden, 2001), allantoinase (Fig. 5) (Kim, Kim, Chung, Ahn, & Rhee, 2009), hydantoinase (Cheon et al., 2002), and imidase (Hung et al., 2010). Although the metal content of these metal-dependent enzymes are somehow different, they possess similar active site architectures (Fig. 6) and catalyze the hydrolysis of the cyclic amide bond of each substrate in the metabolism of purines and pyrimidines (Tables 1–3). Given their involvement in the key reactions of nucleotide synthesis and degradation, some of these amidohydrolases are suggested as chemotherapeutic targets for anticancer (Huang, Lien, Chen, Lin, & Huang, 2020; Lee et al., 2018; Rabinovich et al., 2015), antimicrobial (Huang, 2015;

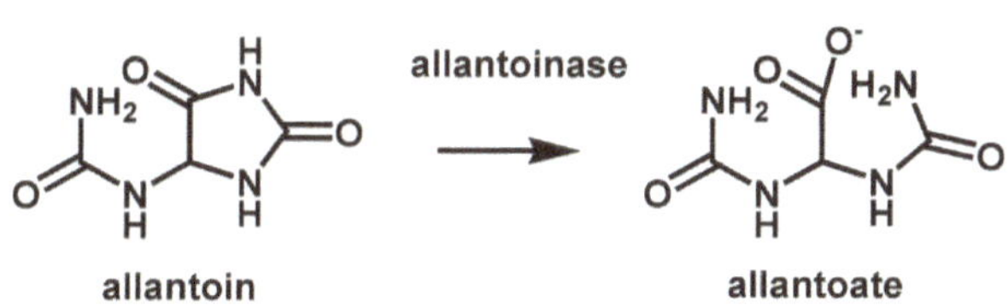


**Fig. 5** The chemical reaction catalyzed by allantoinase.

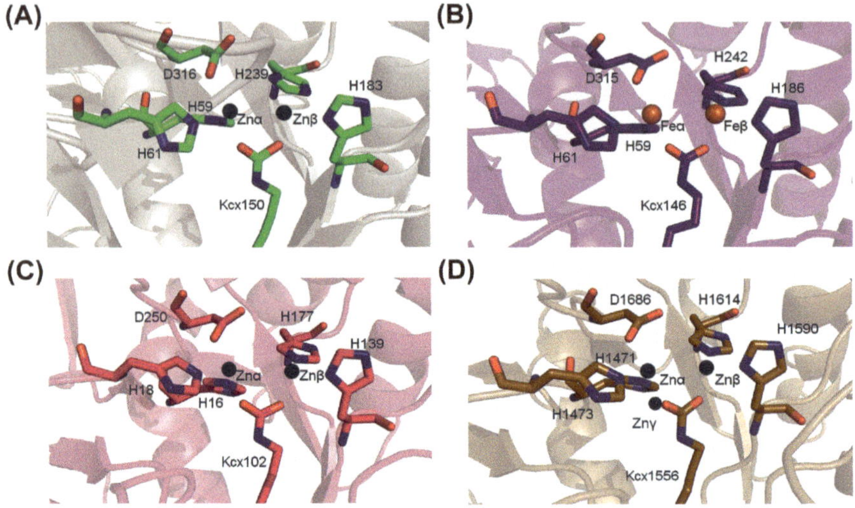


**Fig. 6** The active site of members of the cyclic amidohydrolase family. Their active sites contain four His, one Asp, and one post-carbamylated Lys (Kcx) residues, which are required for metal binding and catalytic activity. (A) *Pseudomonas aeruginosa* dihydropyrimidinase (PDB entry 5E5C). (B) *Escherichia coli* allantoinase (PDB entry 3E74). Note that the active site contains two Fe metal ions. (C) *E. coli* dihydroorotase (PDB entry 2EG6). (D) The human dihydroorotase domain of CAD protein (PDB entry 4C6C). Note that the active site contains three Zn metal ions.

Huang et al., 2020; Peng & Huang, 2014; Verrier et al., 2020), and antimalarial drug developments (Lee et al., 2007; Seymour, Lyons, Phillips, Rieckmann, & Christopherson, 1994).

Dihydropyrimidinase deficiency is an autosomal recessive disease characterized by a large accumulation of dihydrouracil and dihydrothymine in urine, blood, and cerebrospinal fluid (van Kuilenburg et al., 2010). Dihydropyrimidinase is a component in the chain for the biosynthesis of the neurotransmitter β-alanine, and its deficiency results in neurological abnormalities, including severe delay in speech development and white matter abnormalities (Henderson, Ward, Simmonds, Duley, & Davies, 1993; Tsuchiya et al., 2019; van Kuilenburg et al., 2007). Severe developmental delay is also found in infants with dihydropyrimidinase deficiency. Dihydropyrimidinase is the second enzyme in the catabolism of the anticancer drug 5-fluorouracil (5-FU), and 5-FU-associated toxicity is reported in asymptomatic patients with dihydropyrimidinase deficiency who underwent anticancer therapy (Sumi et al., 1998). Fig. 7 shows that 5-FU can bind to the active site of dihydropyrimidinase (Huang, Ning, & Huang, 2019). Many dihydropyrimidinase-like

**Table 1** List of the structures of dihydropyrimidinase/hydantoinase available in Protein Data Bank.

| PDB ID | Structure | Unique ligand | Metal α binding site | | | Bridge | Metal β binding site | |
|---|---|---|---|---|---|---|---|---|
| 1GKP | *Thermus* sp. D-hydantoinase | Zn(αβ) | H59 | H61 | D315 | Kcx150 | H183 | H239 |
| 1GKQ | *Thermus* sp. D-hydantoinase | Zn(αβ) | H59 | H61 | D315 | Kcx150 | H183 | H239 |
| 1GKR | *Arthrobacter aurescens* L-hydantoinase | Zn(αβ) | H60 | H62 | D312 | Kcx147 | H183 | H239 |
| 1NFG | *Ralstonia pickettii* D-hydantoinase | Zn(αβ) | H57 | H69 | D313 | Kcx148 | H181 | H237 |
| 1YNY | *Bacillus* sp. AR9 D-hydantoinase | Mn(αβ) | H58 | H60 | D315 | (K150) | H183 | H239 |
| 2FTW | *Dictyostelium discoideum* dihydropyrimidinase | Zn(αβ) | H65 | H68 | D325 | Kcx158 | H191 | H247 |
| 2FTY | *Saccharomyces kluyveri* dihydropyrimidinase | Zn(αβ) | H62 | H64 | D358 | Kcx167 | H199 | H255 |
| 2FVK | *Saccharomyces kluyveri* dihydropyrimidinase | Zn(αβ), dihydrouracil | H62 | H64 | D358 | Kcx167 | H199 | H255 |
| 2FVM | *Saccharomyces kluyveri* dihydropyrimidinase | Zn(αβ), NCBA | H62 | H64 | D358 | Kcx167 | H199 | H255 |
| 2VR2 | Human dihydropyrimidinase | Zn(αβ) | H67 | H69 | D326 | (K159) | H192 | H248 |
| 3DC8 | *Sinorhizobium meliloti* dihydropyrimidinase | Zn(αβ) | H56 | H58 | D313 | Kcx147 | H180 | H236 |
| 3SFW | *Brevibacillus agri* NCHU1002 dihydropyrimidinase | Zn(αβ) | H58 | H60 | D315 | Kcx150 | H182 | H239 |
| 4GZ7 | Apo-dihydropyrimidinase from *Tetraodon nigroviridis* | No metal | | | | | | |
| 4H00 | Mono-Zn dihydropyrimidinase from *Tetraodon nigroviridis* | Znα | H63 | H65 | D322 | Kcx155 | | |
| 4H01 | Di-Zn dihydropyrimidinase from *Tetraodon nigroviridis* | Zn(αβ) | H63 | H65 | D322 | Kcx155 | H188 | H244 |

*(Continued)*

**Table 1** List of the structures of dihydropyrimidinase/hydantoinase available in Protein Data Bank.—cont'd

| PDB ID | Structure | Unique ligand | Metal α binding site | | | Bridge | Metal β binding site | |
|---|---|---|---|---|---|---|---|---|
| 4LCQ | Di-Zn dihydropyrimidinase from *Tetraodon nigroviridis* in complex with NCBI | Zn(αβ), NCBI | H63 | H65 | D322 | Kcx155 | H188 | H244 |
| 4LCR | Di-Zn dihydropyrimidinase from *Tetraodon nigroviridis* in complex with NCBA | Zn(αβ), NCBA | H63 | H65 | D322 | Kcx155 | H188 | H244 |
| 4LCS | Di-Zn dihydropyrimidinase from *Tetraodon nigroviridis* in complex with hydantoin | Zn(αβ), hydantoin | H63 | H65 | D322 | Kcx155 | H188 | H244 |
| 4TQT | *Brucella suis* dihydropyrimidinase | Zn(αβ) | H61 | H63 | D317 | Kcx151 | H184 | H240 |
| 5E5C | *Pseudomonas aeruginosa* PAO1 dihydropyrimidinase | Zn(αβ) | H59 | H61 | D316 | Kcx150 | H183 | H239 |
| 5YKD | *Pseudomonas aeruginosa* PAO1 dihydropyrimidinase | Zn(αβ) | H59 | H61 | D316 | Kcx150 | H183 | H239 |
| 6AJD | Mono-Zn dihydropyrimidinase from *Pseudomonas aeruginosa* PAO1 | Znα | H59 | H61 | D316 | (K150) | | |
| 6KLK | *Pseudomonas aeruginosa* dihydropyrimidinase in complex with 5-FU | Zn(αβ), 5-FU | H59 | H61 | D316 | Kcx150 | H183 | H239 |

**Table 2** List of the structures of dihydroorotase available in Protein Data Bank.

| PDB ID | Structure | Unique ligand | Metal α binding site | | | Bridge | Metal β binding site | |
|---|---|---|---|---|---|---|---|---|
| 1J79 | *Escherichia coli* dihydroorotase | Zn(αβ), orotic acid, carbamoyl aspartic acid | H16 | H18 | D250 | Kcx102 | H139 | H177 |
| 1XGE | *Escherichia coli* dihydroorotase | Zn(αβ), dihydroorotate, carbamoyl aspartic acid | H16 | H18 | D250 | Kcx102 | H139 | H177 |
| 1XRF | *Aquifex aeolicus* dihydroorotase | Znα | H61 | H63 | D305 | D153 | | |
| 1XRT | *Aquifex aeolicus* dihydroorotase | Zn(αβ) | H61 | H63 | D305 | D153 | H208 | H232 |
| 2E25 | The T109S mutant of *Escherichia coli* dihydroorotase in complex with 5FOA | Zn(αβ), 5FOA | H16 | H18 | D250 | Kcx102 | H139 | H177 |
| 2EG6 | *Escherichia coli* dihydroorotase | Zn(αβ) | H16 | H18 | D250 | Kcx102 | H139 | H177 |
| 2EG7 | *Escherichia coli* dihydroorotase in complex with HDDP | Zn(αβ), HDDP | H16 | H18 | D250 | Kcx102 | H139 | H177 |
| 2EG8 | *Escherichia coli* dihydroorotase in complex with 5FOA | Zn(αβ), 5FOA | H16 | H18 | D250 | Kcx102 | H139 | H177 |
| 2GWN | *Porphyromonas gingivalis* dihydroorotase | Zn(αβ) | Q63 | H65 | D313 | Kcx149 | H181 | H239 |
| 2OGJ | *Agrobacterium fabrum* dihydroorotase | Zn(αβ) | H77 | H79 | D291 | Kcx175 | H208 | H231 |
| 2Z00 | *Thermus thermophiles* dihydroorotase | Zn(αβ) | H55 | H57 | D306 | D147 | H174 | H233 |
| 2Z24, 2Z25, 2Z26, 2Z27, 2Z28, 2Z29, 2Z2A | Thr110Ser, Thr110Val, Thr110Ala, Thr109Ser, Thr109Val, Thr109Ala and Thr109Gly dihydroorotase from *Escherichia coli* | Zn(αβ), dihydroorotate | H16 | H18 | D250 | Kcx102 | H139 | H177 |

*(Continued)*

**Table 2** List of the structures of dihydroorotase available in Protein Data Bank.—cont'd

| PDB ID | Structure | Unique ligand | Metal α binding site | | | Bridge | Metal β binding site | |
|---|---|---|---|---|---|---|---|---|
| 2Z2B | Deletion 107–116 mutant of dihydroorotase from *Escherichia coli* | Zn(αβ) | H16 | H18 | D250 | Kcx102 | H139 | H177 |
| 3CJP | Amidohydrolase CAC3332 from *Clostridium acetobutylicum* | Zn(αβ) | H8 | H10 | D250 | E121 | H139 | H177 |
| 3D6N | *Aquifex* dihydroorotase activated by Aspartate Transcarbamoylase | Zn(αβ) | H61 | H63 | D305 | D153 | H208 | H232 |
| 3GRI | *Staphylococcus aureus* dihydroorotase | Zn(αβ) | H58 | H60 | D303 | D150 | H177 | H230 |
| 3JZE | *S. typhimurium* LT2 dihydroorotase | Zn(αβ) | H17 | H19 | D250 | Kcx103 | H140 | H178 |
| 3MJM | His257Ala dihydroorotase from *Escherichia coli* | Zn(αβ), dihydroorotate | H16 | H18 | D250 | Kcx102 | H139 | H177 |
| 3MPG | *Bacillus anthracis* dihydroorotase | Zn(αβ) | H59 | H61 | D304 | D151 | H178 | H231 |
| 3PNU | *Campylobacter jejuni* dihydroorotase | Zn(αβ) | H10 | H12 | D237 | Kcx92 | H131 | H165 |
| 4BJH | The *Aquifex reactor* complex formed by dihydroorotase (H180A, H232A) with dihydroorotate and aspartate transcarbamoylase with PALA | Znα, PALA, dihydroorotate | H61 | H63 | D305 | D153 | | |
| 4BY3 | The dihydroorotase domain of human CAD in apo- form obtained recombinantly from *E. coli* | Zn(αβγ) | H1471 | H1473 | D1686 | Kcx1556 | H1590 | H1614 |

| | | | | | | | | |
|---|---|---|---|---|---|---|---|---|
| 4C6B | The dihydroorotase domain of human CAD with incomplete active site, obtained recombinantly from *E. coli* | No metal | | | | | | |
| 4C6C | The dihydroorotase domain of human CAD obtained recombinantly from HEK293 cells | Zn(αβγ) | H1471 | H1473 | D1686 | Kcx1556 | H1590 | H1614 |
| 4C6D, 4C6F, 4C6I, 4C6J | The dihydroorotase domain of human CAD bound to substrate at pH 6.0, 6.5, 7.0, and 7.5 | Zn(αβγ), dihydroorotate, carbamoyl aspartic acid | H1471 | H1473 | D1686 | Kcx1556 | H1590 | H1614 |
| 4C6E, 4C6K | The dihydroorotase domain of human CAD bound to substrate at pH 5.5 and 8.0 | Zn(αβ), dihydroorotate, carbamoyl aspartic acid | H1471 | H1473 | D1686 | Kcx1556 | H1590 | H1614 |
| 4C6L, 4C6M | The dihydroorotase domain of human CAD bound to the inhibitor fluoroorotate at pH 6.0 and 7.0 | Zn(αβγ), 5FOA | H1471 | H1473 | D1686 | Kcx1556 | H1590 | H1614 |
| 4C6N | The dihydroorotase domain of human CAD E1637T mutant bound to substrate at pH 6.0 | Zn(αβ), carbamoyl aspartic acid | H1471 | H1473 | D1686 | Kcx1556 | H1590 | H1614 |
| 4C6O | The dihydroorotase domain of human CAD C1613S mutant in apo-form at pH 6.0 | Zn(αβ) | H1471 | H1473 | D1686 | Kcx1556 | H1590 | H1614 |

*(Continued)*

**Table 2** List of the structures of dihydroorotase available in Protein Data Bank.—cont'd

| PDB ID | Structure | Unique ligand | Metal α binding site | | | Bridge | Metal β binding site | |
|---|---|---|---|---|---|---|---|---|
| 4C6P | The dihydroorotase domain of human CAD C1613S mutant in apo-form at pH 7.0 | Zn(αβγ) | H1471 | H1473 | D1686 | Kcx1556 | H1590 | H1614 |
| 4C6Q | The dihydroorotase domain of human CAD C1613S mutant bound to substrate at pH 7.0 | Zn(αβ), carbamoyl aspartic acid | H1471 | H1473 | D1686 | Kcx1556 | H1590 | H1614 |
| 4LFY | *Burkholderia cenocepacia* dihydroorotase | Zn(αβ) | H31 | H33 | D268 | Kcx117 | H154 | H192 |
| 4YIW | *Bacillus anthracis* dihydroorotase | Zn(αβ), carbamoyl aspartic acid | H59 | H61 | D304 | D151 | H178 | H231 |
| 5NNL | Inactive dihydroorotase-like domain of *Chaetomium thermophilum* CAD-like multifunctional protein | No metal | | | | | | |
| 5VGM | *Vibrio cholera* dihydroorotase | Zn(αβ) | H13 | H15 | D246 | Kcx98 | H135 | H173 |
| 5YNZ | The dihydroorotase domain (K1556A) of human CAD | Znα | H1471 | H1473 | D1686 | (A1556) | | |
| 6CTY | *Yersinia pestis* dihydroorotase | Zn(αβ) | H17 | H19 | D251 | Kcx104 | H140 | H176 |
| 6GDD, 6GDF | Dihydroorotase from *aquifex aeolicus* under 1200 and 600 bar of hydrostatic pressure | Znα | H61 | H63 | D305 | D153 | | |
| 6GDE | Dihydroorotase from *aquifex aeolicus* standard (P,T) | Znα | H61 | H63 | D305 | D153 | | |

| | | | | | | | | |
|---|---|---|---|---|---|---|---|---|
| 6HFD, 6HFE, 6HFF, 6HFI | Human dihydroorotase mutant F1563L, F1563T, F1563Y, and F1563A apo structure | Zn(αβγ) | H1471 | H1473 | D1686 | Kcx1556 | H1590 | H1614 |
| 6HFH, 6HFL, 6HFP, 6HFR | Human dihydroorotase mutant F1563A, F1563L, F1563T and F1563Y co-crystallized with carbamoyl aspartate at pH 7.0 | Zn(αβγ), dihydroorotate | H1471 | H1473 | D1686 | Kcx1556 | H1590 | H1614 |
| 6HFJ, 6HFN, 6HFQ, 6HFU | Human dihydroorotase mutant F1563A, F1563L, F1563T and F1563Y co-crystallized with carbamoyl aspartate at pH 7.5 | Zn(αβγ), dihydroorotate | H1471 | H1473 | D1686 | Kcx1556 | H1590 | H1614 |
| 6HFK, 6HFS, | Human dihydroorotase mutant F1563L and F1563Y co-crystallized with carbamoyl aspartate at pH 6.5 | Zn(αβγ), dihydroorotate | H1471 | H1473 | D1686 | Kcx1556 | H1590 | H1614 |
| 6HG1 | Hybrid dihydroorotase domain of human CAD with *E. coli* flexible loop in apo state | Zn(αβ) | H1471 | H1473 | D1688 | Kcx1556 | H1592 | H1616 |
| 6HG2 | Hybrid dihydroorotase domain of human CAD with *E. coli* flexible loop, bound to 5FOA | Zn(αβ), 5FOA | H1471 | H1473 | D1688 | Kcx1556 | H1592 | H1616 |
| 6HG2 | Hybrid dihydroorotase domain of human CAD with *E. coli* flexible loop, bound to dihydroorotate | Zn(αβ), dihydroorotate | H1471 | H1473 | D1688 | Kcx1556 | H1592 | H1616 |

**Table 3** List of the structures of allantoinase available in Protein Data Bank.

| PDB ID | Structure | Unique ligands | Metal α binding site | | | Bridge | Metal β binding site | |
|---|---|---|---|---|---|---|---|---|
| 3CL6 | *Pseudomonas fluorescens* allantoinase | No metal | | | | | | |
| 3CL7 | *Pseudomonas fluorescens* allantoinase in complex with hydantoin | No metal, hydantoin | | | | | | |
| 3CL8 | *Pseudomonas fluorescens* allantoinase in complex with ACA | No metal, ACA | | | | | | |
| 3E74 | *Escherichia coli* allantoinase with iron ions at the metal center | Fe(αβ) | H59 | H61 | D315 | Kcx146 | H186 | H242 |
| 3HM7 | Mono-Zn allantoinase from *Bacillus halodurans* C-125 | Znα | H63 | H65 | D315 | (K150) | | |

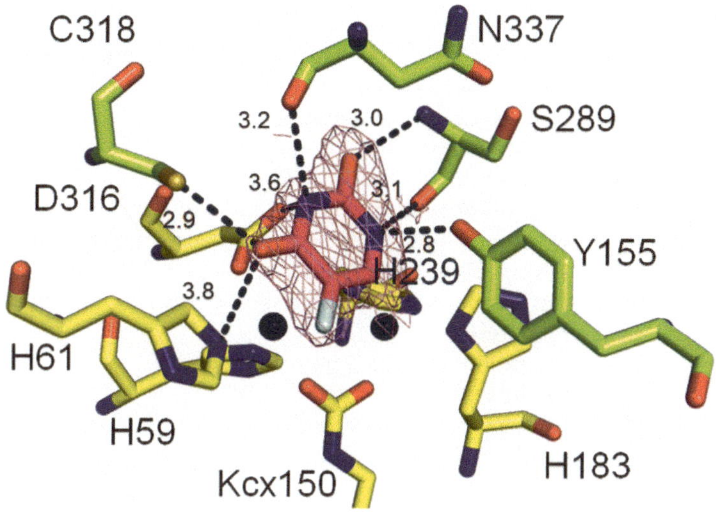


**Fig. 7** 5-FU binding mode of *P. aeruginosa* dihydropyrimidinase. His61, Tyr155, Asp316, Cys318, Ser289 and Asn337 of *P. aeruginosa* dihydropyrimidinase are involved in 5-FU binding. His61 and Asp316 (in yellow) are essential metal-binding sites of *P. aeruginosa* dihydropyrimidinase. 5-FU (in hotpink) interacts with the main chains of residues Ser289 and Asn337 and the side chains of residues Tyr155 and Cys318 (in limon). Ser289, Asn337, and Tyr155 are crucial for substrate binding.

proteins such as collapsing response-mediator proteins, which highly resemble dihydropyrimidinase in terms of sequence and structure, are discovered (Table 4). Human dihydropyrimidinase shares an identity with dihydropyrimidinase-related proteins 1, 2, 3, 4, and 5 by 61%, 60%, 60%, 59%, and 58%, respectively and is higher than the *Pseudomonas aeruginosa* counterpart that shares amino acid residue identity with human dihydropyrimidinase by 51% (Fig. 8). Interestingly, these dihydropyrimidinase-related proteins are involved in neuronal development without the catalytic activity of dihydropyrimidinase (Wang & Strittmatter, 1997). The active site residues in dihydropyrimidinase are not conserved in human dihydropyrimidinase-related proteins, resulting in the absence of sites for metal binding and enzyme activity (Fig. 9).

Many studies on the biochemical properties and structural comparison with the mechanistic data of dihydropyrimidinase before 2009 have been thoroughly reviewed (Hsu, Lu, & Yang, 2010; Schnackerz & Dobritzsch, 2008). Over the last decade, investigations on the structure of

**Table 4** List of the structures of dihydropyrimidinase-related protein available in Protein Data Bank.

| PDB ID | Structure |
|---|---|
| 2GSE | Crystal structure of human dihydropyrimidinease-like 2 |
| 2VM8 | Human CRMP-2 crystallized in the presence of Mg |
| 4B90 | Crystal structure of WT human CRMP-5 |
| 4B91 | Crystal structure of truncated human CRMP-5 |
| 4B92 | Crystal structure of truncated human CRMP-5 soaked with Zn |
| 4BKN | Human dihydropyrimidinase-related protein 3 (DPYSL3) |
| 4B3Z | Structure of the human collapsin response mediator protein-1, a lung cancer suppressor |
| 4CNS | Crystal structure of truncated human CRMP-4 |
| 4CNT | Crystal structure of WT human CRMP-4 |
| 4CNU | Crystal structure of WT human CRMP-4 from lattice translocation |
| 5LXX | High-resolution structure of human collapsin response mediator protein 2 |
| 5MKV | Crystal structure of human dihydropyrimidinase-like 2 (DPYSL2A)/ collapsin response mediator protein (CRMP2) residues 13-516 |
| 5MLE | Crystal structure of human dihydropyrimidinease-like 2 (DPYSL2A)/ collapsin response mediator protein (CRMP2 13–516) mutant Y479E/Y499E |
| 5UQC | Crystal structure of mouse CRMP-2 |
| 5X1A | Crystal structure of human CRMP-2 |
| 5X1C | Crystal structure of human CRMP-2 without C-terminal tail |
| 5X1D | Crystal structure of T246A-N247A human CRMP-2 mutant |
| 5NKS | Human dihydropyrimidinase-related protein 4 (DPYSL4, CRMP3, ULIP-4) |
| 5YZ5 | Crystal structure of human CRMP-2 with T509D-T514D-S518D-S522D mutations |
| 5YZA | Crystal structure of human CRMP-2 with S522D mutation |
| 5YZB | Crystal structure of human CRMP-2 with S522D-T509D-T514D-S518D mutations crystallized with GSK3b |
| 6JV9 | Crystal structure of human CRMP2 1–532, unmodified |
| 6JVB | Crystal structure of human CRMP2 1–532, AGE-modified |

dihydropyrimidinase from different species, as well as the inhibition modes, have provided considerable insights into the structure–function relationships of dihydropyrimidinase. In this chapter, we review articles from 2010 to 2020 and summarize and discuss current knowledge and recent advances on the structure, catalytic mechanism, posttranslational lysine carbamylation, and inhibition of dihydropyrimidinases.

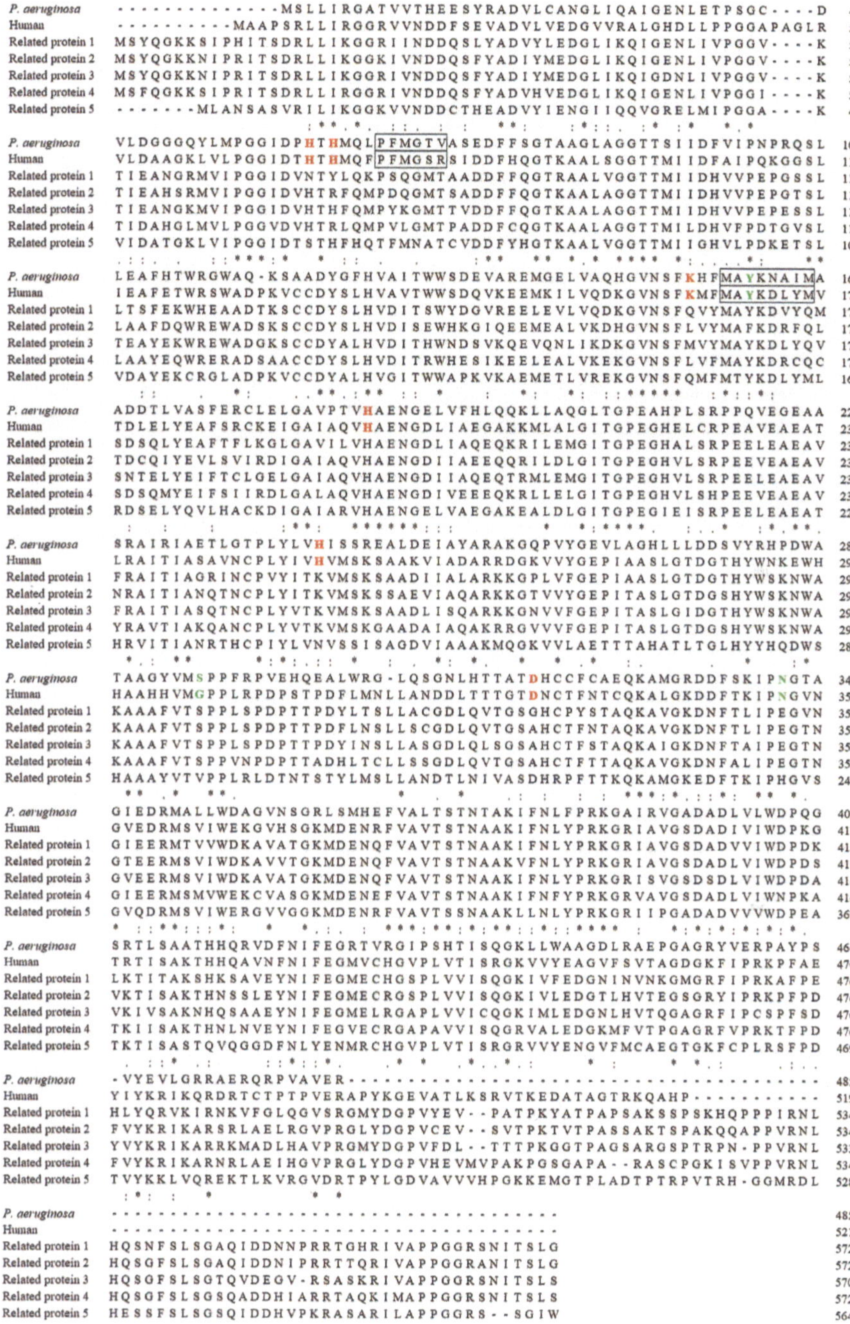

**Fig. 8** Multiple amino acid sequence alignment of dihydropyrimidinases and the related proteins. The amino acids in dihydropyrimidinases required for metal and substrate binding are shown in red and green, respectively. The two dynamic loops are boxed. Human dihydropyrimidinase shares an identity with dihydropyrimidinase-related proteins 1, 2, 3, 4, and 5 by 61%, 60%, 60%, 59%, and 58%, respectively and is higher than the *P. aeruginosa* counterpart that shares amino acid residue identity by 51%.

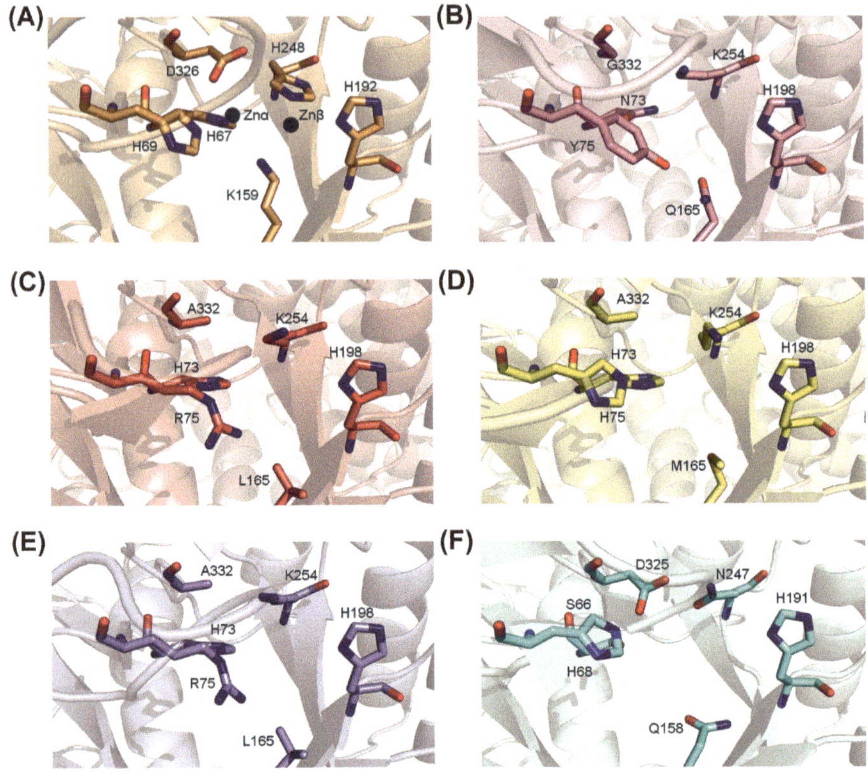


**Fig. 9** The active site residues in human dihydropyrimidinase are not conserved in dihydropyrimidinase-related proteins. (A) The active site of human dihydropyrimidinase (PDB entry 2VR2). Note that the Lys159 is uncarbamylated. The corresponding residues in (B) human dihydropyrimidinase-related protein 1 (PDB entry 4B3Z), (C) human dihydropyrimidinase-related protein 2 (PDB entry 2VM8), (D) human dihydropyrimidinase-related protein 3 (PDB entry 4BKN), (E) human dihydropyrimidinase-related protein 4 (PDB entry 5NKS), and (F) human dihydropyrimidinase-related protein 5 (PDB entry 4B90) are shown. Notably, the active site residues in dihydropyrimidinase are not conserved in human dihydropyrimidinase-related proteins, resulting in the absence of sites for metal binding and enzyme activity.

## 2. Structural mechanisms of Lys carbamylation in dihydropyrimidinase

The metal center in the amidohydrolase superfamily is vital for enzymatic activities (Seibert & Raushel, 2005). Members of this family contain either a mononuclear metal center or a binuclear metal center (Seibert & Raushel, 2005). As a member of the amidohydrolase superfamily, dihydropyrimidinase/hydantoinase from different organisms contains one or two metal ions per subunit. The crystal structures of dihydropyrimidinase/hydantoinase from *Thermus* sp. (Abendroth et al., 2002), *Burkholderia pickettii* (Xu et al., 2003), *Bacillus stearothermophilus* (Cheon et al., 2002), *Saccharomyces kluyveri* (Lohkamp, Andersen, Piskur, & Dobritzsch, 2006), *Dictyostelium discoideum* (Lohkamp et al., 2006), *Sinorhizobium meliloti* CECT4114 (Martinez-Rodriguez et al., 2010), and *P. aeruginosa* PAO1 (Tzeng, Huang, & Huang, 2016) are resolved. A carbamylated Lys located within the active site bridges the two metal ions (Fig. 4B). The crystal structures of human dihydropyrimidinase (Fig. 9A) and *Bacillus* sp. AR9 hydantoinase (Radha Kishan et al., 2005) are also determined, but the usual carbamylation on the active-site residue Lys does not happen. However, dihydropyrimidinase purified from bovine (Brooks, Jones, Kim, & Sander, 1983), rat (Kikugawa et al., 1994), pig (Huang & Yang, 2002), and fish liver (Huang & Yang, 2003) contains only a single divalent metal ion in the active site. Even in monometal-containing enzymes, the sites for dimetal binding in all these dihydropyrimidinases/hydantoinases are perfectly conserved.

The structures of apo-, monometal-, and dimetal-containing dihydropyrimidinase from *Tetraodon nigroviridis* (Fig. 10) reveal four steps in the assembly of the holoprotein with the carbamylated Lys and two metal ions (Hsieh et al., 2013). Structural and functional analyses indicate that only one metal (Zn$\alpha$) is sufficient for stabilizing carbamylation on the active-site residue Lys and for the maximal enzymatic activity of this fish dihydropyrimidinase (Hsieh et al., 2013). The structures of another monometallic dihydropyrimidinase from *P. aeruginosa* and the human dihydroorotase domain K1556A mutant further reveal that carbamylation on Lys is not required for Zn$\alpha$ binding (Cheng, Huang, Lin, & Huang, 2018b). The presence or absence of Lys carbamylation does not affect Zn$\alpha$ binding to the active site of dihydropyrimidinase. Thus, the stable carbamylation on the Lys in dihydropyrimidinase may be a Zn$\alpha$-mediated process. Although Lys carbamylation in dihydropyrimidinase does not need an additional enzyme to achieve covalent modification and is a spontaneous and reversible process, Zn$\alpha$ may play a role in stabilizing the developing negative charge of carbamate (Fig. 10). When Lys is carbamylated, the negative charge of $O^-$ facilitates additional metal coordination (Zn$\beta$) for the regulation of specific enzymatic activities by controlling the conformations of two dynamic loops for the substrate entrance. This phenomenon may explain the ability of carboxylic acids to facilitate the assembly of the binuclear metal center only partially and temporarily (Ho, Huang, & Huang, 2013; Huang, Hsu,

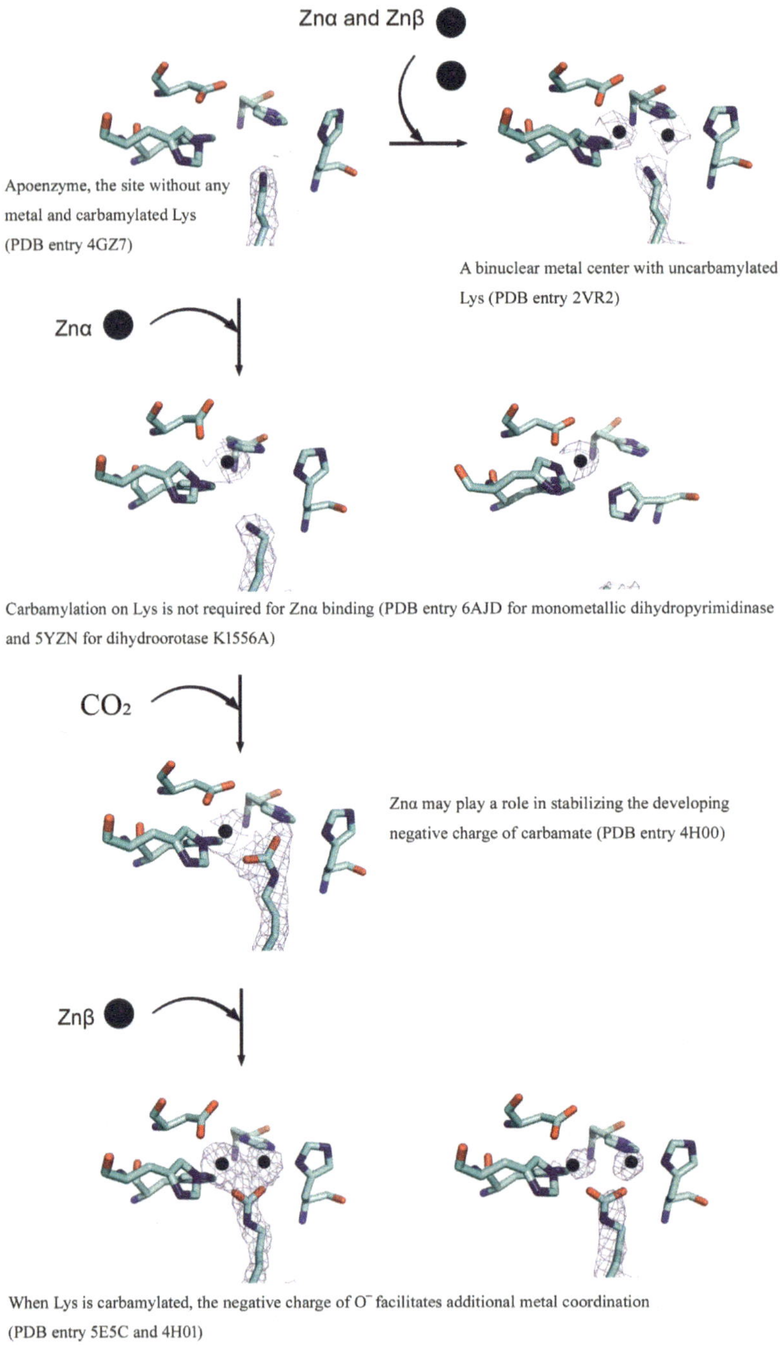


**Fig. 10** Structural mechanisms of Lys carbamylation in dihydropyrimidinases.

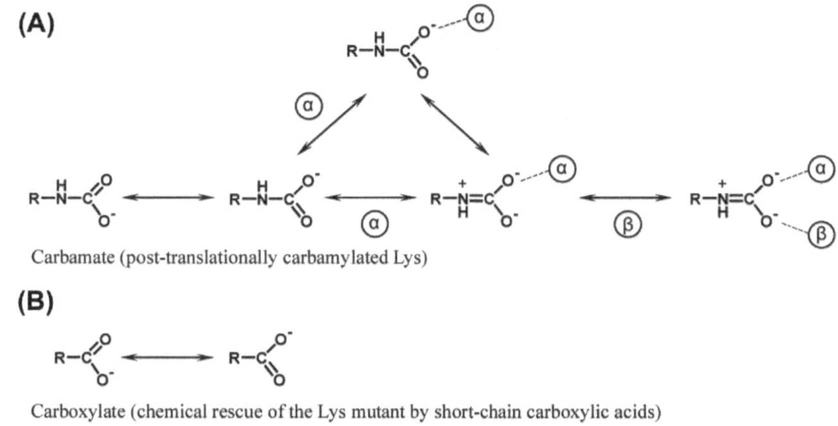


**Fig. 11** Different chemical properties between carbamate and carboxylate. (A) Resonance forms for carbamate of carboxylated Lys. The function of the carbamylated Lys in dihydropyrimidinases is to coordinate metal ion(s) and maintain a suitable active site environment. (B) Resonance forms for carboxylate. The activity of the Lys mutant can be chemically rescued only partially by adding metal ions and short-chain carboxylic acids.

Chen, & Yang, 2009). Owing to different chemical properties between carbamate and carboxylate, only a resonance structure of carbamate can result in oxygen atoms forming negative charges (Fig. 11) that may promote di-metal assembly in dihydropyrimidinase. This phenomenon may also increase the nucleophilicity of the hydroxide in the active site of dihydropyrimidinase for catalysis (Kumar, Saxena, Sarma, & Radha Kishan, 2011).

The active site of dihydropyrimidinase may be grouped into six types: (I) a binuclear metal center with carbamylated Lys (Fig. 12A), (II) a binuclear metal center with uncarbamylated Lys (Fig. 12B), (III) a mononuclear metal center with carbamylated Lys (Fig. 12C), (IV) a mononuclear metal center with uncarbamylated Lys (Fig. 12D), (V) the site without any metal and carbamylated Lys (Fig. 12E), and (VI) the site without metal but with carbamylated Lys (Fig. 12F). With the analysis of enzyme activity as basis, the active site types I, II, and III are the catalytically active forms of dihydropyrimidinases found in dihydropyrimidinase/hydantoinase from *Agrobacterium radiobacter* (Huang et al., 2009), human (PDB entry 2VR2) and *T. nigroviridis* (Hsieh et al., 2013). At present, the active site type VI (no metal but with carbamylated Lys) is never found in any dihydropyrimidinase and in other cyclic amidohydrolases. Possibly, the spontaneous carbamylation on Lys must be stabilized by a metal ion within the active site of dihydropyrimidinase; otherwise, it cannot be maintained stably for a long time.

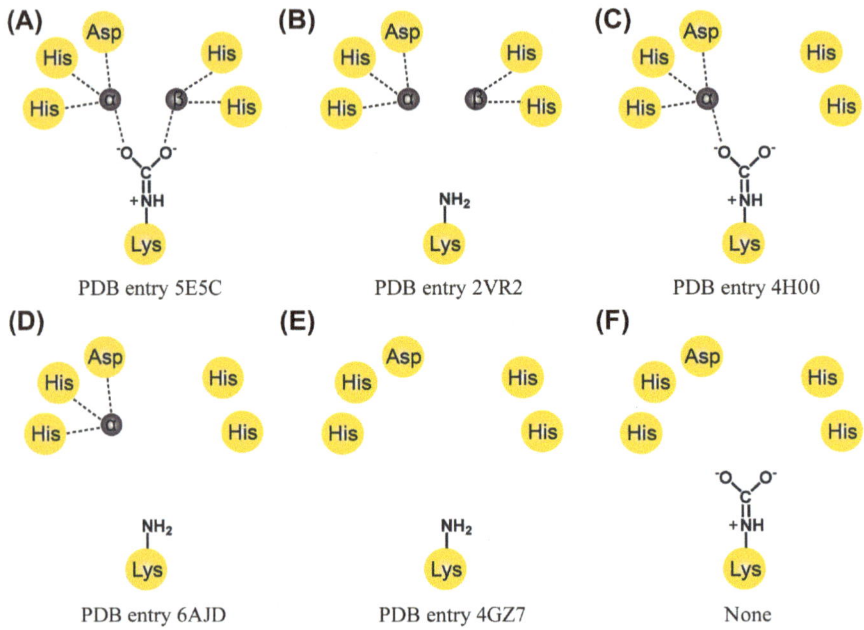


**Fig. 12** Six active site types of dihydropyrimidinase. (A) A binuclear metal center with carbamylated Lys. (B) A binuclear metal center with uncarbamylated Lys. (C) A mononuclear metal center with carbamylated Lys. (D) A mononuclear metal center with uncarbamylated Lys. (E) The site without any metal and carbamylated Lys. (F) The site without metal but with carbamylated Lys. With the analysis of enzyme activity as basis, the active site types I, II, and III are the catalytically active forms of dihydropyrimidinases.

## 3. Molecular basis of dimer or tetramer formation: different contact residues

The global architecture of dihydropyrimidinase monomer consists of two domains: a large domain with a classic $(\beta/\alpha)_8$-barrel structure core embedding the catalytic dimetal center and a small β-sandwich domain. The length of a long C-terminal tail in dihydropyrimidinase varies among different species (Fig. 13A–F). Structural and functional analyses have indicated that all known dihydropyrimidinases/hydantoinases are tetramers except pseudomonal enzymes (Kim & Kim, 1998a; Niu, Zhang, Shi, & Yuan, 2007; Zhang, Niu, Shi, & Yuan, 2008). *P. aeruginosa* dihydropyrimidinase forms a dimer, rather than a tetramer, in the crystalline state and in the solution (Tzeng et al., 2016). The overall structure of each *P. aeruginosa* dihydropyrimidinase subunit consists of 17 α-helices and 19 β-sheets. The

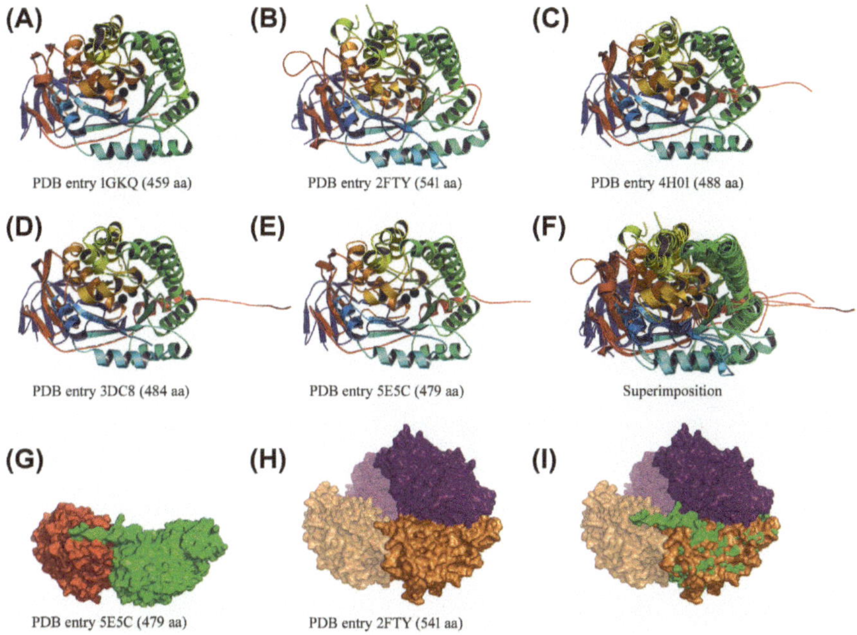


**Fig. 13** Comparison of the C-terminal tail of dihydropyrimidinases. The structures of dihydropyrimidinase monomer from (A) *Thermus* sp., (B) *S. kluyveri*, (C) *T. nigroviridis*, (D) *S. meliloti*, and (E) *P. aeruginosa* are shown. (F) The superimposed structures of these dihydropyrimidinases. Their C-terminal tails differ in length. (G) The C-terminal tails of each *P. aeruginosa* dihydropyrimidinase subunit extend toward another monomer in a swapping-like manner. The C-terminal tail contributes to the main interactions for maintaining the dimeric state of the protein. (H) *S. kluyveri* tetramer. The C-terminal tails of each *S. kluyveri* dihydropyrimidinase subunit do not significantly extend toward another monomer. (I) Comparison by superimposition of dihydropyrimidinases from *S. kluyveri* and *P. aeruginosa*. The length and conformation of C-terminal tails significantly differ between *S. kluyveri* (tetramer) and *P. aeruginosa* (dimer) dihydropyrimidinases.

C-terminal tails of each subunit extend toward another monomer in a swapping-like manner (Fig. 13G), which may be involved in dihydropyrimidinase tetramerization (Lohkamp et al., 2006; Martinez-Rodriguez et al., 2010) and dimerization (Hsieh et al., 2013; Tzeng et al., 2016). Although *P. aeruginosa* dihydropyrimidinase has an extension C-terminal tail, *P. aeruginosa* dihydropyrimidinase forms a dimer (Fig. 13H and I). If any, contact is minimal, and many important and conserved residues are lacking in the dimer—dimer interface of *P. aeruginosa* dihydropyrimidinase and are not enough to form a tetramer. However, the oligomerization of *P. aeruginosa* dihydropyrimidinase is a pH-dependent process (Cheng, Huang, Huang,

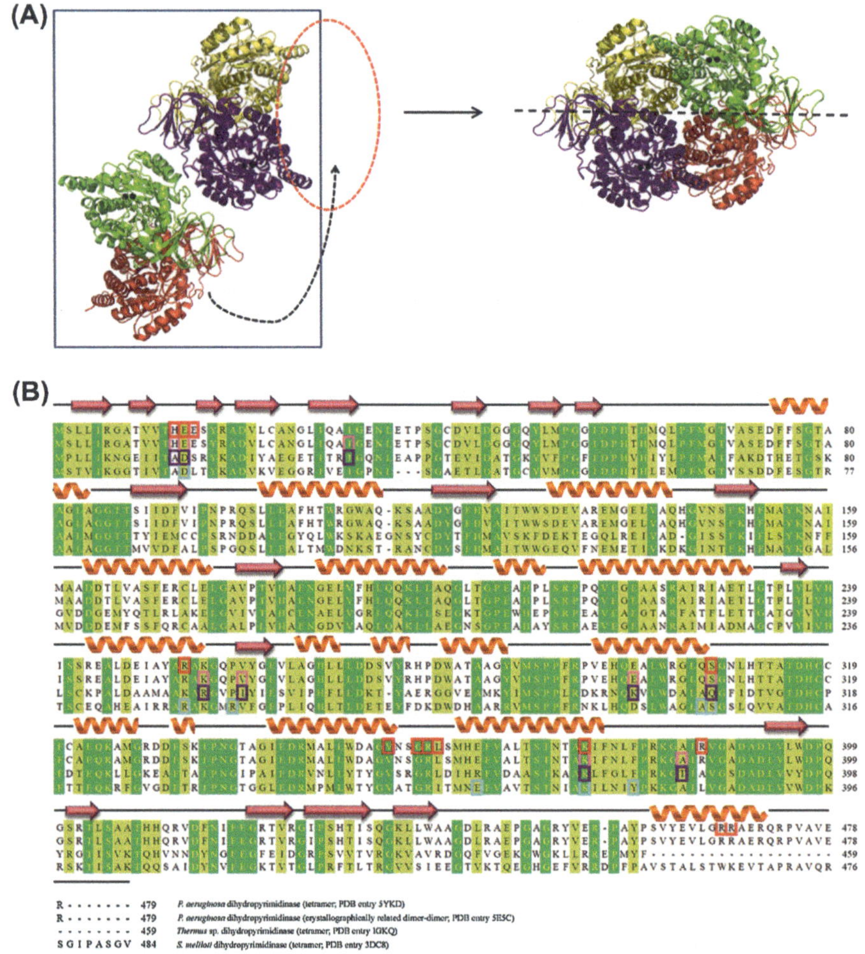


**Fig. 14** The dimer-dimer interface of dihydropyrimidinases. (A) The structure of *P. aeruginosa* dihydropyrimidinase tetramer. An asymmetric unit contains four crystallographically independent *P. aeruginosa* dihydropyrimidinase monomers. (B) Sequence alignment of dihydropyrimidinases from *P. aeruginosa*, *Thermus* sp., and *S. meliloti* dihydropyrimidinases. The amino acids that are involved in dimer-dimer interface are boxed, respectively. The crystallographically related dimer-dimer interface of *P. aeruginosa* dihydropyrimidinase is also compared.

& Huang, 2018a). A mixture of dimers and tetramers of *P. aeruginosa* dihydropyrimidinase (Fig. 14A) is found at pH 5.9 rather than at neutral pH (Cheng et al., 2018a). Residues crucial to forming a tetramer among *P. aeruginosa* (Cheng et al., 2018a), *Thermus* sp., and *S. meliloti* dihydropyrimidinases differ (Fig. 14B).

## 4. Catalytic mechanism: dynamic loops, tunnel bottleneck, and metal content of dihydropyrimidinases

Basing on their similar binuclear metal active site architecture and post-carbamylation, dihydropyrimidinase/hydantoinase and dihydroorotase share several enzymatic properties and may use a similar mechanism for catalysis (Gojkovic et al., 2003). For *S. kluyveri* dimetal-containing dihydropyrimidinase, two amino acids, Asn392 and Ser331, are directly involved in substrate and product binding. The binding of this enzyme to the reaction product proceeds in a manner similar to that of an enzyme and a substrate (Schnackerz & Dobritzsch, 2008). Notably, no solvent molecules are involved in ligand binding. A narrow channel probably used for substrate entry is found from the protein surface on the active site of *S. kluyveri* dimetal-containing dihydropyrimidinase. Tyr172 located within a dynamic loop in this enzyme may play an essential role in the stabilization of the tetrahedral transition state during hydrolysis of the substrate, collapse of the transition state, product formation, and product release (Cheon et al., 2002; Lohkamp et al., 2006).

Two dynamic loops (segments Ala69–Arg74 and Met158–Met165) and the C-terminal tail are important for the substrate entrance of *T. nigroviridis* dihydropyrimidinase. Tyr160 that structurally corresponds to the essential Tyr172 residue in *S. kluyveri* dihydropyrimidinase is also found in *T. nigroviridis* dihydropyrimidinase (Hsieh et al., 2013). Three stages are proposed on the basis of the structures of the apo- (the open form), monometal- (the intermediate stage), and dimetal-containing dihydropyrimidinases (the closed form) from *T. nigroviridis*. The cross-section bottlenecks of the active site pocket and the conformations of dynamic loops among the three stages of this enzyme are significantly different. The tunnel bottlenecks in the open and closed forms of *T. nigroviridis* dihydropyrimidinase are approximately 10 and 3 Å, respectively. Given that the tunnel bottleneck in the closed form (dimetallic enzyme) is narrower than the average size of substrates (6 Å), the bottleneck size may indicate a substrate entrance. However, the dynamic process for substrate diffusion and different catalytic mechanisms make whether this assumption is applicable to cases of dimetallic dihydropyrimidinase unclear (Fig. 15). For example, the bottleneck sizes of the tunnel in the monometallic (Cheng et al., 2018b) and dimetallic *P. aeruginosa* dihydropyrimidinase (Tzeng et al., 2016) are 4.5 and 4.0 Å, respectively. Both values

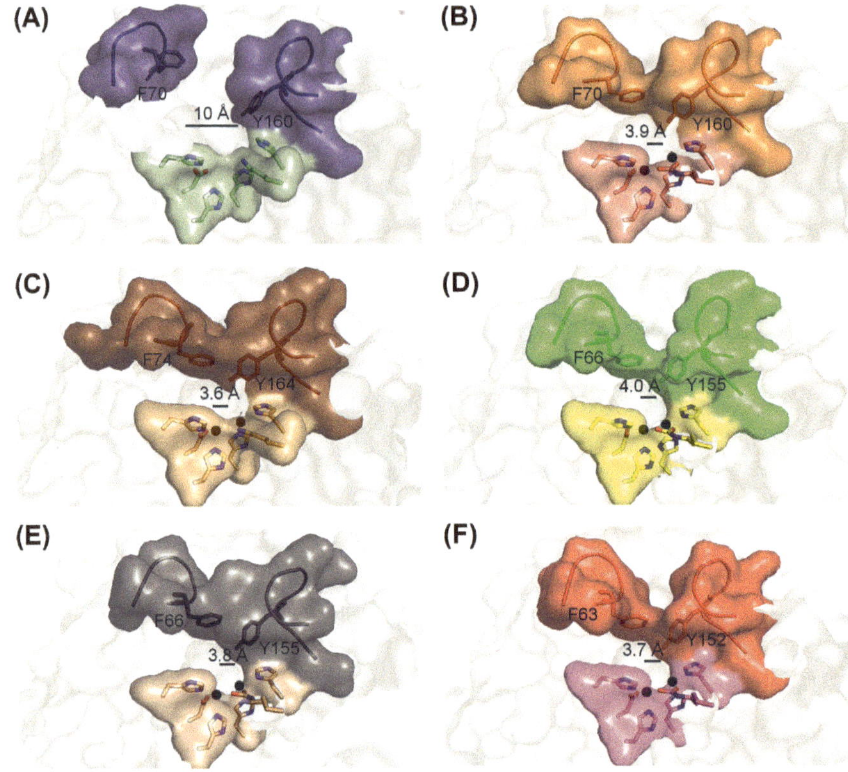


**Fig. 15** Bottleneck size of the tunnel in dihydropyrimidinases. Two dynamic loops in dihydropyrimidinase are important for the substrate entrance. The active site with the dynamic loops in (A) the open form of *T. nigroviridis* dihydropyrimidinase, (B) the close form of *T. nigroviridis* dihydropyrimidinase, (C) human dihydropyrimidinase, (D) dimetal-containing *P. aeruginosa* dihydropyrimidinase, (E) *Thermus* sp. dihydropyrimidinase, and (F) *S. meliloti* dihydropyrimidinase is shown.

suggest a closed form, but the dimetallic *P. aeruginosa* dihydropyrimidinase is catalytically active.

The active form of animal dihydropyrimidinases is a monometallic enzyme (Brooks et al., 1983; Huang & Yang, 2002, 2003; Kikugawa et al., 1994). In addition to structural analysis, the linear titration curve for apo-form enzyme reactivation with zinc ions indicates that only one metal ion is required for the active *T. nigroviridis* dihydropyrimidinase (Fig. 16A) (Hsieh et al., 2013). This phenomenon may explain why the metal ion becomes an inhibitor as soon as the first metal binding site is occupied (Hsieh et al., 2013). Binding of the second zinc ion to *T. nigroviridis*

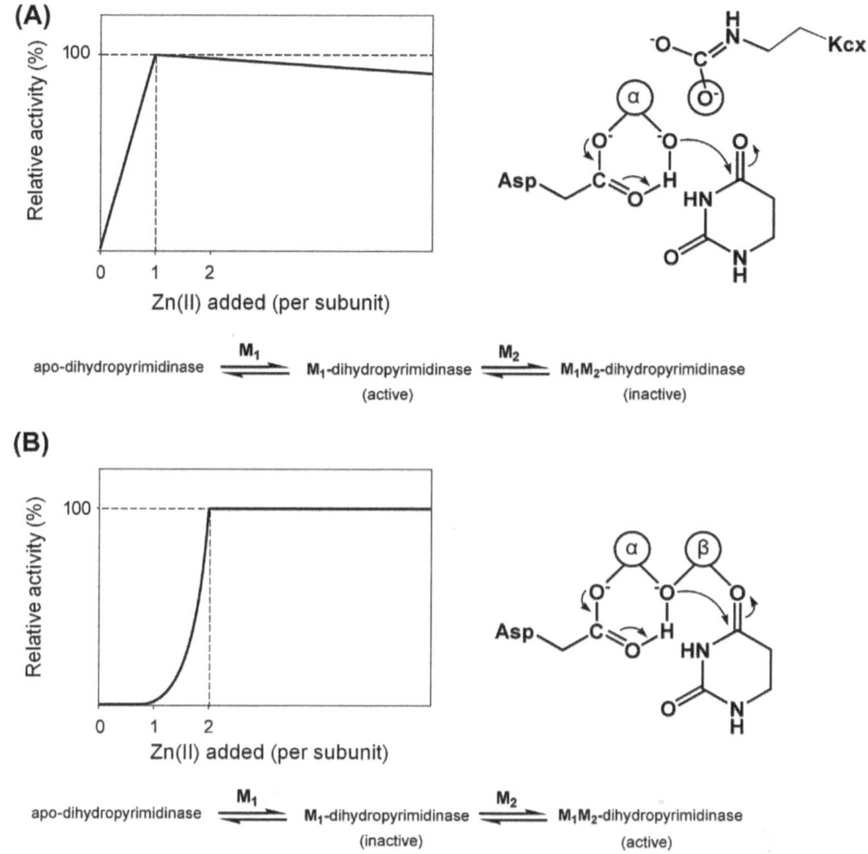


**Fig. 16** Proposed kinetic models for the metal-activated dihydropyrimidinase. (A) The kinetic model of the monometallic dihydropyrimidinase. Only one metal ion is required for this active dihydropyrimidinase. (B) The kinetic model of the dimetallic dihydropyrimidinase. A nonlinear increase in the activity of apo-form enzyme suggests that a kinetic model in which only the form of dihydropyrimidinase with a fully assembled binuclear metal center ($M_1M_2$-dihydropyrimidinase) is catalytically active.

dihydropyrimidinase induces a conformational change on the bottleneck of the active site tunnel, thereby blocking the site for substrate binding as also seen in thermolysin (Holland, Hausrath, Juers, & Matthews, 1995). The structural and biochemical analyses demonstrate that the catalytic mechanism of *T. nigroviridis* dihydropyrimidinase somehow differs from that of the dimetallic dihydropyrimidinases and dihydroorotases (Fig. 16B) (Gojkovic et al., 2003) and provide a distinctive mechanistic view currently held in the field. The human dihydropyrimidinase is a dimetal-containing enzyme with uncarbamylated Lys, but whether this form is catalytically active or is a (second) metal ion-inhibited form remains unclear (Fig. 9A). Perhaps, the titration curve for reactivating the apo-form enzyme with zinc ions should be plotted to determine the active form of human dihydropyrimidinase. A metal titration curve showing a nonlinear increase in the activity of apo-form enzyme suggests that a kinetic model in which only dihydropyrimidinase with a fully assembled binuclear metal center ($M_1M_2$-dihydropyrimidinase) is catalytically active. The first metal ($M_1$) is preferentially bound to the binuclear center (Fig. 16B), as in the case of *A. radiobacter* hydantoinase (Huang et al., 2009).

Although human dihydropyrimidinase and dihydropyrimidinase-related proteins share a high level of sequence homology (with 58%—61% identity and 73%—77% similarity), dihydropyrimidinase-related proteins are not the enzymes responsible for dihydrouracil and dihydrothymine hydrolysis. The active site metal-binding residues in dihydropyrimidinase are not conserved in human dihydropyrimidinase-related proteins, leading to the absence of a catalytic activity (Fig. 9). However, the substrate-binding residue (Tyr164), as well as the dynamic loop (segment Met162—Met169) for substrate diffusion in human dihydropyrimidinase, is somehow similar in these neuroproteins. Whether the big cavity in these dihydropyrimidinase-related proteins is responsible for some unknown molecule accommodation and interaction is not clear (Fig. 9B—F). For example, the essential Tyr residue in human dihydropyrimidinase is almost conserved in dihydropyrimidinase-related proteins (Fig. 8). Whether these neuroproteins can bind dihydrouracil and the reaction product (e.g., *N*-carbamoyl-β-alanine) remains unknown. β-alanine is a small molecule neurotransmitter widely distributed in the central nervous system and the reductive degradation of uracil in mammals is the sole source of β-alanine. Whether the empty cavity in these dihydropyrimidinase-related proteins is associated with binding or transporting dihydrouracil via the dynamic loop should be further studied via structural and functional identification.

## 5. Competitive inhibitors of dihydropyrimidinase

Considering that dihydropyrimidinase is a component in the chain of pyrimidine catabolism required for metabolizing the DNA base, blocking its activity may be useful to limit bacterial and cancer cell growth and survival.

Substrate analogs for any enzyme are usually potential inhibitors, but this common rule is not applicable to dihydropyrimidinase. Possibly due to the broad substrate specificity of dihydropyrimidinase, inhibitors that target its active site are not enough to strongly suppress its activity. Although some chelators, such as 8-hydroxyquinoline-5-sulfonic acid (Brooks et al., 1983), inhibit dihydropyrimidinase, they are harmful to humans. Reaction products, such as *N*-carbamyl-β-alanine (with $K_i$ value of 0.68 mM), are inhibitors of dihydropyrimidinase (Kikugawa et al., 1994). Product analogs glutarate monoamide and 4-ureidobutyrate with $K_i$ values of 0.21 and 0.99 mM, respectively, inhibit dihydropyrimidinase (Kikugawa et al., 1994). High concentrations of dihydroorotate, 5-hydantoin acetic acid, and acetohydroxamate (within the millimolar range) are required to greatly inhibit dihydropyrimidinase (Huang, 2015). Thus, substrate/product analogs do not work as potent inhibitors on dihydropyrimidinase.

Given the poor inhibition ability of substrate analogs, screening new dihydropyrimidinase inhibitors from natural products may be a good strategy. Plumbagin isolated from the extract of the carnivorous plant *Nepenthes miranda* is an inhibitor of *P. aeruginosa* dihydropyrimidinase with a half maximal inhibitory concentration ($IC_{50}$) value of 58 μM (Huang et al., 2020). The flavonoid dihydromyricetin significantly inhibits human (Basbous et al., 2020) and *P. aeruginosa* dihydropyrimidinase activity (Huang, 2015) with $IC_{50}$ values of approximately 6 and 48 μM, respectively. These naturally occurring compounds seem to inhibit dihydropyrimidinase through similar but different modes by competing with dihydrouracil for the active sites of dihydropyrimidinase. According to the docking models of *P. aeruginosa* dihydropyrimidinase (Fig. 17), dihydromyricetin interacts with Ile95, Ser289, and Asp316 (Huang, 2015), whereas plumbagin interacts with Tyr155, Lys156, Val190, Gln194, Arg212, and Asn337 (Huang et al., 2020). Residues Tyr155, Ser289, and Asn337 are crucial for substrate binding, and Asp316 has essential dual roles in metal Znα binding and in catalysis in *P. aeruginosa* dihydropyrimidinase. Thus, these compounds partially occupy and block the active site of dihydropyrimidinase for inhibition. In addition, Tyr155 is the most important residue within the dynamic loop involved in stabilizing the intermediate state during catalysis (Hsieh et al., 2013). Further research can focus on designing compounds that target the dynamic loops and catalytic sites in dihydropyrimidinase for inhibitor optimization and drug development (Huang et al., 2020).

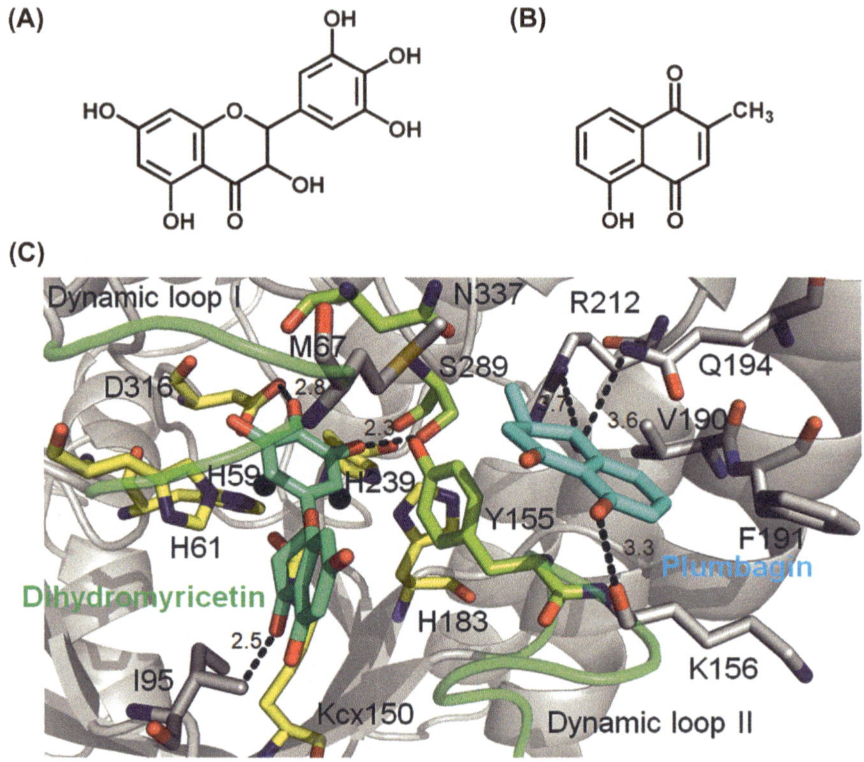


**Fig. 17** Docking models of dihydropyrimidinase-inhibitor complex. (A) Molecular structure of dihydromyricetin. (B) Molecular structure of plumbagin. (C) Structural model of *P. aeruginosa* dihydropyrimidinase in complex with dihydromyricetin and plumbagin. This model is directly constructed by superimposing the two docking models of dihydromyricetin and plumbagin to dihydropyrimidinase. It can be possible that dihydromyricetin and plumbagin bind simultaneously to the active site of dihydropyrimidinase. However, this speculation needs to be confirmed by further biochemical and cellular experiments.

## 6. Conclusions

Over the last several years, enzymes involved in pyrimidine nucleotide metabolism as targets for chemotherapy have been intensively studied. Many pyrimidine analogs, such as azidothymidine and 5-FU, have been used as antiviral and anticancer chemotherapeutic drugs. Regulation of dihydropyrimidinase activity may induce different effects on living cells. Recently, crystal structures for monometallic dihydropyrimidinase, as well as the complex with 5-FU, are determined. The elucidation of different crystal structures of dihydropyrimidinase, together with biochemical studies, will be helpful in providing new knowledge of their structure–function relationships. Some natural products (but not substrate analogs) as inhibitors on dihydropyrimidinase are discovered. Designing and finding potent inhibitors remain a challenge. Further studies are necessary for inhibition and drug optimization on dihydropyrimidinase and finding promising drugs for suppressing pyrimidine metabolism.

## Acknowledgements

This research was supported by a grant from the Ministry of Science and Technology, Taiwan (MOST 108-2320-B-040-010 to C.Y. Huang).

## References

Abendroth, J., Niefind, K., & Schomburg, D. (2002). X-ray structure of a dihydropyrimidinase from Thermus sp. at 1.3 A resolution. *Journal of Molecular Biology, 320*(1), 143–156.

Altenbuchner, J., Siemann-Herzberg, M., & Syldatk, C. (2001). Hydantoinases and related enzymes as biocatalysts for the synthesis of unnatural chiral amino acids. *Current Opinion in Biotechnology, 12*(6), 559–563.

Basbous, J., Aze, A., Chaloin, L., Lebdy, R., Hodroj, D., Ribeyre, C., et al. (2020). Dihydropyrimidinase protects from DNA replication stress caused by cytotoxic metabolites. *Nucleic Acids Research, 48*(4), 1886–1904.

Brooks, K. P., Jones, E. A., Kim, B. D., & Sander, E. G. (1983). Bovine liver dihydropyrimidine amidohydrolase: Purification, properties, and characterization as a zinc metalloenzyme. *Archives of Biochemistry and Biophysics, 226*(2), 469–483.

Cheng, J. H., Huang, C. C., Huang, Y. H., & Huang, C. Y. (2018a). Structural basis for pH-dependent oligomerization of dihydropyrimidinase from *Pseudomonas aeruginosa* PAO1. *Bioinorganic Chemistry and Applications, 2018*, 9564391.

Cheng, J. H., Huang, Y. H., Lin, J. J., & Huang, C. Y. (2018b). Crystal structures of monometallic dihydropyrimidinase and the human dihydroorotase domain K1556A mutant reveal no lysine carbamylation within the active site. *Biochemical and Biophysical Research Communications, 505*(2), 439–444.

Cheon, Y. H., Kim, H. S., Han, K. H., Abendroth, J., Niefind, K., Schomburg, D., et al. (2002). Crystal structure of D-hydantoinase from Bacillus stearothermophilus: Insight into the stereochemistry of enantioselectivity. *Biochemistry, 41*(30), 9410–9417.

Clemente-Jimenez, J. M., Martinez-Rodriguez, S., Rodriguez-Vico, F., & Heras-Vazquez, F. J. (2008). Optically pure alpha-amino acids production by the "Hydantoinase Process". *Recent Patents on Biotechnology, 2*(1), 35–46.

Del Cano-Ochoa, F., Moreno-Morcillo, M., & Ramon-Maiques, S. (2019). CAD, A multienzymatic protein at the head of de Novo pyrimidine biosynthesis. *Subcellular Biochemistry, 93*, 505–538.

Eadie, G. S., Bernheim, F., & Bernheim, M. L. (1949). The partial purification and properties of animal and plant hydantoinases. *Journal of Biological Chemistry, 181*(2), 449–458.

Evans, D. R., & Guy, H. I. (2004). Mammalian pyrimidine biosynthesis: Fresh insights into an ancient pathway. *Journal of Biological Chemistry, 279*(32), 33035–33038.

Gerlt, J. A., & Babbitt, P. C. (2001). Divergent evolution of enzymatic function: Mechanistically diverse superfamilies and functionally distinct suprafamilies. *Annual Review of Biochemistry, 70*, 209–246.

Gojkovic, Z., Rislund, L., Andersen, B., Sandrini, M. P., Cook, P. F., Schnackerz, K. D., et al. (2003). Dihydropyrimidine amidohydrolases and dihydroorotases share the same origin and several enzymatic properties. *Nucleic Acids Research, 31*(6), 1683–1692.

Grande-Garcia, A., Lallous, N., Diaz-Tejada, C., & Ramon-Maiques, S. (2014). Structure, functional characterization, and evolution of the dihydroorotase domain of human CAD. *Structure, 22*(2), 185–198.

Hayaishi, O., & Kornberg, A. (1952). Metabolism of cytosine, thymine, uracil, and barbituric acid by bacterial enzymes. *Journal of Biological Chemistry, 197*(2), 717–732.

Henderson, M. J., Ward, K., Simmonds, H. A., Duley, J. A., & Davies, P. M. (1993). Dihydropyrimidinase deficiency presenting in infancy with severe developmental delay. *Journal of Inherited Metabolic Disease, 16*(3), 574–576.

Ho, Y. Y., Huang, Y. H., & Huang, C. Y. (2013). Chemical rescue of the post-translationally carboxylated lysine mutant of allantoinase and dihydroorotase by metal ions and short-chain carboxylic acids. *Amino Acids, 44*(4), 1181–1191.

Holland, D. R., Hausrath, A. C., Juers, D., & Matthews, B. W. (1995). Structural analysis of zinc substitutions in the active site of thermolysin. *Protein Science, 4*(10), 1955–1965.

Hsieh, Y. C., Chen, M. C., Hsu, C. C., Chan, S. I., Yang, Y. S., & Chen, C. J. (2013). Crystal structures of vertebrate dihydropyrimidinase and complexes from *Tetraodon nigroviridis* with lysine carbamylation: Metal and structural requirements for post-translational modification and function. *Journal of Biological Chemistry, 288*(42), 30645–30658.

Hsu, C. C., Lu, L. Y., & Yang, Y. S. (2010). From sequence and structure of sulfotransferases and dihydropyrimidinases to an understanding of their mechanisms of action and function. *Expert Opinion on Drug Metabolism and Toxicology, 6*(5), 591–601.

Huang, C. Y. (2015). Inhibition of a putative dihydropyrimidinase from *Pseudomonas aeruginosa* PAO1 by flavonoids and substrates of cyclic amidohydrolases. *PLoS One, 10*(5), e0127634.

Huang, C. Y., Hsu, C. C., Chen, M. C., & Yang, Y. S. (2009). Effect of metal binding and posttranslational lysine carboxylation on the activity of recombinant hydantoinase. *Journal of Biological Inorganic Chemistry, 14*(1), 111–121.

Huang, Y. H., Lien, Y., Chen, J. H., Lin, E. S., & Huang, C. Y. (2020). Identification and characterization of dihydropyrimidinase inhibited by plumbagin isolated from Nepenthes miranda extract. *Biochimie, 171–172*, 124–135.

Huang, Y. H., Ning, Z. J., & Huang, C. Y. (2019). Crystal structure of dihydropyrimidinase in complex with anticancer drug 5-fluorouracil. *Biochemical and Biophysical Research Communications, 519*(1), 160–165.

Huang, C. Y., & Yang, Y. S. (2002). The role of metal on imide hydrolysis: Metal content and pH profiles of metal ion-replaced mammalian imidase. *Biochemical and Biophysical Research Communications, 297*(4), 1027–1032.

Huang, C. Y., & Yang, Y. S. (2003). A novel cold-adapted imidase from fish *Oreochromis niloticus* that catalyzes hydrolysis of maleimide. *Biochemical and Biophysical Research Communications, 312*(2), 467–472.

Huang, C. Y., & Yang, Y. S. (2005). Discovery of a novel N-iminylamidase activity: Substrate specificity, chemicoselectivity and catalytic mechanism. *Protein Expression and Purification, 40*(1), 203–211.

Hung, C. L., Lee, C., Lin, C. Y., Chang, C. H., Chung, Y. C., & Yi Tang, C. (2010). Feature amplified voting algorithm for functional analysis of protein superfamily. *BMC Genomics, 11*(Suppl. 3), S14.

Kautz, J., & Schnackerz, K. D. (1989). Purification and properties of 5,6-dihydropyrimidine amidohydrolase from calf liver. *European Journal of Biochemistry, 181*(2), 431–435.

Kikugawa, M., Kaneko, M., Fujimoto-Sakata, S., Maeda, M., Kawasaki, K., Takagi, T., et al. (1994). Purification, characterization and inhibition of dihydropyrimidinase from rat liver. *European Journal of Biochemistry, 219*(1–2), 393–399.

Kim, G. J., & Kim, H. S. (1998a). C-terminal regions of D-hydantoinases are nonessential for catalysis, but affect the oligomeric structure. *Biochemical and Biophysical Research Communications, 243*(1), 96–100.
Kim, G. J., & Kim, H. S. (1998b). Identification of the structural similarity in the functionally related amidohydrolases acting on the cyclic amide ring. *Biochemical Journal, 330*(Pt 1), 295–302.
Kim, K., Kim, M. I., Chung, J., Ahn, J. H., & Rhee, S. (2009). Crystal structure of metal-dependent allantoinase from *Escherichia coli*. *Journal of Molecular Biology, 387*(5), 1067–1074.
van Kuilenburg, A. B., Dobritzsch, D., Meijer, J., Meinsma, R., Benoist, J. F., Assmann, B., et al. (2010). Dihydropyrimidinase deficiency: Phenotype, genotype and structural consequences in 17 patients. *Biochimica et Biophysica Acta, 1802*(7–8), 639–648.
van Kuilenburg, A. B., Meijer, J., Dobritzsch, D., Meinsma, R., Duran, M., Lohkamp, B., et al. (2007). Clinical, biochemical and genetic findings in two siblings with a dihydropyrimidinase deficiency. *Molecular Genetics and Metabolism, 91*(2), 157–164.
Kumar, V., Saxena, N., Sarma, M., & Radha Kishan, K. V. (2011). Carboxylated lysine is required for higher activities in Hydantoinases. *Protein and Peptide Letters, 18*(7), 663–669.
Lee, J. S., Adler, L., Karathia, H., Carmel, N., Rabinovich, S., Auslander, N., et al. (2018). Urea cycle dysregulation generates clinically relevant genomic and biochemical signatures. *Cell, 174*(6), 1559–1570.e22.
Lee, M., Chan, C. W., Graham, S. C., Christopherson, R. I., Guss, J. M., & Maher, M. J. (2007). Structures of ligand-free and inhibitor complexes of dihydroorotase from *Escherichia coli*: Implications for loop movement in inhibitor design. *Journal of Molecular Biology, 370*(5), 812–825.
Lohkamp, B., Andersen, B., Piskur, J., & Dobritzsch, D. (2006). The crystal structures of dihydropyrimidinases reaffirm the close relationship between cyclic amidohydrolases and explain their substrate specificity. *Journal of Biological Chemistry, 281*(19), 13762–13776.
Martinez-Rodriguez, S., Martinez-Gomez, A. I., Clemente-Jimenez, J. M., Rodriguez-Vico, F., Garcia-Ruiz, J. M., Las Heras-Vazquez, F. J., et al. (2010). Structure of dihydropyrimidinase from Sinorhizobium meliloti CECT4114: New features in an amidohydrolase family member. *Journal of Structural Biology, 169*(2), 200–208.
Niu, L., Zhang, X., Shi, Y., & Yuan, J. (2007). Subunit dissociation and stability alteration of D hydantoinase deleted at the terminal amino acid residue. *Biotechnology Letters, 29*(2), 303–308.
O'Donovan, G. A., & Neuhard, J. (1970). Pyrimidine metabolism in microorganisms. *Bacteriological Reviews, 34*(3), 278–343.
Ogawa, J., Soong, C. L., Honda, M., & Shimizu, S. (1997). Imidase, a dihydropyrimidinase-like enzyme involved in the metabolism of cyclic imides. *European Journal of Biochemistry, 243*(1–2), 322–327.
Pardee, A. B., & Yates, R. A. (1956). Pyrimidine biosynthesis in *Escherichia coli*. *Journal of Biological Chemistry, 221*(2), 743–756.
Peng, W. F., & Huang, C. Y. (2014). Allantoinase and dihydroorotase binding and inhibition by flavonols and the substrates of cyclic amidohydrolases. *Biochimie, 101*, 113–122.
Rabinovich, S., Adler, L., Yizhak, K., Sarver, A., Silberman, A., Agron, S., et al. (2015). Diversion of aspartate in ASS1-deficient tumours fosters de novo pyrimidine synthesis. *Nature, 527*(7578), 379–383.
Radha Kishan, K. V., Vohra, R. M., Ganesan, K., Agrawal, V., Sharma, V. M., & Sharma, R. (2005). Molecular structure of D-hydantoinase from Bacillus sp. AR9: Evidence for mercury inhibition. *Journal of Molecular Biology, 347*, 95–105.

Schnackerz, K. D., & Dobritzsch, D. (2008). Amidohydrolases of the reductive pyrimidine catabolic pathway purification, characterization, structure, reaction mechanisms and enzyme deficiency. *Biochimica et Biophysica Acta, 1784*(3), 431–444.

Schoemaker, H. E., Mink, D., & Wubbolts, M. G. (2003). Dispelling the myths–biocatalysis in industrial synthesis. *Science, 299*(5613), 1694–1697.

Seibert, C. M., & Raushel, F. M. (2005). Structural and catalytic diversity within the amidohydrolase superfamily. *Biochemistry, 44*(17), 6383–6391.

Seymour, K. K., Lyons, S. D., Phillips, L., Rieckmann, K. H., & Christopherson, R. I. (1994). Cytotoxic effects of inhibitors of de novo pyrimidine biosynthesis upon Plasmodium falciparum. *Biochemistry, 33*(17), 5268–5274.

Sumi, S., Imaeda, M., Kidouchi, K., Ohba, S., Hamajima, N., Kodama, K., et al. (1998). Population and family studies of dihydropyrimidinuria: Prevalence, inheritance mode, and risk of fluorouracil toxicity. *American Journal of Medical Genetics, 78*(4), 336–340.

Thoden, J. B., Phillips, G. N., Jr., Neal, T. M., Raushel, F. M., & Holden, H. M. (2001). Molecular structure of dihydroorotase: A paradigm for catalysis through the use of a binuclear metal center. *Biochemistry, 40*(24), 6989–6997.

Tsuchiya, H., Akiyama, T., Kuhara, T., Nakajima, Y., Ohse, M., Kurahashi, H., et al. (2019). A case of dihydropyrimidinase deficiency incidentally detected by urine metabolome analysis. *Brain & Development, 41*(3), 280–284.

Tzeng, C. T., Huang, Y. H., & Huang, C. Y. (2016). Crystal structure of dihydropyrimidinase from *Pseudomonas aeruginosa* PAO1: Insights into the molecular basis of formation of a dimer. *Biochemical and Biophysical Research Communications, 478*(3), 1449–1455.

Verrier, E. R., Weiss, A., Bach, C., Heydmann, L., Turon-Lagot, V., Kopp, A., et al. (2020). Combined small molecule and loss-of-function screen uncovers estrogen receptor alpha and CAD as host factors for HDV infection and antiviral targets. *Gut, 69*(1), 158–167.

Vogels, G. D., & Van der Drift, C. (1976). Degradation of purines and pyrimidines by microorganisms. *Bacteriological Reviews, 40*(2), 403–468.

Wallach, D. P., & Grisolia, S. (1957). The purification and properties of hydropyrimidine hydrase. *Journal of Biological Chemistry, 226*(1), 277–288.

Wang, L. H., & Strittmatter, S. M. (1997). Brain CRMP forms heterotetramers similar to liver dihydropyrimidinase. *Journal of Neurochemistry, 69*(6), 2261–2269.

Xu, Z., Liu, Y., Yang, Y., Jiang, W., Arnold, E., & Ding, J. (2003). Crystal structure of D-hydantoinase from Burkholderia pickettii at a resolution of 2.7 angstroms: Insights into the molecular basis of enzyme thermostability. *Journal of Bacteriology, 185*(14), 4038–4049.

Yang, Y. S., Ramaswamy, S., & Jakoby, W. B. (1993). Rat liver imidase. *Journal of Biological Chemistry, 268*(15), 10870–10875.

Yang, Y. L., Ramaswamy, S. G., & Jakoby, W. B. (1998). Enzymatic hydrolysis of organic cyclic carbonates. *Journal of Biological Chemistry, 273*(14), 7814–7817.

Zhang, X. Y., Niu, L. X., Shi, Y. W., & Yuan, J. M. (2008). The flexibility of the non-conservative region at the C terminus of D-hydantoinase from Pseudomonas putida YZ-26 is extremely limited. *Applied Biochemistry and Biotechnology, 144*(3), 237–247.

CHAPTER FOUR

# Catalytic activity regulation through post-translational modification: the expanding universe of protein diversity

**Michael Kokkinidis**[a,b], **Nicholas M. Glykos**[c,*] **and Vasiliki E. Fadouloglou**[c,*]
[a]Institute of Molecular Biology and Biotechnology, Heraklion, Crete, Greece
[b]Department of Biology, University of Crete, Voutes University Campus, Heraklion, Crete, Greece
[c]Department of Molecular Biology and Genetics, Democritus University of Thrace, Dragana University Campus, Alexandroupolis, Evros, Greece
[*]Corresponding authors: E-mail: glykos@mbg.duth.gr; fadoulog@mbg.duth.gr

## Contents



## Abstract

Protein composition is restricted by the genetic code to a relatively small number of natural amino acids. Similarly, the known three-dimensional structures adopt a limited number of protein folds. However, proteins exert a large variety of functions and show a remarkable ability for regulation and immediate response to intracellular and extracellular stimuli. To some degree, the wide variability of protein function can be attributed

*Advances in Protein Chemistry and Structural Biology*, Volume 122
ISSN 1876-1623
https://doi.org/10.1016/bs.apcsb.2020.05.001

to the post-translational modifications. Post-translational modifications have been observed in all kingdoms of life and give to proteins a significant degree of chemical and consequently functional and structural diversity. Their importance is partly reflected in the large number of genes dedicated to their regulation. So far, hundreds of post-translational modifications have been observed while it is believed that many more are to be discovered along with the technological advances in sequencing, proteomics, mass spectrometry and structural biology. Indeed, the number of studies which report novel post translational modifications is getting larger supporting the notion that their space is still largely unexplored. In this review we explore the impact of post-translational modifications on protein structure and function with emphasis on catalytic activity regulation. We present examples of proteins and protein families whose catalytic activity is substantially affected by the presence of post translational modifications and we describe the molecular basis which underlies the regulation of the protein function through these modifications. When available, we also summarize the current state of knowledge on the mechanisms which introduce these modifications to protein sites.

## 1. Introduction

Post-translational modifications (PTMs) are the chemical changes occurring at the side chains or the main chain of a protein after the completion of its translation in the ribosome. These modifications involve different types of chemical reactions and their final products may include addition of new chemical moieties, double bond formation and formation of cross-linked products, L-to D-amino acid conversion, heteroatom substitution and proteolytic cleavages (Bischoff & Schlüter, 2012; Lin, Du, & Jiang, 2008; Müller, 2018; Okeley & Van Der Donk, 2000). Addition of small or larger chemical moieties to the side chains are by far the most common PTMs in proteins. Hydroxylation, methylation, acetylation, phosphorylation and glycosylation are some of the commonest and best studied PTMs as far as the source of the prosthetic group, the type of reaction and the biological impact is concerned (Lin, Du, & Jiang, 2008). Ubiquitylation is another major modification which involves the addition of the whole ubiquitin protein at specific sites of another protein to target it for degradation. Similar to the addition of chemical groups to the side chains, small groups usually hydroxyl or methyl moieties can also be added to the protein backbone (Müller, 2018). Cross linking of spatially adjacent side chains is generally a rare modification. However, the disulphide bond which is one of the commonest and more frequent PTM is formed by the cross linking of side chains. Likewise, a side chain could react with an adjacent backbone atom and form a ring product or a new residue. Deamidation of asparagine provides a characteristic example which proceeds through a cyclic succinimide intermediate. Dependent on the local environment succinimide may be the final product (Kumar et al., 2016), even though it is usually unstable and opens to isoAsp which has a shortened side chain compared to Asp and an one-carbon atom longer backbone compared to usual residues (Bischoff & Schlüter, 2012; Ravikiran & Mahalakshmi, 2014; Müller, 2018). Heteroatom substitution, as for example thioamide formation and Cα-Cβ double bond formation are two unusual PTMs with impact on backbone chemistry and flexibility. On the other hand, the backbone hydrolytic cleavage is a well-established and common modification.

PTMs can have multiple effects on a protein such as, for example, they can change the physicochemical properties, increase the steric hindrances or change the chemical reactivity of a specific site. Thus, a prosthetic group might substantially change hydrophobicity/hydrophilicity, charge and hydrogen bonding potential of the site where it is introduced. For instance, histone acetylation activates transcription, in part, by relaxation of chromatin structure through masking the lysine's positive charge and thus, attenuating interactions with the negatively charged DNA. Likewise, histone methylation influences the histone-DNA interactions by the addition of a hydrophobic group. The prosthetic group might impose steric hindrances and restrict accessibility to a biological important site i.e. an active, binding or recognition site. For example, such a role is usually assigned to AMPylations, which restrict access to significant sites of the modified protein. Moreover, a prosthetic group may induce structural changes with impact on the local or overall structure. Typically, backbone modifications influence structural and conformational stability and flexibility with a local or long-range effect (Craveur, Narwani, Rebehmed, & de Brevern, 2019; Müller, 2018). In addition, a PTM may provide a novel functional group which complement the chemistry of the site where it is introduced and potentially generate novel cofactors (Okeley & Van Der Donk, 2000).

PTMs may be reversible or irreversible, enzymatically installed or spontaneous, strictly regulated by cell mechanisms or not under any cell control. They may be dependent on the concentration of metabolism intermediate products and environmental factors such as oxidation stress and adaptation pressure. Thus, PTMs have been associated with feedback mechanisms of metabolism, regulation of biochemical pathways and often reflect a cellular ability for rapid and local response to a sharp stimulus. Sometimes PTMs are the result of chemical or radiation damage due to exposure to stressful conditions or due to the natural accumulation of deleterious factors during the cell's lifespan. In these cases, they are associated with aging and age-related disorders and can be utilized as biomarkers or as tools to guide drug design procedures.

Current literature affords many excellent reviews which thoroughly and systematically cover specific and general topics related to PTMs. Here, our aim is not to provide an exhaustively detailed overview of the field. Instead, we anthologize and briefly present a few examples from the very recent literature which either describe a novel modification or provide new findings which expand our knowledge and understanding of how PTMs regulate protein function and especially how they modify enzymatic activity and consequently control biological pathways. These examples show the chemical and structural complexity involved in such processes and underline the still-to-be-discovered biological potential of PTMs.

## 2. Post-translational modifications at protein backbones

### 2.1 A backbone proline hydroxylation promotes active-site maturation

Proline hydroxylation, primarily at Cγ and secondarily at Cβ atoms (Bischoff & Schlüter, 2012), is a common enzymatic post-translational modification mostly catalyzed by prolyl hydroxylases. For example, Cγ-hydroxylated proline is a recurrent feature of collagen with a profound structural role at the stabilization of the triple helix (Myllyharju & Kivirikko, 2001). Hydroxyproline is also present at the plant cell wall and in proteins related to hypoxia response (Lamport, Kieliszewski, Chen, & Cannon, 2011; Palmer & Clegg, 2014). On the other hand, proline hydroxylation is largely unexplored in bacteria. Although a putative bacterial peptidyl-propyl hydroxylase has been detected, its specific substrate and biochemical pathway remain unknown (Culpepper, Scott, & Limburg, 2010).

In this context, it was of great surprise the discovery of a modified proline hydroxylated at its Cα atom in a bacterial enzyme named Bc0361 (Fadouloglou et al., 2013). Bc0361, is a *Bacillus cereus* polysaccharide deacetylase from CE4 family (Fig. 1A) and the modified residue is a conserved proline lying in the active site (Balomenou & Arnaouteli, 2015). The modification was first detected when during refinement of the crystal structure extra electron density attached to that proline was best modeled as Cα-hydroxylated-L-proline (Fadouloglou et al., 2017). The modification was then confirmed by tandem mass spectrometry. In addition, it was detected in many other

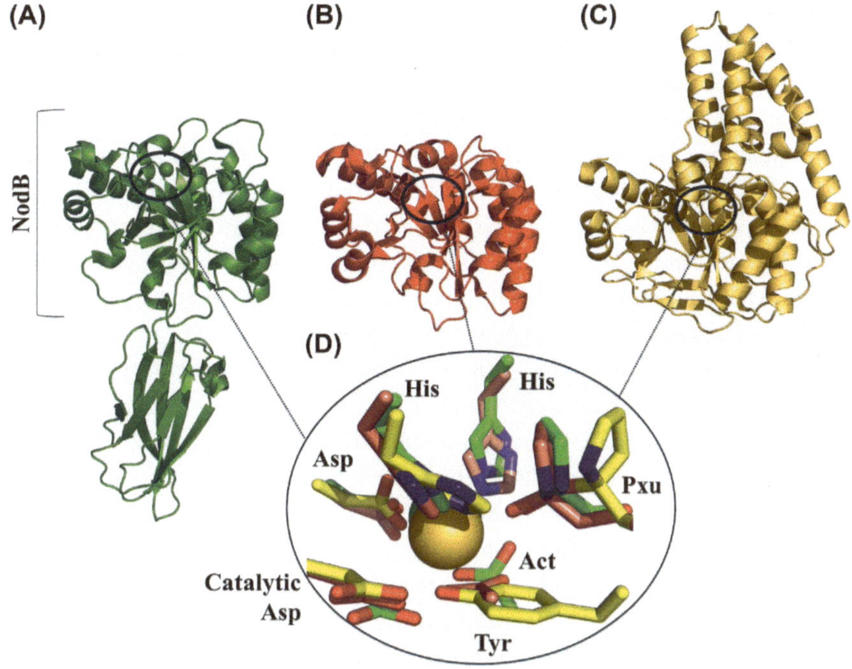


**Fig. 1** Crystal structures of three CE4 deacetylases whose active site proline is hydroxylated at the Cα (Pxu). For comparison the molecules are shown at the same orientation. Ribbon representations of the (A) *Bacillus cereus* Bc0361 (PDB 4HD5), (B) *B. cereus* Bc1960 (PDB 4L1G) and (C) *Bdellovibrio bacteriovorus* Bd3279 (PDB 5JP6). (D) Close-up view of the superimposed active sites. The following residues are shown (coloring as in panels (A)–(C)): Cα hydroxylated Pro (Pxu); His-His-Asp metal coordination triad; catalytic Asp; the fourth ion-ligand is an acetate molecule (Act) for Bc0361 and Bc1960 and a Tyr residue for Bd3279. The ion is shown as sphere (it is only present in the structures of Bc0361 and Bd3279).

CE4 family polysaccharide deacetylases from the gram positive bacteria *B. cereus* and *B. anthracis* (Fadouloglou et al., 2017). Indeed, mass spectrometric (MS/MS) characterization of the hydroxylation levels at the CE4 family of deacetylases showed significant percent occurrence at the conserved proline position for many members of the family i.e. Ba1961, Ba0330, Ba3942, Bc1960, produced by different expression systems and under different conditions (Fadouloglou et al., 2017). In addition to that, Cα-hydroxylation of the conserved proline has been observed in the crystal structures of *B. anthracis* Ba0330 (Arnaouteli et al., 2015), *B. cereus* Bc1960 (Fadouloglou et al., 2017) and *Bdellovibrio bacteriovorus* Bd3279 (Lambert et al., 2016) polysaccharide deacetylases (Fig. 1, A–C).

In the CE4 family of polysaccharide deacetylases, catalysis is performed by a conserved NodB domain, which folds as a distorted TIM-like barrel (Fig. 1, A—C; Balomenou & Arnaouteli, 2015). The active site contains a divalent ion coordinated by a His-His-Asp triad and it is complemented by a catalytic Asp and a His/Asp electron relay (Fig. 1D; Fadouloglou et al., 2013). The modified proline is located at the one side of the active site and the prosthetic hydroxyl group points toward the metal ion and the substrate binding site (Fig. 1D). Although the mechanism of hydroxylation is still elusive, recent findings suggest a non-enzymatic hydroxylation implicating at least one of the deacetylation reaction catalytic residues i.e. the catalytic Asp. Moreover, the origin of the hydroxyl moiety appears to be molecular oxygen rather than water (Fadouloglou et al., 2017).

Measurement of the deacetylase activities of modified and unmodified Bc1960 shows that the modified enzyme is more active by a factor of 10. Inspection of the structure and a recent computational analysis (Prejanò, Romeo, Sgrizzi, Russo, & Marino, 2019) underline that the proline hydroxylation substantially extends the hydrogen-bonding network in the active site and contributes to the stability of the intermediate state. Prejanò et al. performed a quantum mechanistics (QM) cluster model study on the reaction mechanism of Bc1960 considering the canonical and Cα-hydroxylated active site proline and showed that the hydroxyl moiety on the modified Cα establishes an extended interaction network with proline's closest neighbors. The result is a substantial stabilization of the transition state and a decrease in the activation energy. Thus, the addition of such a small chemical moiety in the active site can cause significant improvements in enzymatic activity and contributes to active site maturation.

Collectively, Cα-hydroxylation at a strictly conserved, active-site proline residue contributes to the active site maturation by enhancing the hydrogen-bonding network and thus stabilizing the transition state.

## 2.2 Asparagine deamidation to isoAspartate impairs binding interactions with ligand

Asparagine deamidation is a commonly occurring spontaneous modification which starts with the nucleophilic attack of a backbone amide nitrogen to the side chain amide of an adjacent Asn to form a cyclic succinimide intermediate (Bischoff & Schlüter, 2012). Succinimide is usually unstable and is hydrolyzed to either Asp or isoAsp. Asn deamidation is usually accumulated in proteins with time, therefore it has been strongly correlated with protein aging such as, for example, calmodulin, crystallin, β-amyloid and human growth hormone aging as well as age-related disorders such as cataract and neurodegenerative diseases (Fujii, Takata, Fujii, & Aki, 2016; Johnson et al., 1989; Moro et al., 2018; Müller, 2018; Potter, Henzel, & Aswad, 1993; Reissner & Aswad, 2003). However, their involvement to other biological processes is increasingly reported (Eschenburg & Schönbrunn, 2000) revealing a functional potential which extends from protein thermostability (Kumar et al., 2016) to viral pathogenicity (Mallagaray et al., 2019).

Asn deamidation to isoAsp was recently detected on the glycan-binding site of norovirus capsid protein by triple-resonance 3D NMR experiments and verified by crystallography and hydrogen/deuterium exchange (HDX) MS (Mallagaray et al., 2019). The dense layer of glycans which cover the surface of red cell membrane and consist the major part of histo-blood group antigens (HBGAs) can be exploited by microorganisms and viruses to be attached to the host-cell membrane (Heggelund, Varrot, Imberty, & Krengel, 2017; Shanker et al., 2017). Such a virus is the human norovirus, the causative agent of acute gastroenteritis, which recognizes and binds HBGAs to achieve infection (Taube, Mallagaray, & Peters, 2018). The norovirus capsid protein comprises a shell domain and a protruding (P) domain which exists as a dimer (Fig. 2A) and is responsible for host-cell interactions and strain diversity (Singh, Leuthold, & Hansman, 2015).

The virus-glycan interaction implicates a HBGA binding pocket on the surface of the P domain dimer. Mutations highlighting the evolution of P protein among the viral variants are mainly located on the protein surface except for the binding pocket beneath the HBGA and the surrounding regions which show little variation. In particular, on the rim of the binding pocket is found a common set of conserved residues which form numerous interactions with divergent HBGA types (Singh et al., 2015). One of them is Asp374 immediately after Asn373, the residue which is irreversible modified to isoAsp373 (Fig. 2; Mallagaray et al., 2019).

The formation of isoAsp dramatically changes the binding potential of P protein for HBGAs and causes a loss of affinity of approximately one order of magnitude (Mallagaray et al., 2019). It has been shown that isoAsp can affect protein in two ways which could explain the observed decrease in HBGA binding affinity. Firstly, there is a local effect on recognition and binding of glycans. It has been shown that Asp374 forms two direct hydrogen bonds with a fucose residue of the glycan ligand. Inspection of the structure reveals that in the deamidated P-protein the side chain of Asp374 adopts a different rotamer (Fig. 2B)-probably induced by the backbone changes occurring on the adjacent residue 373- which results in loss of one hydrogen bond (Fig. 2C).

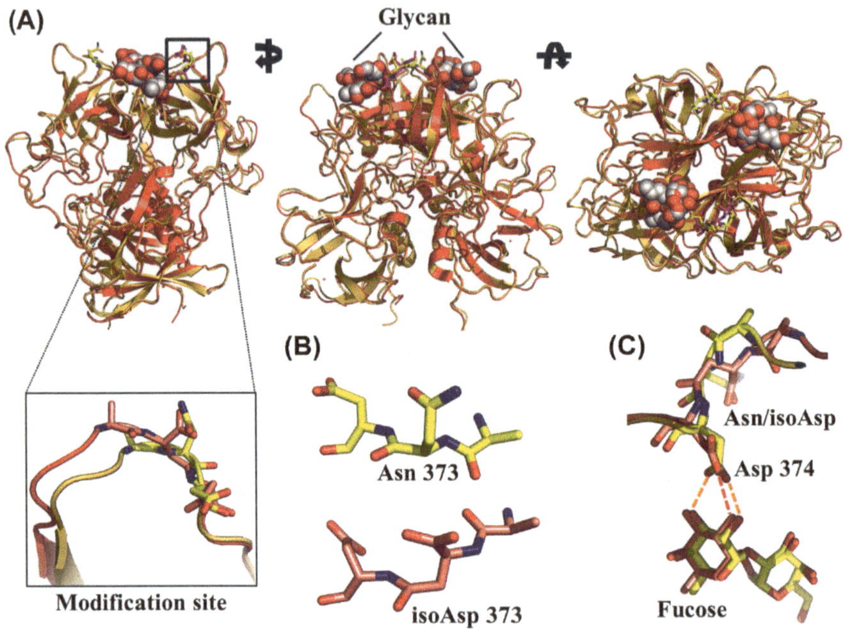


**Fig. 2** Crystal structure of norovirus P domain dimer. (A) Superimposed ribbon diagrams of P domain dimers containing the unmodified Asn 373 (yellow, PDB 4X06) and the modified isoAsp 373 (red, PDB 6H9V). Bound glycan is shown with space filling model. The modification site is shown with sticks and highlighted in the insertion. (B) Comparison of the conformations adopted by the peptides which contain unmodified Asn 373 and modified isoAsp 373. (C) Close-up view of the P domain-glycan complex structure. The fucose moiety of the glycan is held by two direct hydrogen bonds from the side chain of Asp 374 when 373 is unmodified Asn. Instead, only one hydrogen bond remains when 373 is converted to isoAsp.

Secondly, isoAsp formation has a long-range effect on protein dynamics, which is known to have significant impact on the ability of P-dimers to efficiently recognize and bind divergent HBGAs (Singh et al., 2015). NMR and HDX MS data agree that in the deamidated protein certain regions of P dimer display higher flexibility compared to the native protein independently on the presence of glycans. Interestingly, the elevated flexibility is not detected in the peptide which contains the isoAsp.

The deamidation of Asn373 is an irreversible modification and depends on the temperature (Mallagaray et al., 2019). It was shown that at room temperature or lower the modification occurs slowly displaying a half-life of several days or weeks. Conversely, in body temperature the half-time of modification is just 1–2 days. Moreover, it is shown that the Asn373-Asp374 motif (Fig. 2) as well as the conformation of the loop which incorporates the motif are well conserved. Thus, the Asn deamidation may represent a conserved viral mechanism for modulating recognition and binding of glycans in the course of infection.

Collectively, Asn deamidation to isoAsp occurs on an area which is critically involved in glycan binding. In addition, the modification induces changes in overall protein dynamics and protein's binding potential. Accordingly, the modification causes a significant loss of binding affinity for glycans.

## 3. Metabolism regulation by post-translational modifications

### 3.1 Multiple post-translational modifications in a single active site

Methyl coenzyme M reductase (MCR) is universally used by methanogenic archaea to catalyze both the final step of methane-forming reaction and the first step of the anaerobic oxidation of methane (Wongnate & Ragsdale, 2015). Protein homologs share a common overall structure and a highly conserved active site. They are assembled in hexamers of three different subunits ($\alpha\beta\gamma$). The subunits are organized in a dimer of trimers forming two equivalent active sites (Fig. 3A; Ermler, Grabarse, Shima, Goubeaud, & Thauer, 1997; Grabarse et al., 2001). The active site pocket incorporates a redox-sensitive nickel tetrahydrocorphinoid (coenzyme F430) and a coenzyme B. The enzyme needs both these factors to catalyze the reduction of methyl coenzyme M to produce methane (Fig. 3A; Grabarse et al., 2001). Interestingly, the $\alpha$ subunit of MCR features several, mostly unusual but well conserved post-translational modifications located within the active site (Fig. 3A). The modifications, which have been confirmed by numerous crystal structures and mass spectrometric data (Ermler et al., 1997; Grabarse, Mahlert, Shima, Thauer, & Ermler, 2000; Grabarse et al., 2001) include a group of methylated residues i.e. methyl-histidine, methyl-arginine, methyl-cysteine and methyl-glutamine, a thio-glycine, a hydroxyl-tryptophan and a didehydro-aspartic acid (Fig. 3B).

Besides a heavily modified active site, MCR incorporates quite uncommon modifications. Arginine is modified at C-5 instead of the guanidino nitrogen atom where arginine methylation usually occurs (Fig. 3B). Histidine is methylated at N$\pi$ which is a common modification site for this residue. However, histidine methylation is an unusual modification. Methyl-His and thio-Gly are strictly conserved in all MCRs examined to date (Mahanta, Liu, Dong, Nair, & Mitchell, 2018; Nayak, Mahanta, Mitchell, & Metcalf,

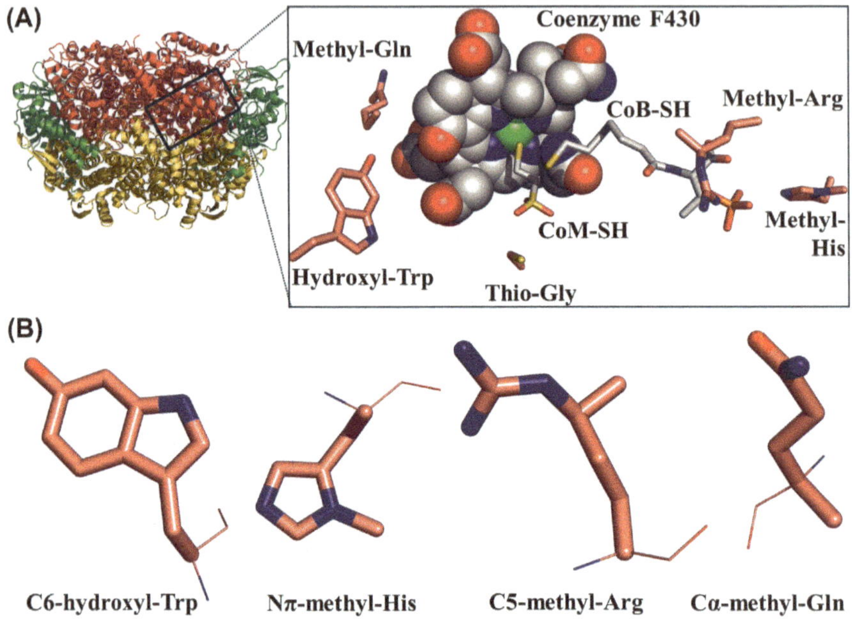


**Fig. 3** Crystal structure of Methyl coenzyme M reductase (MCR). (A) Ribbon representation of the *Methanothermococcus thermolithotrophicus* MCR III (PDB 5N1Q). The different colors represent each of the α, β and γ subunits. The insert is a close-up view of the active site which shows the arrangement of the modified residues. The tetrahydrocorphinoid (coenzyme F430) is represented as a space-filling model. CoB-SH and CoM-SH are also shown. (B) Stick models of modified residues commonly found in the MCR active site.

2017). Thio-Gly is a highly unusual protein modification, where the peptide amide bond is converted to thioamide i.e. the carbonyl oxygen has been replaced by sulfur. As a result, a higher multiple bond character is observed along the C—N bond and consequently a larger rotational barrier. Unlike the other methylated residues methyl-Gln is methylated at the Cα atom instead of the side chain. The didehydro-Asp contains a double bond between Cα and Cβ atoms (Siodłak, 2015; Wagner, Kahnt, Ermler, & Shima, 2016) and it is present only in a subgroup of MCR isoenzymes (Wagner, Wegner, Kahnt, Ermler, & Shima, 2017). Likewise, the 6-hydroxyl-Trp, which is a novel modification is not highly conserved among MCRs (Wagner et al., 2017).

Although post-translational modifications in the MCR have been known for many years, the elucidation of their exact impact on catalytic mechanism and the identification of their source remain largely unresolved.

The newly discovered Trp hydroxylation to C-6 introduces an additional polar group and therefore increases the hydrogen bonding network in the active site (Wagner et al., 2017). Indeed, inspection of the crystal structures demonstrates that the hydroxy-group is in hydrogen-bonding distance with main chain and side chain atoms. On the other hand, the double bond between Cα and Cβ in the didehydro aspartic acid causes a local backbone distortion which also affects the interaction distances of the polar side chain group with adjacent residues. Recent findings show that the modification is reversible since it is easily reduced to aspartate (Wagner et al., 2016). However, no evidence exists for the biochemical pathway and the mechanism which generates this modification.

The didehydro-Asp is adjacent to the thio-Gly and both residues are found on a bend structure. Indeed, the functional role which has been proposed for thio-Gly is stabilization of the protein secondary structure in the active site (Nayak et al., 2017). The formation of thio-Gly is enzymatic and at least two enzymes i.e. YcaO and TfuA have been identified to be involved in its biosynthesis (Mahanta et al., 2018). Thio-Gly is not essential for MCR activity but it does affect methanotrophic archaea growth under specific conditions of nutrient and temperature (Nayak et al., 2017). Furthermore, thio-Gly does not influence the MCR global stability since the thermal stability of the protein is not significantly affected by the presence of this modification. Accordingly, it has been proposed that the role of the thioamide is to maintain the architecture of the active site and the conformations of the surrounding amino acids.

Methylation of methyl-modified residues in MCR has long been presumed to be catalyzed by S-adenosylmethionine-dependent methyltransferases (Selmer et al., 2000). However, only recently a study was able to correlate the MCR methyl-Arg to MmpO protein, which is a S-adenosylmethionine methylase (Lyu et al., 2020; Radle, Miller, Laremore, & Booker, 2019). The study not only demonstrates that MmpO and methyl-Arg are correlated in a highly specific manner but also shows that arginine methylation has an impact in methanogenesis and cell growth (Lyu et al., 2020).

MCR serves a chemically challenging and complex catalytic mechanism. PTMs complement the active site with important chemical features and largely contribute to the necessary fine-tuning (Nayak et al., 2020). Thus, PTMs optimize the catalysis in such a way that they are indispensable.

## 3.2 Lysine acylations: novel discoveries of PTMs of a commonly modified residue

Lysine is commonly modified at the side-chain nitrogen atom by alkylation and acylation reactions. Mono-, di- and tri-methylations as well as acetylation, formylation and ubiquitinylation belong to the most well-known lysine modifications (Zee & Garcia, 2012) with significant impact on biological systems. Particularly, lysine methylation and acetylation have attracted much attention mainly because of their relation to gene transcription regulation through histone modifications (Karve & Cheema, 2011; Lin, Du, & Jiang, 2008). Histone acetylation is primarily correlated with transcription activation through a relaxation of chromatin packing on the nucleosome. In addition, acetylated Lys is able to recruit other proteins such as bromodomain containing proteins which help with gene transcription activation. On the other hand, histone methylation is associated with both transcription activation and inactivation. Histone acetylases and methylases catalyze these modifications during reactions which chiefly use acetyl-CoA and S-adenylosyl-methionine (SAM) as acetylating and methylating agents, respectively. Likewise, removal of these modifications is enzymatically catalyzed by histone deacetylases and demethylases. At least two types of histone deacetylases are known, the zinc-dependent hydrolases and the NAD-dependent sirtuins. It is interesting that sirtuins can couple NAD-degradation to Lys-deacetylation by sensing the cell metabolic state i.e. sirtuins sense the NAD concentration in the cell and regulate Lys deacetylation accordingly (Imai, Armstrong, Kaeberlein, & Guarente, 2000; Sauve, Wolberger, Schramm, & Boeke, 2006).

In the last decade a growing number of new, non-enzymatic Lys-acylations has been reported and their significance as metabolism regulators has been demonstrated (Lin, Su, & He, 2012; Harmel & Fiedler, 2018). The acylation group typically originates from a cellular metabolite which is converted to acyl-CoA that is the most common acylating agent. This kind of spontaneous modifications are usually enzymatically reversible, occur to metabolic enzymes and regulate their function -activate or inactivate them (Lin et al., 2012). For instance, it was shown that lysine acetylation, apart from its well established role in histone regulation, is also present in enzymes that catalyze metabolic processes including glycolysis, gluconeogenesis, glycogen and fatty acid metabolism, urea and tricarboxylic acid cycle (Zhao et al., 2010). The acetylation state of the enzymes is greatly dependent on the concentration of metabolites such as glucose, amino acids and fatty acids. Moreover, it was shown that Lys-acetylation can cause different effects i.e. activation, deactivation or destabilization on different enzymes.

Another example is the selective modification of lysine residues in the active site of glycolytic enzymes by the glycolytic intermediate 1,3-bisphosphoglycerate. The formation of 3-phosphoglyceryl-lysines results in enzymatic inhibition which is believed that is a feedback mechanism for redirection of glycolytic intermediates to other biosynthetic processes (Moellering & Cravatt, 2013).

In 2014, Tan and workers detected for the first time a lysine glutarylation in metabolic enzymes and mitochondrial proteins (Tan et al., 2014). In particular, it was shown that the carbamoyl phosphate synthase 1, an enzyme important for ammonia detoxification in the urea cycle, is Lys-glutarylated and that the effect of this modification is enzyme inhibition. Moreover, it was demonstrated that the modification is reversible and that lysine deglutarylation is performed by sirtuin 5, an enzyme previously annotated as deacetylase (Tan et al., 2014).

The latest addition to the list of acylated lysines is the lactoylysine which was reported recently by two different groups. Gaffney and co-workers (Gaffney et al., 2020) showed that the new lysine modification is derived from methylglyoxal, a glycolytic by-product. According to the study, methylglyoxal is rapidly conjugated to glutathione to produce lactoylglutathione. Enzymatic hydrolysis of lactoylglutathione produces glutathione and lactate. Lactate can then modify lysine residues mainly of glycolytic enzymes to generate the lactoylysine modification which result in a decreased glycolytic output. Therefore, it was proposed that lysine lactoylation serves as a regulatory feedback mechanism for glycolysis. The lactoylysine levels are directly dependent on the levels of lactoylglutathione and its hydrolytic enzyme. Moreover, findings support a non-enzymatic acyl transfer of the lactoyl-group similar to the non-enzymatic acyl-transfer mechanism of lysine acetylation from acetyl-CoA and acetyl-glutathione.

The existence and biological relevance of lysine lactoylation was further verified by another group which reported a histone lactoylysine modification (Zhang et al., 2019), derived by a lactoyl-CoA intermediate. Zhang and co-workers showed that lactate production and histone lysine lactoylation levels were induced by glucose and regulated by glycolysis. Use of a non-metabolizable glucose analog resulted in a decrease of both lactate production and histone lysine lactoylation. Moreover, histone lysine lactoylation by lactate was proposed to serve as an epigenetic mechanism which directly stimulates gene transcription.

Collectively, a growing number of new Lys-acylations has been discovered the last decade. The source of these modifications are metabolic intermediate products and their targets are primarily enzymes involved in metabolism. Glucose is the major source of intracellular lactate and lactate is used to modify Lys to lactoylated-Lys which is the newest reported Lys-acylation and plays a role as regulator of glucose metabolism and stimulator of gene transcription.

## 4. AMPylation: new insights to an emerging modification

AMPylation is the transfer of AMP to a protein substrate during an enzymatically catalyzed reaction called adenylylation (Casey & Orth, 2018). AMPylases (or AMPylators) have the same substrate as kinases i.e. an ATP molecule and primarily modify Ser, Thr and Tyr residues. Two classes of AMPylases have been identified so far, those containing an adenyl transferase (ATase) domain and those containing a Fic domain. AMPylation is an emerging modification which influences the enzymatic activity of the modified proteins or alters the pattern of interactions with molecular partners. In this way, for instance, it is used for the rapid regulation of a protein folding chaperon in response to its substrate fluctuations or it is exploited by bacterial pathogens to manipulate the host cellular machinery.

### 4.1 AMPylation regulates protein folding

BiP is an Hsp70 protein folding chaperone located at the endoplasmatic reticulum (ER). The protein comprises a nucleotide binding domain (NBD) which is responsible for ATP binding and hydrolysis, a substrate binding domain (SBD) which binds unfolded protein and a linker between them (Fig. 4). It is known that in the course of BiP-mediated protein folding, the chaperon binds the unfolded client-protein in a process that is coupled to ATP hydrolysis and also involves major conformational changes. When BiP is in the ADP state (and in the apo form) adopts an extended, domain-undocked, conformation (Fig. 4B) and binds with high affinity the exposed hydrophobic residues of unfolded or misfolded proteins (Fig. 4C; Bertelsena, Chang, Gestwicki, & Zuiderweg, 2009). ATP binding, on the other hand, induce a conformational change to a more compact, domain-docked, structure (Fig. 4A), which results in release of the client protein (Yan, Rato, Rohland, Preissler, & Ron, 2019).

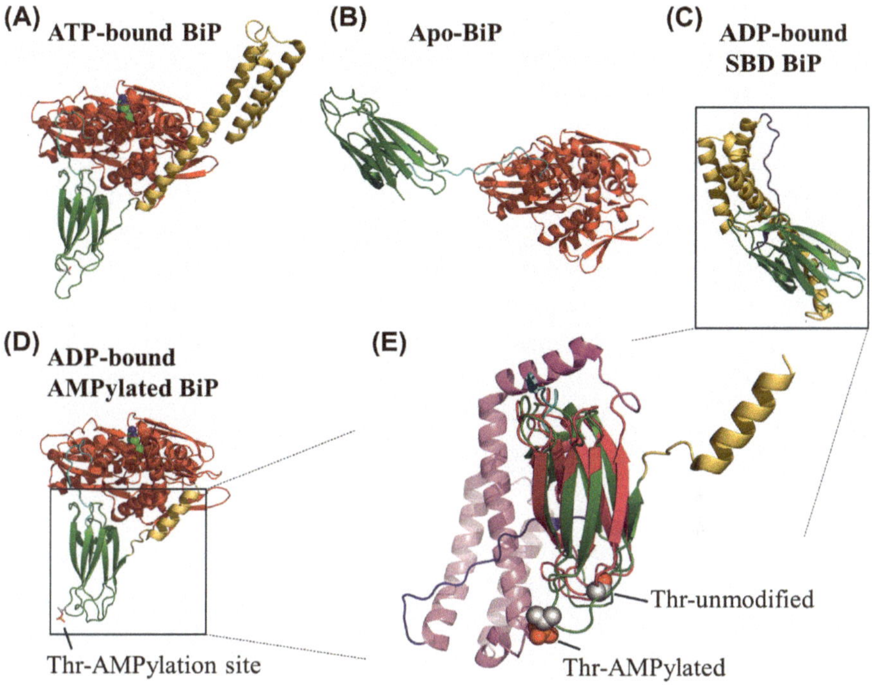


**Fig. 4** Crystal structures of BiP protein in different conformations. In panels (A)—(D) the color scheme is as follows: the nucleotide binding domain (NBD) is shown in red, the substrate binding domain (SBD) is shown in green, the C-terminal helical extension of SBD is shown in yellow. The structures are superimposed on their NBDs. (A) Ribbon representation of BiP in complex with ATP (space filling model). (B) Apo BiP which is structurally equivalent to the ADP-bound conformation. (C) SDB and its C-terminal helical extension. A peptide from the C-terminus occupies the peptide binding pocket and it is shown in blue. (D) Ribbon representation of AMPylated BiP in complex with ADP (space filling model). (E) Comparison of the SDBs and their C-terminal helical extensions from panels (C) and (D). Here, the structure from (C) panel is shown in magenta for clarity.

Because the load of unfolded proteins is continually fluctuating, the levels of active BiP must be accordingly regulated by a rapid and highly effective mechanism. It has long been speculated that such a mechanism should be regulated at the post-translational level. Indeed, recent evidence supports a BiP regulation mechanism which is mediated by BiP AMPylation through the FICD enzyme of the ER. The mechanism is regulated as follows: when the load of unfolded proteins is low, FICD transfers an AMP moiety to a specific Thr residue of BiP and AMPylates it (Fig. 4D and E). This modification inactivates BiP since the modified BiP binds less firmly peptide-substrates than the unmodified or it has an impaired response in particular stimuli. Notably, when there is a high load of unfolded proteins, FICD activates a reverse mechanism and deAMPylates Thr (Perera et al., 2019).

Crystal structure determination of the AMPylated BiP complexed with ADP revealed that modification favors a domain-docked BiP conformation (Fig. 4D) similar to that adopted by the ATP-bound chaperone (Fig. 4A) instead of the one adopted by the ADP-bound unmodified BiP (Fig. 4B; Preissler et al., 2017). This conformation corresponds to the low affinity binding of BiP for client-proteins (Fig. 4E). Moreover, it was shown that apart from the consequences that AMPylation has on the substrate binding, it also affects the interactions of BiP on J-domain co-chaperones. J-domain are non-client bound proteins which assist BiP by stimulating the ATPase activity.

## 4.2 AMPylation mediates bacterial pathogenicity and host-defense mechanisms

*Legionella pneumophila* is a gram-negative bacterium, the causative agent of Legionnaires' disease. During infection, *L. pneumophila* is internalized into the host cells and forms a protective membrane-vacuole where replicates. To achieve replication, the bacterium secretes to the host an arsenal of effector proteins which modulate host-cell functions. DrrA and SdeA are two *L. pneumophila* effector proteins whose activity is related to AMPylation and they will be discussed in detail below.

*L. pneumophila* DrrA protein AMPylates the eukaryotic host's Rab1b GTPase therefore provides an example of AMPylation performed by a bacterial protein to a eukaryotic one. Crystal structure determination of DrrA revealed structural similarities with nucleotidyl transferases. In particular, the N-terminal domain of DrrA folds similarly to the C-terminal domain of glutamine synthetase adenylyl transferase (GS-ATPase) and contains a motif similar to the GS-ATPase catalytic motif. Initially, these observations raised speculations for the AMPylase activity of DrrA and led to the experiments which confirmed this activity. Subsequently, it was shown that the high cytotoxicity of DrrA for mammalian cells is associated with its activity to AMPylate the Rab1b GTPase (Marcucci et al., 2010).

In particular, tandem MS and site-directed mutagenesis demonstrated that the AMPylation site is a surface Tyr residue on the switch II region of Rab1b GTPase (Fig. 5A; Marcucci et al., 2010). Switch II region is known to be involved in binding to effector and GTPase-activating proteins

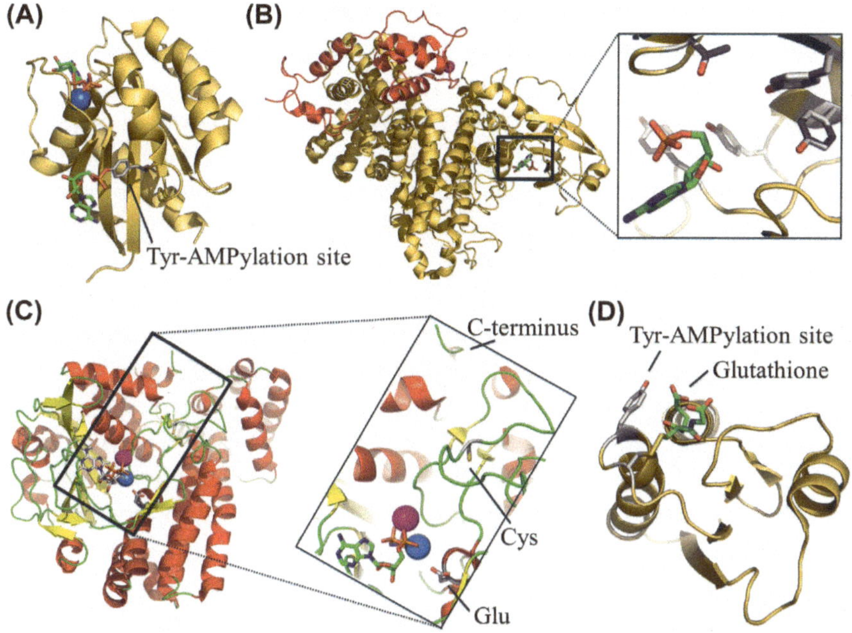


**Fig. 5** Crystal structures of proteins associated with AMPylation. Nucleotides and analogs are shown as green sticks, side chains as gray sticks and metal ions as spheres (blue for Mg, magenta for Ca). (A) Ribbon representation of the human Rab 1b covalently modified with AMP at Tyr 77 (PDB 3NKV). It is shown (i) a GppNHp and a magnesium ion (blue sphere) bound in the GTP-binding site (ii) the AMPylated Tyr on the switch II area of the protein. (B) Ribbon representation of *Legionella pneumophila* SidJ (yellow) in complex with CaM (red) and AMP (stick model) (PDB 6K4R) (PDB 6K4R). The insertion is a close-up view of the AMP-binding pocket. Several residues within the pocket could be possible AMPylation sites (sticks in gray). (C) Ribbon representation of the *Pseudomonas syringae* SelO in complex with AMP-PNP (PDB 6EAC). Insertion is a close-up view of the nucleotide binding site. It is shown a Glu residue with a possible catalytical role and the Cys in the enzyme's activating loop which forms a disulfide bridge with the C-terminal Sec/Cys. (D) Ribbon representation of the *E.coli* Glutaredoxin (PDB 1GRX). The AMPylation site (Tyr) and the Cys which interacts with glutathione are shown in stick representation.

(GAPs). Structural comparison of unmodified and modified Rab1b GTPases showed that no significant structural changes are induced from the AMPylation consistent with the location of AMP group to the exterior of the structure (Fig. 5A). On the other hand, AMPylation significantly inhibits GAPs stimulating GTP hydrolysis and prolongs Rab1b GTP state presumably by restricting the access of GAPs to switch II region. Likewise, AMPylation selectively affects interactions with effector proteins. For instance, it was shown that AMPylation abolish binding to the mammalian effector MICAL-3 but not to the bacterial LidA (Marcucci et al., 2010).

*L. pneumophila* SdeA is a mono-ADP ribosyltransferase involved in ubiquitination of host proteins associated with the ER. Bacterial dependent ubiquitination of host proteins starts with the ADP-ribosylation of specific ubiquitin sites, an activity catalyzed by SdeA. The modified ubiquitin impairs the function of eukaryotic cells by inhibiting canonical ubiquitination. Thus, SdeA is an important bacterial virulence factor. The deleterious function of SdeA for the host is inhibited by the post-translational modification of a catalytically important residue. Activation of this residue for modification is achieved through an AMPylated intermediate (Gan et al., 2019). Especially, it was shown that SdeA toxicity can be suppressed by another bacterial protein named SidJ. SidJ inhibits the mono-ADP ribosyltransferase activity of SdeA through glutamylation of a catalytically important Glu residue. SidJ exerts the glutamylase activity and modifies Glu by adding one or more glutamate moieties (Gan et al., 2019).

Interestingly, the inhibitory effects of SidJ to SdeA were only evident when the proteins were expressed in eukaryotic cells supporting the hypothesis that the SidJ/SdeA interaction is mediated by a eukaryotic factor. Indeed, it was found that SdeA can only be modified in the presence of calmodulin (CaM) and that CaM/SidJ/L-glutamate complex suppressed SdeA-mediated ubiquitination (Fig. 5B; Gan et al., 2019).

Further examination of the SidJ catalyzed reaction revealed that glutamylation mechanism involves ATP hydrolysis to AMP (Fig. 5B). Moreover, when the reaction cannot be productive i.e. in the absence of substrate (SdeA) or glutamate, SidJ undergoes self-AMPylation. Taking together these findings it was proposed that SdeA-Glu glutamylation from SidJ proceeds through an AMPylated-Glu intermediate as follows: First, SidJ activates its target i.e. SdeA-Glu by its adenylation (AMPylation). Then, the glutamate amino group attacks the activated carbonyl group of the unstable intermediate Glu-AMP to form a Glu-glutamate final modification (Gan et al., 2019).

Collectively, bacterial SidJ is a CaM-dependent glutamylase that catalyzes the ligation of glutamates to a catalytically important Glu of bacterial SdeA. As a result, SdeA is inactivated and its eukaryotic toxicity is abolished. The mechanism of inhibition includes an intermediate state where a transient Glu-AMP is formed. In addition, when SidJ cannot catalyze a productive reaction undergoes self-AMPylation.

## 4.3 The SelO pseudokinase is actually an AMPylase

Perhaps the most exciting example of AMPylation comes from an evolutionary conserved seleno pseudokinase (SelO). Members of SelO family have been characterized as pseudokinases because they share sequence similarity with kinases and they lack amino acids which are important for catalysis. Interestingly, SelO homologs from higher eukaryotes incorporate the uncommon 21st genetically encoded amino acid namely Sec. Sec is structurally equivalent to Cys except for the sulfur atom which is replaced by a selenium. Despite their similarities, Sec has a lower pKa than Cys and it is deprotonated at physiological pH; therefore, it is a stronger nucleophile with higher oxidoreductase efficiency. There is a single Sec at the C-terminus of the Sec-containing SelO homologs, while in lower eukaryotes and prokaryotes the equivalent position is invariably occupied by a Cys residue (Sreelatha et al., 2018).

Crystal structure determination of the *Pseudomonas syringae* SelO homolog in complex with ATP (Fig. 5C) confirmed previous predictions that the protein adopts a protein kinase fold and revealed an unexpected flipped binding orientation for ATP (Sreelatha et al., 2018). This accidental observation raised questions about a possible previously overlooked protein functionality and triggered a series of experiments. It was demonstrated that SelO can transfer AMP from ATP to protein substrates, therefore it is an AMPylase (Sreelatha et al., 2018). SelO activity depends on the presence of divalent cations in the active site (Fig. 5C), similarly to kinases. Interestingly, SelO activity also depends on the formation of a disulfide bond between the conserved Cys/Sec at the C-terminus and a Cys in the enzyme's activation loop (Fig. 5C). Furthermore, the structural inspection of SelO suggests that the originally considered as impaired active site, because of the lack of a catalytically important Asp, may not be actually impaired. The apparently absent catalytic Asp could be afforded by a different position which better suits the new, flipped substrate orientation.

Two proteins have been identified as direct protein targets for SelO namely Glutaredoxin (GlxA) and the SucA component of the α-ketoglutarate dehydrogenase. Both, SucA and GlxA are involved in redox processes i.e. the first is a redox sensor and the latter removes glutathione from cysteines in an oxidative stress defense mechanism conserved in eukaryotes and prokaryotes. It was shown that SelO AMPylates a Tyr residue of the highly conserved redox active site of GlxA (Fig. 5D). AMPylation inactivates GlxA and protect yeast cells from oxidative stress. It was proposed that Tyr-AMPylation imposes steric hindrance which renders the active site inaccessible to glutathione and inhibits GlxA activity.

In conclusion, SelO-mediated GlxA AMPylation regulates protein glutathionylation and protects against oxidative stress.

## 5. Stress responsive post-translational modifications: cysteine oxidation and S-nitrosylation

Human DJ-1 is a small (21 kDa) ubiquitously expressed redox protein associated with many neurodegenerative diseases and the cellular oxidative stress response (Fig. 6A; Antipova & Bandopadhyay, 2017). DJ-1 has been associated with neurodegenerative disorders because deletion of the DJ-1 gene or loss-of-function mutations causes early-onset Parkinson's disease. Other studies implicate DJ-1 to oxidative stress response as, for instance,

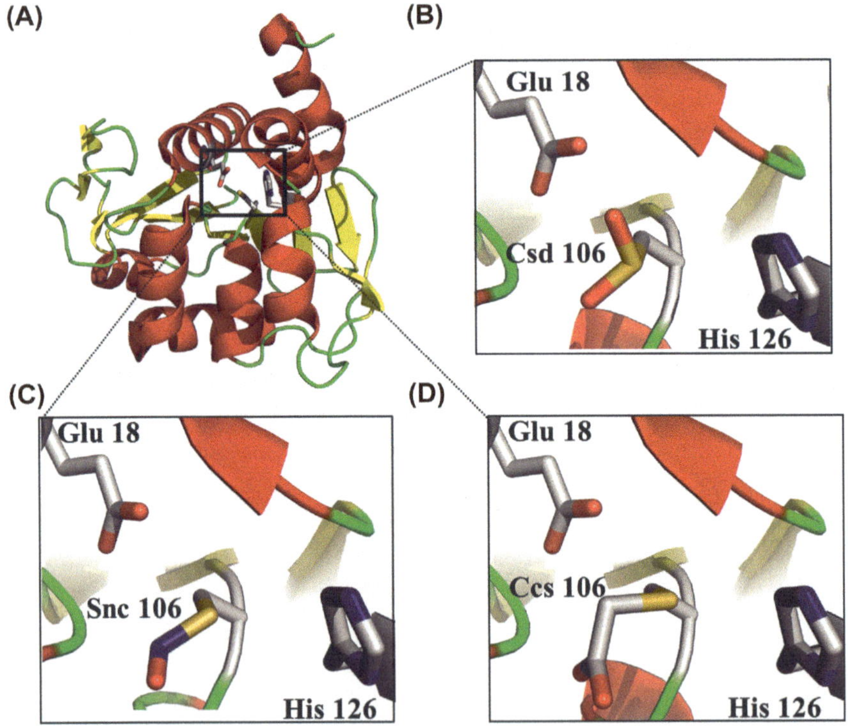


**Fig. 6** Crystal structure of DJ-1 protein. (A) Ribbon diagram of the human DJ-1 (PDB 1P5F). (B)–(C) Close-up views of the modified Cys in three different cases: (B) Sulfinic-Cys (Csd; PDB 1SOA), (C) S-nitrosylated-Cys (Snc; PDB 4RKY), (D) Carboxymethylated- Cys (Ccs; PDB 6E5Z).

that activates SOD1 (superoxide dismutase) by acting as a metallochaperone which transfers copper to SOD1 (Girotto et al., 2014). DJ-1 shows analogies with thiol peroxidases and sulfoxide reductases. There are studies which attribute to DJ-1 a translational/transcriptional regulation role or even a function outside the cell (Tashiro, Caaveiro, Wu, Hoang, & Tsumoto, 2014). Although the precise biochemical function remains still elusive, accumulating evidence correlates DJ-1 function with the ability of a specific Cys to undergo post-transcriptional modifications.

An acidic isoform of DJ-1 is formed *in vivo* in response to oxidative stress suggesting the presence of a sensor on the protein for reactive oxygen species. Indeed, a conserved Cys residue, extremely sensitive to radiation damage during DJ-1 crystal structure determination (Wilson, Collins, Hod, Ringe, & Petsko, 2003), was identified as the preferred target for post-translational modifications by reactive oxygen species generated by oxidative stress. This Cys (Cys 106), which is surrounded by a cluster of conserved acidic residues, is a catalytic residue in homologs and systematically adopts an unfavorable backbone conformation in the Ramachandran plot. The sulfinic modification of Cys 106 was directly observed by X-ray crystallography (Fig. 6B; Canet-Avilés et al., 2004), was also confirmed by mass spectrometry and it is consistent with the formation of an acidic DJ-1 isoform whose accumulation is observed under oxidative stress. Moreover, it was shown that oxidation of Cys 106 is reversible and signals the DJ-1 re-localization to mitochondria where exerts a protective activity against mitochondrial damage. For example, toxicity induced by the MPP + toxin was eliminated by overexpression of wt DJ-1 but not by Cys106Ala variant (Blackinton et al., 2009; Canet-Avilés et al., 2004). Apart from the mitochondrial localization, the sulfinic Cys106 modification is implicated in neuroprotection by at least two more mechanisms i.e. regulating a redox-regulated chaperone activity and directly scavenging reactive oxygen species (Andres-Mateos et al., 2007; Girotto et al., 2014). Furthermore, the reduced structure of DJ-1 -stabilized by a Glu-Leu motif of a peptide bound to a site close to Cys106-was captured and showed no structural differences compared to the oxidized protein (Premkumar, Dobaczewska, & Riedl, 2011).

An unusually low pKa (5.4) has been recorded spectrophotometrically for Cys106 and has been attributed to the formation of a thiolate anion which is substantially stabilized by the local protein environment (Witt et al., 2008). Structural inspection reveals a Cys106-Glu18 interaction which is conserved in every structurally characterized member of the DJ-1 superfamily (Fig. 6). The significance of this interaction was demonstrated by site-directed mutagenesis and shows that even minor substitutions significantly influence the pKa value of C106. A bond length analysis, after unrestrained bond length refinement of a high-resolution structure (1.2 Å) of DJ-1 shows that Glu18 is protonated (and so uncharged) and able to donate a hydrogen bond to Cys106 and thus to stabilize the thiolate anion.

Similar to oxidative stress, nitrosative stress, namely overproduction of reactive NO species, causes neuron death and it is associated with many neurodegenerative disorders. On the other hand, small concentrations of NO play multiple regulatory roles in cell pathways. S-nitrosylation is a post-translational modification often involved in processes that regulate the biological functions of NO. It is known that the phosphatase activity of PTEN (phosphatase and tensin homolog) causes neuronal death and that DJ-1 is a negative regulator of PTEN. Choi and co-workers showed that under mild-NO stress DJ-1 is post-translationally S-nitrosylated at Cys106 (Fig. 6C) and this modification has a neuroprotective effect through regulation of the PTEN pathway (Choi et al., 2014). In particular, S-nitrosylated DJ-1 transfers the NO group to PTEN (transnitrosylation) and thus inhibits the phosphatase activity of PTEN and suppresses the PTEN-dependent cell death.

Recent findings imply that Cys106 may undertake even more modifications such as a transient carboxymethylation (Fig. 6D) mediating an, as yet, uncharacterized metabolic pathway (Mussakhmetov et al., 2018). Collectively, oxidative/nitrosative stress contribute to the pathogenesis of neurodegenerative diseases, whereas detoxification of the reactive species is neuroprotective. DJ-1 is a conserved multifunctional protein which is implicated in a number of cellular pathways. The ability of DJ-1 to respond to oxidative/nitrosative stress mitigates cellular toxicity and results in neuroprotection. Cys106 is a redox-active site of oxidation and S-nitrosylation which trigger different processes of neuroprotection.

## 6. Conclusions

Post-translational modifications are used by all kingdoms of life to generate molecular diversity and afford biological macromolecules a huge potential for functional and structural flexibility and adaptation. Based on chemistry rather than genetics, this mechanism expands, modifies and regulates the physiological roles of proteins so that they are able for quick and selective responses to numerous stimuli. Thus, PTMs can generate permanent or transient alterations, in the whole or in part of the population of a protein by using either enzymatic or spontaneous processes.

Introducing PTMs to enzymes is a highly versatile tool for controlling catalytic activity. In this way, it is possible to accurately and promptly regulate a wide range of processes such as protein folding, cellular metabolism, stress response, defense against pathogens and pathogenicity itself. It has long been known that pathogens can use PTMs as a virulence mechanism to manipulate the host's cellular machinery (Ribet & Cossart, 2010). Numerous bacterial pathogens including *Streptococcus, Staphylococcus, Yersinia, Pseudomonas, Salmonella and Legionella* species use, in part, PTMs to achieve survival and replication within the host. In these cases, the virulence factors namely the virulence proteins produced by pathogens and mediating the host-pathogen interactions can use PTMs to target host's signaling pathways and evade immune responses (Chiang & Gack, 2017; Ravikumar, Jers, & Mijakovic, 2015; Salomon & Orth, 2013). Likewise, viral pathogens can utilize PTMs in order to overcome the host defense mechanisms and hijack its cellular machinery for its own replication and propagation (Loboda, Soond, Piacentini, & Barlev, 2019). It is well established, for example, that structural and non-structural proteins from the coronavirus family of RNA viruses are extensively post-translationally modified by a variety of modifications. These modifications facilitate utilization of the host's protein synthesis machinery for viral replication and promote pathogenesis (Fung & Liu, 2018).

Besides the infectious diseases, aging and age-related diseases have often been related to PTMs (Santos & Lindner, 2017). Such PTMs are usually those which are no-under cellular control, for example those which are accumulated in proteins during the cell lifespan or those which are generated as the result of oxidative stress (Banks & Andersen, 2019; Svistunova et al., 2019). Moreover, disturbances in the pattern of PTMs affect the cellular control pathways and causes abnormalities. Neurodegenerative and autoimmune diseases as well as heart failure and cancer have been associated with PTMs. Thus, PTMs are often used as biomarkers, indicators of abnormalities and attractive targets for vaccines (Brentville, Vankemmelbeke, Metheringham, & Durrant, 2020; Carubbi, Alunno, Gerli, & Giacomelli, 2019; Didonna & Benetti, 2016; Toya et al., 2020).

Although PTMs have long been considered as potential drug targets (Parekh & Rohlff, 1997), it is the progress of the last decade that has given new impetus to the discovery of medicine based on them. Our better understanding of the mechanisms that control PTMs and the mechanisms through which PTMs regulate interaction networks into the cell makes possible the identification of relevant molecular targets and the design of specific drugs. Thus, PTMs especially those which affect catalytic activity emerge as a highly promising field for investigation and development of new therapeutics for a wide range of infectious and non-infectious diseases.

## References

Andres-Mateos, E., Perier, C., Zhang, L., Blanchard-Fillion, B., Greco, T. M., Thomas, B., et al. (2007). DJ-1 gene deletion reveals that DJ-1 is an atypical peroxiredoxin-like peroxidase. *Proceedings of the National Academy of Sciences of the United States of America, 104*(37), 14807–14812. http://dx.doi.org/10.1073/pnas.0703219104.

Antipova, D., & Bandopadhyay, R. (2017). Expression of DJ-1 in neurodegenerative disorders. *Advances in Experimental Medicine & Biology, 1037*, 25–43. http://dx.doi.org/10.1007/978-981-10-6583-5_3.

Arnaouteli, S., Giastas, P., Andreou, A., Tzanodaskalaki, M., Aldridge, C., Tzartos, S. J., et al. (2015). Two putative polysaccharide deacetylases are required for osmotic stability and cell shape maintenance in Bacillus anthracis. *Journal of Biological Chemistry, 290*(21), 13465–13478. http://dx.doi.org/10.1074/jbc.M115.640029.

Balomenou, S., & Arnaouteli, S. (2015). Polysaccharide deacetylases: New antibacterial drug targets. *Frontiers in Anti-Infective Drug Discovery*, 68–130. http://dx.doi.org/10.2174/9781681080826115040005.

Banks, C. J., & Andersen, J. L. (2019). Mechanisms of SOD1 regulation by post-translational modifications. *Redox Biology, 26*(May), 101270. http://dx.doi.org/10.1016/j.redox.2019.101270.

Bertelsena, E. B., Chang, L., Gestwicki, J. E., & Zuiderweg, E. R. P. (2009). Solution conformation of wild-type *E. coli* Hsp70 (DnaK) chaperone complexed with ADP and substrate. *Proceedings of the National Academy of Sciences of the United States of America, 106*(21), 8471–8476. http://dx.doi.org/10.1073/pnas.0903503106.

Bischoff, R., & Schlüter, H. (2012). Amino acids: Chemistry, functionality and selected non-enzymatic post-translational modifications. *Journal of Proteomics, 75*(8), 2275–2296. http://dx.doi.org/10.1016/j.jprot.2012.01.041.

Blackinton, J., Lakshminarasimhan, M., Thomas, K. J., Ahmad, R., Greggio, E., Raza, A. S., et al. (2009). Formation of a stabilized cysteine sulfinic acid is critical for the mitochondrial function of the parkinsonism protein DJ-1. *Journal of Biological Chemistry, 284*(10), 6476–6485. http://dx.doi.org/10.1074/jbc.M806599200.

Brentville, V. A., Vankemmelbeke, M., Metheringham, R. L., & Durrant, L. G. (2020). Post-translational modifications such as citrullination are excellent targets for cancer therapy. *Seminars in Immunology, 47*(January), 101393. http://dx.doi.org/10.1016/j.smim.2020.101393.

Canet-Avilés, R. M., Wilson, M. A., Miller, D. W., Ahmad, R., McLendon, C., Bandyopadhyay, S., et al. (2004). The Parkinson's disease DJ-1 is neuroprotective due to cysteine-sulfinic acid-driven mitochondrial localization. *Proceedings of the National Academy of Sciences of the United States of America, 101*(24), 9103–9108. http://dx.doi.org/10.1073/pnas.0402959101.

Carubbi, F., Alunno, A., Gerli, R., & Giacomelli, R. (2019). Post-translational modifications of proteins: Novel insights in the autoimmune response in rheumatoid arthritis. *Cells, 8*(7), 657. http://dx.doi.org/10.3390/cells8070657.

Casey, A. K., & Orth, K. (2018). Enzymes involved in AMPylation and deAMPylation. *Chemical Reviews, 118*(3), 1199–1215. http://dx.doi.org/10.1021/acs.chemrev.7b00145. American Chemical Society.

Chiang, C., & Gack, M. U. (2017). Post-translational control of intracellular pathogen sensing pathways. *Trends in Immunology, 38*(1), 39–52. http://dx.doi.org/10.1016/j.it.2016.10.008.

Choi, M. S., Nakamura, T., Cho, S. J., Han, X., Holland, E. A., Qu, J., et al. (2014). Trans-nitrosylation from DJ-1 to PTEN attenuates neuronal cell death in Parkinson's disease models. *Journal of Neuroscience, 34*(45), 15123–15131. http://dx.doi.org/10.1523/JNEUROSCI.4751-13.2014.

Craveur, P., Narwani, T. J., Rebehmed, J., & de Brevern, A. G. (2019). Investigation of the impact of PTMs on the protein backbone conformation. *Amino Acids, 51*(7), 1065–1079. http://dx.doi.org/10.1007/s00726-019-02747-w.

Culpepper, M. A., Scott, E. E., & Limburg, J. (2010). Crystal structure of prolyl 4-hydroxylase from Bacillus anthracis. *Biochemistry, 49*(1), 124–133. http://dx.doi.org/10.1021/bi901771z.

Didonna, A., & Benetti, F. (2016). Post-translational modifications in neurodegeneration. *AIMS Biophysics, 3*(1), 27–49. http://dx.doi.org/10.3934/biophy.2016.1.27.

Ermler, U., Grabarse, W., Shima, S., Goubeaud, M., & Thauer, R. K. (1997). Crystal structure of methyl-coenzyme M reductase: The key enzyme of biological methane formation. *Science, 278*(5342), 1457–1462. http://dx.doi.org/10.1126/science.278.5342.1457.

Eschenburg, S., & Schönbrunn, E. (2000). Comparative X-ray analysis of the un-liganded fosfomycin-target MurA. *Proteins: Structure, Function, and Genetics, 40*(2), 290–298. http://dx.doi.org/10.1002/(SICI)1097-0134(20000801)40:2<290::AID-PROT90>3.0.CO;2-0.

Fadouloglou, V. E., Balomenou, S., Aivaliotis, M., Kotsifaki, D., Arnaouteli, S., Tomatsidou, A., et al. (2017). Unusual α-carbon hydroxylation of proline promotes active-site maturation. *Journal of the American Chemical Society, 139*(15), 5330–5337. http://dx.doi.org/10.1021/jacs.6b12209.

Fadouloglou, V. E., Kapanidou, M., Agiomirgianaki, A., Arnaouteli, S., Bouriotis, V., Glykos, N. M., et al. (2013). Structure determination through homology modelling and torsion-angle simulated annealing: Application to a polysaccharide deacetylase from Bacillus cereus. *Acta Crystallographica Section D Biological Crystallography, 69*(2), 276–283. http://dx.doi.org/10.1107/S0907444912045829.

Fujii, N., Takata, T., Fujii, N., & Aki, K. (2016). Isomerization of aspartyl residues in crystallins and its influence upon cataract. *Biochimica et Biophysica Acta (BBA) - General Subjects, 1860*(1), 183–191. http://dx.doi.org/10.1016/j.bbagen.2015.08.001.

Fung, T. S., & Liu, D. X. (2018). Post-translational modifications of coronavirus proteins: Roles and function. *Future Virology, 13*(6), 405–430. http://dx.doi.org/10.2217/fvl-2018-0008.

Gaffney, D. O., Jennings, E. Q., Anderson, C. C., Marentette, J. O., Shi, T., Schou Oxvig, A. M., et al. (2020). Non-enzymatic lysine lactoylation of glycolytic enzymes. *Cell Chemical Biology, 27*(2), 206–213.e6. http://dx.doi.org/10.1016/j.chembiol.2019.11.005.

Gan, N., Zhen, X., Liu, Y., Xu, X., He, C., Qiu, J., et al. (2019). Regulation of phosphoribosyl ubiquitination by a calmodulin-dependent glutamylase. *Nature, 572*(7769), 387–391. http://dx.doi.org/10.1038/s41586-019-1439-1.

Girotto, S., Cendron, L., Bisaglia, M., Tessari, I., Mammi, S., Zanotti, G., et al. (2014). DJ-1 Is a copper chaperone acting on SOD1 activation. *Journal of Biological Chemistry, 289*(15), 10887–10899. http://dx.doi.org/10.1074/jbc.M113.535112.

Grabarse, W., Mahlert, F., Duin, E. C., Goubeaud, M., Shima, S., Thauer, R. K., et al. (2001). On the mechanism of biological methane formation: Structural evidence for conformational changes in methyl-coenzyme M reductase upon substrate binding. *Journal of Molecular Biology, 309*(1), 315–330. http://dx.doi.org/10.1006/jmbi.2001.4647.

Grabarse, W., Mahlert, F., Shima, S., Thauer, R. K., & Ermler, U. (2000). Comparison of three methyl-coenzyme M reductases from phylogenetically distant organisms: Unusual amino acid modification, conservation and adaptation. *Journal of Molecular Biology, 303*(2), 329–344. http://dx.doi.org/10.1006/jmbi.2000.4136.

Harmel, R., & Fiedler, D. (2018). Features and regulation of non-enzymatic post-translational modifications. *Nature Chemical Biology, 14*(3), 244–252. http://dx.doi.org/10.1038/nchembio.2575.

Heggelund, J. E., Varrot, A., Imberty, A., & Krengel, U. (2017). Histo-blood group antigens as mediators of infections. *Current Opinion in Structural Biology, 44*, 190–200. http://dx.doi.org/10.1016/j.sbi.2017.04.001. Elsevier Ltd.

Imai, S. I., Armstrong, C. M., Kaeberlein, M., & Guarente, L. (2000). Transcriptional silencing and longevity protein Sir2 is an NAD-dependent histone deacetylase. *Nature, 403*(6771), 795–800. http://dx.doi.org/10.1038/35001622.

Johnson, B. A., Shirokawa, J. M., Hancock, W. S., Spellman, M. W., Basa, L. J., & Aswad, D. W. (1989). Formation of isoaspartate at two distinct sites during in vitro aging of human growth hormone. *Journal of Biological Chemistry, 264*(24), 14262–14271. http://www.ncbi.nlm.nih.gov/pubmed/2760065.

Karve, T. M., & Cheema, A. K. (2011). Small changes huge impact: The role of protein post-translational modifications in cellular homeostasis and disease. *Journal of Amino Acids, 2011*, 1–13. http://dx.doi.org/10.4061/2011/207691.

Kumar, S., Prakash, S., Gupta, K., Dongre, A., Balaram, P., & Balaram, H. (2016). Unexpected functional implication of a stable succinimide in the structural stability of Methanocaldococcus jannaschii glutaminase. *Nature Communications, 7*. http://dx.doi.org/10.1038/ncomms12798.

Lambert, C., Lerner, T. R., Bui, N. K., Somers, H., Aizawa, S. I., Liddell, S., et al. (2016). Interrupting peptidoglycan deacetylation during Bdellovibrio predator-prey interaction prevents ultimate destruction of prey wall, liberating bacterial-ghosts. *Scientific Reports, 6*(May), 1–19. http://dx.doi.org/10.1038/srep26010.

Lamport, D. T. A., Kieliszewski, M. J., Chen, Y., & Cannon, M. C. (2011). Role of the extensin superfamily in primary cell wall architecture. *Plant Physiology, 156*(1), 11–19. http://dx.doi.org/10.1104/pp.110.169011.

Lin, H., Du, J., & Jiang, H. (2008). Post-translational modifications to regulate protein function. *Wiley Encyclopedia of Chemical Biology*. http://dx.doi.org/10.1002/9780470048672.wecb467.

Lin, H., Su, X., & He, B. (2012). Protein lysine acylation and cysteine succination by intermediates of energy metabolism. *ACS Chemical Biology, 7*(6), 947–960. http://dx.doi.org/10.1021/cb3001793. American Chemical Society.

Loboda, A. P., Soond, S. M., Piacentini, M., & Barlev, N. A. (2019). Lysine-specific post-translational modifications of proteins in the life cycle of viruses. *Cell Cycle, 18*(17), 1995–2005. http://dx.doi.org/10.1080/15384101.2019.1639305.

Lyu, Z., Shao, N., Chou, C. W., Shi, H., Patel, R., Duin, E. C., et al. (2020). Posttranslational methylation of arginine in methyl coenzyme M reductase has a profound impact on both methanogenesis and growth of Methanococcus maripaludis. *Journal of Bacteriology, 202*(3), 1–18. http://dx.doi.org/10.1128/JB.00654-19.

Mahanta, N., Liu, A., Dong, S., Nair, S. K., & Mitchell, D. A. (2018). Enzymatic reconstitution of ribosomal peptide backbone thioamidation. *Proceedings of the National Academy of Sciences of the United States of America, 115*(12), 3030–3035. http://dx.doi.org/10.1073/pnas.1722324115.

Mallagaray, A., Creutznacher, R., Dülfer, J., Mayer, P. H. O., Grimm, L. L., Orduña, J. M., et al. (2019). A post-translational modification of human Norovirus capsid protein attenuates glycan binding. *Nature Communications, 10*(1). http://dx.doi.org/10.1038/s41467-019-09251-5.

Marcucci, G., Bloomfield, C. D., Caligiuri, M. A., Tachibana, K. E., Laskey, R. A., Coleman, N., et al. (2010). The Legionella effector protein DrrA AMPylates the membrane traffic regulator Rab 1b. *Science, 329*(August), 946–950.

Moellering, R. E., & Cravatt, B. F. (2013). Functional lysine modification by an intrinsically reactive primary glycolytic metabolite. *Science, 341*(6145), 549–553. http://dx.doi.org/10.1126/science.1238327.

Moro, M. L., Phillips, A. S., Gaimster, K., Paul, C., Mudher, A., Nicoll, J. A. R., et al. (2018). Pyroglutamate and Isoaspartate modified Amyloid-Beta in ageing and Alzheimer's disease. *Acta Neuropathologica Communications, 6*(1), 3. http://dx.doi.org/10.1186/s40478-017-0505-x.

Müller, M. M. (2018). Post-translational modifications of protein backbones: Unique functions, mechanisms, and challenges. *Biochemistry, 57*(2), 177–185. http://dx.doi.org/10.1021/acs.biochem.7b00861.

Mussakhmetov, A., Shumilin, I. A., Nugmanova, R., Shabalin, I. G., Baizhumanov, T., Toibazar, D., et al. (2018). A transient post-translational modification of active site cysteine alters binding properties of the parkinsonism protein DJ-1. *Biochemical and Biophysical Research Communications, 504*(1), 328–333. http://dx.doi.org/10.1016/j.bbrc.2018.08.190.

Myllyharju, J., & Kivirikko, K. I. (2001). Collagens and collagen-related diseases. *Annals of Medicine, 33*(1), 7–21. http://dx.doi.org/10.3109/07853890109002055. Royal Society of Medicine Press Ltd.

Nayak, D. D., Liu, A., Agrawal, N., Rodriguez-Carerro, R., Dong, S. H., Mitchell, D. A., et al. (2020). Functional interactions between posttranslationally modified amino acids of methyl-coenzyme M reductase in Methanosarcina acetivorans. *PLoS Biology, 18*(2). http://dx.doi.org/10.1371/journal.pbio.3000507.

Nayak, D. D., Mahanta, N., Mitchell, D. A., & Metcalf, W. W. (2017). Post-translational thioamidation of methyl-coenzyme M reductase, a key enzyme in methanogenic and methanotrophic archaea. *ELife, 6*(I), 1–18. http://dx.doi.org/10.7554/eLife.29218.

Okeley, N. M., & Van Der Donk, W. A. (2000). Novel cofactors via post-translational mocifications of enzyme active sites. *Chemistry & Biology, 7*(7), 159–171. http://dx.doi.org/10.1016/S1074-5521(00)00140-X.

Palmer, B. F., & Clegg, D. J. (2014). Oxygen sensing and metabolic homeostasis. *Molecular and Cellular Endocrinology, 397*(1–2), 51–58. http://dx.doi.org/10.1016/j.mce.2014.08.001. Elsevier Ireland Ltd.

Parekh, R. B., & Rohlff, C. (1997). Post-translational modification of proteins and the discovery of new medicine. *Current Opinion in Biotechnology, 8*(6), 718–723. http://dx.doi.org/10.1016/S0958-1669(97)80126-7.

Perera, L. A., Rato, C., Yan, Y., Neidhardt, L., McLaughlin, S. H., Read, R. J., et al. (2019). An oligomeric state-dependent switch in the ER enzyme FICD regulates AMP ylation and de AMP ylation of BiP. *The EMBO Journal, 38*(21). http://dx.doi.org/10.15252/embj.2019102177.

Potter, S. M., Henzel, W. J., & Aswad, D. W. (1993). In vitro aging of calmodulin generates isoaspartate at multiple Asn–Gly and Asp–Gly sites in calcium-binding domains II, III, and IV. *Protein Science, 2*(10), 1648–1663. http://dx.doi.org/10.1002/pro.5560021011.

Preissler, S., Rohland, L., Yan, Y., Chen, R., Read, R. J., & Ron, D. (2017). AMPylation targets the rate-limiting step of BiP's ATPase cycle for its functional inactivation. *ELife, 6*. http://dx.doi.org/10.7554/eLife.29428.

Prejanò, M., Romeo, I., Sgrizzi, L., Russo, N., & Marino, T. (2019). Why hydroxy-proline improves the catalytic power of the peptidoglycan: N -deacetylase enzyme: Insight from theory. *Physical Chemistry Chemical Physics, 21*(42), 23338–23345. http://dx.doi.org/10.1039/c9cp03804c.

Premkumar, L., Dobaczewska, M. K., & Riedl, S. J. (2011). Identification of an artificial peptide motif that binds and stabilizes reduced human DJ-1. *Journal of Structural Biology, 176*(3), 414–418. http://dx.doi.org/10.1016/j.jsb.2011.08.011.

Radle, M. I., Miller, D. V., Laremore, T. N., & Booker, S. J. (2019). Methanogenesis marker protein 10 (Mmp10) from methanosarcina acetivorans is a radical S-adenosylmethionine methylase that unexpectedly requires cobalamin. *Journal of Biological Chemistry, 294*(31). http://dx.doi.org/10.1074/jbc.RA119.007609.

Ravikiran, B., & Mahalakshmi, R. (2014). Unusual post-translational protein modifications: The benefits of sophistication. *RSC Advances, 4*(64), 33958–33974. http://dx.doi.org/10.1039/c4ra04694c. Royal Society of Chemistry.

Ravikumar, V., Jers, C., & Mijakovic, I. (2015). Elucidating host-pathogen interactions based on post-translational modifications using proteomics approaches. *Frontiers in Microbiology, 6*(NOV), 1–7. http://dx.doi.org/10.3389/fmicb.2015.01312.

Reissner, K. J., & Aswad, D. W. (2003). Deamidation and isoaspartate formation in proteins: Unwanted alterations or surreptitious signals? *Cellular and Molecular Life Sciences, 60*(7), 1281–1295. http://dx.doi.org/10.1007/s00018-003-2287-5.

Ribet, D., & Cossart, P. (2010). Pathogen-mediated posttranslational modifications: A re-emerging field. *Cell, 143*(5), 694–702. http://dx.doi.org/10.1016/j.cell.2010.11.019.

Salomon, D., & Orth, K. (2013). What pathogens have taught us about posttranslational modifications. *Cell Host & Microbe, 14*(3), 269–279. http://dx.doi.org/10.1016/j.chom.2013.07.008. Cell Press.

Santos, A. L., & Lindner, A. B. (2017). Protein posttranslational modifications: Roles in aging and age-related disease. *Oxidative Medicine and Cellular Longevity, 2017*. http://dx.doi.org/10.1155/2017/5716409.

Sauve, A. A., Wolberger, C., Schramm, V. L., & Boeke, J. D. (2006). The biochemistry of sirtuins. *Annual Review of Biochemistry, 75*(1), 435–465. http://dx.doi.org/10.1146/annurev.biochem.74.082803.133500.

Selmer, T., Kahnt, J., Goubeaud, M., Shima, S., Grabarse, W., Ermler, U., et al. (2000). The biosynthesis of methylated amino acids in the active site region of methyl-coenzyme M reductase. *Journal of Biological Chemistry, 275*(6). http://dx.doi.org/10.1074/jbc.275.6.3755.

Shanker, S., Hu, L., Ramani, S., Atmar, R. L., Estes, M. K., & Venkataram Prasad, B. V. (2017). Structural features of glycan recognition among viral pathogens. *Current Opinion in Structural Biology, 44*, 211–218. http://dx.doi.org/10.1016/j.sbi.2017.05.007. Elsevier Ltd.

Singh, B. K., Leuthold, M. M., & Hansman, G. S. (2015). Human noroviruses' fondness for histo-blood group Antigens. *Journal of Virology, 89*(4), 2024–2040. http://dx.doi.org/10.1128/jvi.02968-14.

Siodłak, D. (2015). α,β-Dehydroamino acids in naturally occurring peptides. *Amino Acids, 47*(1), 1–17. http://dx.doi.org/10.1007/s00726-014-1846-4.

Sreelatha, A., Yee, S. S., Lopez, V. A., Park, B. C., Kinch, L. N., Pilch, S., et al. (2018). Protein AMPylation by an evolutionarily conserved pseudokinase. *Cell, 175*(3), 809–821. http://dx.doi.org/10.1016/j.cell.2018.08.046. e19.

Svistunova, D. M., Simon, J. N., Rembeza, E., Crabtree, M., Yue, W. W., Oliver, P. L., et al. (2019). Oxidation resistance 1 regulates post-translational modifications of peroxiredoxin 2 in the cerebellum. *Free Radical Biology and Medicine, 130*(July 2018), 151–162. http://dx.doi.org/10.1016/j.freeradbiomed.2018.10.447.

Tan, M., Peng, C., Anderson, K. A., Chhoy, P., Xie, Z., Dai, L., et al. (2014). Lysine glutarylation is a protein posttranslational modification regulated by SIRT5. *Cell Metabolism, 19*(4), 605–617. http://dx.doi.org/10.1016/j.cmet.2014.03.014.

Tashiro, S., Caaveiro, J. M. M., Wu, C. X., Hoang, Q. Q., & Tsumoto, K. (2014). Thermodynamic and structural characterization of the specific binding of Zn(II) to human protein DJ-1. *Biochemistry, 53*(14), 2218–2220. http://dx.doi.org/10.1021/bi500294h.

Taube, S., Mallagaray, A., & Peters, T. (2018). Norovirus, glycans and attachment. *Current Opinion in Virology, 31*, 33–42. http://dx.doi.org/10.1016/j.coviro.2018.04.007. Elsevier B.V.

Toya, T., Ito, K., Kagami, K., Osaki, A., Sato, A., Kimura, T., et al. (2020). Impact of oxidative posttranslational modifications of SERCA2 on heart failure exacerbation in young patients with non-ischemic cardiomyopathy: A pilot study. *IJC Heart and Vasculature, 26*, 100437. http://dx.doi.org/10.1016/j.ijcha.2019.100437.

Wagner, T., Kahnt, J., Ermler, U., & Shima, S. (2016). Didehydroaspartate modification in methyl-coenzyme M reductase catalyzing methane formation. *Angewandte Chemie International Edition, 55*(36), 10630–10633. http://dx.doi.org/10.1002/anie.201603882.

Wagner, T., Wegner, C. E., Kahnt, J., Ermler, U., & Shima, S. (2017). Phylogenetic and structural comparisons of the three types of methyl coenzyme M reductase from Methanococcales. *Journal of Bacteriology, 199*(16), 1–15. http://dx.doi.org/10.1128/JB.00197-17.

Wilson, M. A., Collins, J. L., Hod, Y., Ringe, D., & Petsko, G. A. (2003). The 1.1-Å resolution crystal structure of DJ-1, the protein mutated in autosomal recessive early onset Parkinson's disease. *Proceedings of the National Academy of Sciences of the United States of America, 100*(16), 9256–9261. http://dx.doi.org/10.1073/pnas.1133288100.

Witt, A. C., Lakshminarasimhan, M., Remington, B. C., Hasim, S., Pozharski, E., & Wilson, M. A. (2008). Cysteine pKa depression by a protonated glutamic acid in human DJ-1. *Biochemistry, 47*(28), 7430–7440. http://dx.doi.org/10.1021/bi800282d.

Wongnate, T., & Ragsdale, S. W. (2015). The reaction mechanism of methyl-coenzyme M reductase. *Journal of Biological Chemistry, 290*(15), 9322–9334. http://dx.doi.org/10.1074/jbc.M115.636761.

Yan, Y., Rato, C., Rohland, L., Preissler, S., & Ron, D. (2019). MANF antagonizes nucleotide exchange by the endoplasmic reticulum chaperone BiP. *Nature Communications, 10*(1), 1–15. http://dx.doi.org/10.1038/s41467-019-08450-4.

Zee, B. M., & Garcia, B. A. (2012). Discovery of lysine post-translational modifications through mass spectrometric detection. *Essays in Biochemistry, 52*(1), 147–163. http://dx.doi.org/10.1042/BSE0520147.

Zhang, D., Tang, Z., Huang, H., Zhou, G., Cui, C., Weng, Y., et al. (2019). Metabolic regulation of gene expression by histone lactylation. *Nature, 574*(7779), 575–580. http://dx.doi.org/10.1038/s41586-019-1678-1.

Zhao, S., Xu, W., Jiang, W., Yu, W., Lin, Y., Zhang, T., et al. (2010). Regulation of cellular metabolism by protein lysine acetylation. *Science, 327*(5968), 1000–1004. http://dx.doi.org/10.1126/science.1179689.

CHAPTER FIVE

# Current advances on the development of BET inhibitors: insights from computational methods

**Fernando D. Prieto-Martínez*** **and José L. Medina-Franco**
Department of Pharmacy, School of Chemistry, National Autonomous University of Mexico, Mexico City, Mexico
*Corresponding author: E-mail: ferdpm4@hotmail.com

## Contents



## Abstract

Epigenetics was coined almost 70 years ago for the description of heritable phenotype without altering DNA sequences. Research on the field has uncovered significant roles of such mechanisms, that account for the biogenesis of several diseases. Further studies have led the way for drug development which targets epi-enzymes, mainly for cancer treatment. Of the numerous epi-targets involved with histone acetylation, bromodomains have captured the spotlight of drug discovery focused on novel therapies. However, due to high sequence identity, the development of potent and selective inhibitors poses a significant challenge. Herein, we discuss recent computational developments

*Advances in Protein Chemistry and Structural Biology*, Volume 122
ISSN 1876-1623
https://doi.org/10.1016/bs.apcsb.2020.06.002

on BET inhibitors and other methods that may be applied for drug discovery in general. As a proof-of-concept, we discuss a virtual screening to identify novel BET inhibitors based on coumarin derivatives. From public data, we identified putative structure-activity relationships of coumarin scaffold and propose R-group modifications for BET selectivity. Results showed that the optimization and design of novel coumarins could be further explored.

## 1. Introduction

Chemical modifications of histones (H3 or H4) involve a series of enzymes collectively known as the epigenome. These enzymes may follow three main mechanisms: writing (chemical addition), erasing (chemical removal) or reading functional groups; e.g., methyl, acetyl or ubiquitin (Bredfeldt & Walker, 2010). Thus, epigenetic modifications have a direct impact on histone stability and are crucial for the preservation of cellular identity and development (Hassler & Egger, 2012). In addition to this, it has been shown that mutations on the epigenome play pivotal roles in physiological and pathological states (Ye, Huang, Wang, Luo, & Zhao, 2019).

Histone acetylation has been studied throughly since 1964, then Allfrey et al., hypothesized about the role of these modifications as they happened independently of transcription (McCullough & Marmorstein, 2016). It was later found that the ε-amino group of lysine residues was the target of acetylation (Simon, Robaa, Alhalabi, Sippl, & Jung, 2016). Now it is widely regarded as a crucial epi-modification, because its patterns lead to altered folding of nucleosome fibers, thus participating in chromatin remodeling (Eberharter & Becker, 2002).

There are two main systems responsible for the remodeling cycle: Histone Acetyl Transferases (HATs) and Histone Deacetylases (HDACs), with the latter been linked to cancer development (Robert & Rassool, 2012). However, only a few HDAC inhibitors have been approved by the Food and Drug Administration (FDA) of the United States: Vorinostat, Panobinostat (Guha, 2015), and Romidepsin (VanderMolen, McCulloch, Pearce, & Oberlies, 2011). Moreover, non-selective HDAC inhibitors present pleiotropic effects with no clear mechanism of action (Jain & Barton, 2017). Therefore, research has also been extended to epigenetic readers, associated with acetylation (Filippakopoulos & Knapp, 2014). Currently, only three types of domains have been identified as KAc readers (Gil, Ramírez-Torres, & Encarnación-Guevara, 2017): YEATS (which bind to

H3K9Ac) and associate with HATs; Plant homeodomain (PHD) fingers (which bind to H3 and H4) and bromodomains (BRDs), which play a major role in KAc modulation and are focus of this chapter.

Bromodomains are small finger-proteins first identified on the brahma (*brm*) gene of the *Drosophila* genus, later found as a common motif in most organisms (Yang, 2004). Early studies on these proteins contemplated the promotion of chromatin interactions as their main function. The discovery of a bromodomain-related protein on the Gcn5 family of transcription-linked HATs confirmed this hypothesis (Marmorstein & Berger, 2001). Bromodomain containing proteins are primarily located on the nucleus and cytosol, taking part on multiprotein complexes which serve for signaling and recruitment (Gil et al., 2017). These recognize KAc residues, as demonstrated by spectral analysis in presence of H4 peptides (Dhalluin et al., 1999).

Currently, 61 isoforms have been found in the human proteome (expressed in 46 proteins) and are classified on eight phylogenetic families (Prinjha, Witherington, & Lee, 2012). Of these, family II, or the Bromodomain and Extra-terminal domain protein (BET) family, includes eight isoforms: BRD2, BRD3, BRD4, and BRDT, each with two domains, i.e., BD1 and BD2. BET BRDs have a distinctive structure (Fig. 1). It is based on a bundle of four left-handed helices known as αZ, αA, αB, and αC; this rather unusual trait is the result of the loops which connect them. Due to the length of the ZA loop, the left-handed turn is favored as tertiary structure interactions are expected with the BC loop. This results in the formation of a hydrophobic pocket, which interacts with the acetylated proteins (Smith & Zhou, 2015).

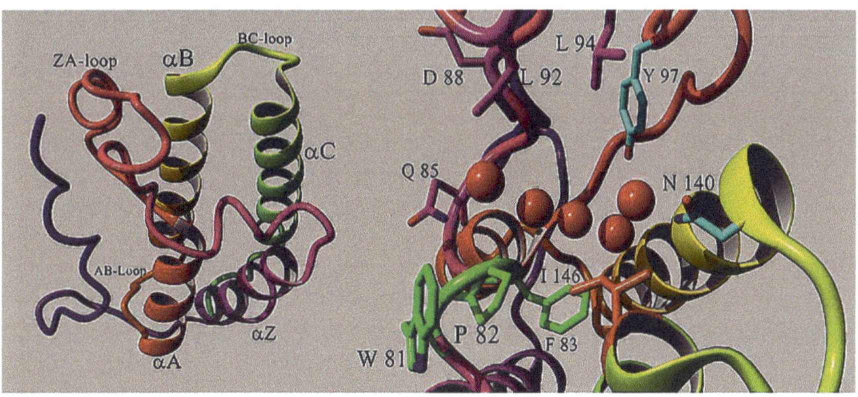


**Fig. 1** Structural features of BET bromodomains (left). Active site view (right) with hotspots highlighted: WPF shelf (green), ZA channel (magenta), Ac-binding tandem (cyan), gatekeeper (orange), and water network (red spheres).

The binding site of BET bromodomains is comprised by well-established hot-spots. These include the Ac-binding motifs (a "tandem" comprised by asparagine and tyrosine residues), the WPF shelf (a unique trait of the BET family), the ZA channel, and a "gatekeeper" residue (usually an isoleucine or histidine). Additionally, the presence of a conserved water network in the ZA channel is crucial for binding (Bharatham, Slavish, Young, & Shelat, 2018).

BET bromodomains play a key role in the cell cycle (see Table 1), acting as growth promoting genes. These proteins dock to chromatin by their bromodomains, while the ET domain acts as a recruiter of transcriptional machinery (Fujisawa & Filippakopoulos, 2017). Each isoform of the BET family has shown specific roles, all related to cell proliferation. For example, BRD2 controls the alternative splicing of multiple genes, some of them involved on MAPK pathways (Wang & Filippakopoulos, 2015). Moreover, its participation in normal development of crucial neural structures has been studied in mice (Taniguchi, 2016). BRD3 takes part in erythropoiesis due to interactions with GATA1 factor (Fujisawa & Filippakopoulos, 2017) and it has shown some functional overlaps with BRD2. As for the most studied isoform of the BET family BRD4, its major interactions are with P-TEFb and NELF (Shi & Vakoc, 2014). Finally, BRDT is mostly expressed in testis (hence the T) and is a main modulator of spermatogenesis during meiotic and post-meiotic stages of differentiation (Gaucher et al., 2012). Notably, BRDT inhibition results on "reversible infertility" which prompts further study of such putative contraceptive effect (Matzuk et al., 2012).

Based on the current literature, there is no doubt on the importance of BETs for major physiological functions. In addition, BETs have also been

**Table 1** Major biological roles currently identified for BETs.

| Isoform | Cells with high expression | Main roles | Potential therapeutic uses |
|---|---|---|---|
| BRD2 | Fetal tissue | Cell proliferation | Disruptor of K-Ras mediated oncogenesis |
| | Reproductive tissue | Promoter of gene transcription | |
| | Blood cells | | |
| BRD3 | Fetal tissue | Promoter of gene transcription | NUT carcinoma |
| | Retina | Erythropoiesis | Cardiac disease |
| | Blood cells | | |
| BRD4 | Blood cells | Cell proliferation | NUT carcinoma |
| | | Gene expression | Antinflammatory |
| | | NF-κB activation | Cardiac disease |
| BRDT | Testis | Spermatogenesis | Contraceptive |

**Table 2** Interacting partners of BETs and their role in several diseases/therapies.

| **Isoform** | **BRD2** | **BRD3** | **BRD4** | **BRDT** |
|---|---|---|---|---|
| Interacting proteins | CK2<br>TFIID<br>TFIIH<br>RNAPII<br>RING3 | p21<br>SWI/SNF<br>TFIID<br>TFIIH<br>RNAPII<br>GATA1 | p21<br>p53<br>p65<br>P-TEFb<br>SWI/SNF<br>NELF | TEFb<br>NELF |
| Diseases | Metabolic syndrome<br>Sepsis<br>Autoinmmune disease<br>Epilepsy | Atheroesclerosis<br>c-Myc related cancers | Psoriasis<br>Diabetes<br>Cancer<br>Sepsis | |
| Possible combination therapies | PPARγ-inhibitors | Statins | HDAC-inhibitors<br>Kinase inhibitors | |
| Potential risk or toxic effects | Theratogenic<br>Metabolic imbalances | Theratogenic<br>Thrombocytopenia | Theratogenic<br>Viral reactivation | Spermatozoa<br>Infertility |

associated with different diseases (see Table 2). One of the promising roles of these bromodomains in disease is their ability to inhibit *MYC*. These are often identified as oncogenes, due to their role on the cell cycle; yet current *MYC* inhibitors have not reached clinical approval (Vita & Henriksson, n.d.). It has been shown that BETs and *MYC* have a quite complex relation, as the inhibitory effect is significant but heterogeneous across tumor types (Doroshow, Eder, & LoRusso, 2017). BET inhibition has also been identified as a novel mechanism for the treatment of other neoplasms such as leukemia and nuclear protein in testis (NUT) midline carcinoma (Basheer & Huntly, 2015). BRD4 has been related to diabetes, and inflammatory processes (Wang & Filippakopoulos, 2015). With the latter gaining more interest, as it establishes the role of epigenetics in the activation of NF-κB (Paneni, Costantino, Volpe, Lüscher, & Cosentino, 2013).

Therefore, BETs and BRD4 (in particular) have been in the spotlight of drug discovery efforts. This has led to the identification of several inhibitory compounds (see Chart 1); with JQ1 and Molibresib being potent pan-BET inhibitors with $IC_{50}$ values in the nanomolar range (Zaware & Zhou, 2017). Recently, it has been confirmed that JQ1 diminishes inflammation in psoriasis (Nadeem et al., 2015) and in pulmonary hypertension models (Mumby

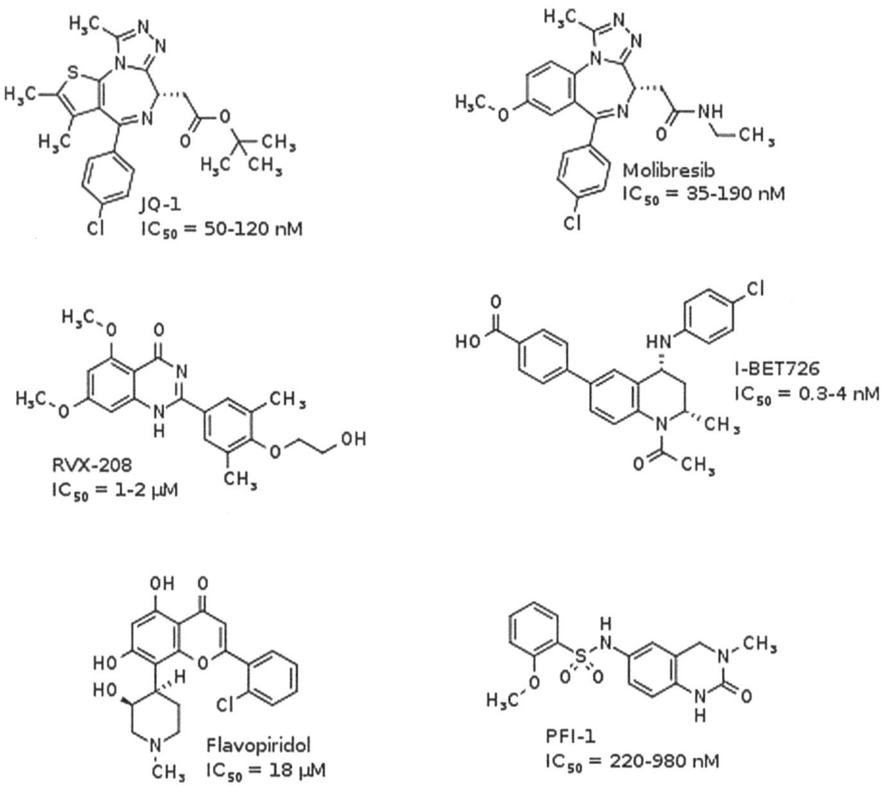


**Chart 1** Representative BRDs inhibitors.

et al., 2017). Such findings are further supported by the reduced lethality of lipopolysaccharide (LPS), followed by Molibresib administration in mice. This suggests that non-histone signaling with BETs plays a crucial role in NF-κB regulation (Hajmirza et al., 2018).

There is additional evidence that links neurological disorders with BET protein dysregulation; such as epilepsy and memory impairment (Padmanabhan, Mathur, Manjula, & Tripathi, 2016). Other studies have identified the role of BETs (through JQ1 inhibition) in the development of L-DOPA induced dyskinesia (Figge & Standaert, 2017). Such pleiotropic effects suggest that BET inhibition may be used as concomitant therapies on chronic disease. For example, clinical studies with Apabetalone (RVX-208) in atherosclerothic patients showed a reduction in cardiovascular events (Schiano et al., 2020). Similarly, combined therapies for autoimmune and hormonal diseases with BET inhibitors have been proposed and remain under clinical study (Bechter & Schöffski, 2020).

Despite the significant progress on BRD drug discovery, the physiology of BRDs and its overall epigenetic scope is mostly unknown, hampering the development of potent, but safe, drug leads (Andrieu, Belkina, & Denis, 2016; Conway, 2012). This may be due to the low scaffold diversity of currently developed inhibitors (Prieto-Martínez, Fernandez-de Gortari, Méndez-Lucio, & Medina-Franco, 2016). Hence, the search for novel ligands should not be limited to potentially therapeutic, but also probe molecules. Therefore, different research groups have taken interest in the description of the physiology of BRDs and development of novel inhibitors. A quick search on primary sources shows a growing tendency for "bromodomain" key word in research papers, during the last decade (Fig. 2).

The aim of this chapter is to discuss general and current developments on the characterization of BET inhibitors, focusing on computational methodologies used for discovery or optimization. The chapter is organized in five major sections. After this introduction, Section 2 discusses general aspects of *in silico* drug discovery and design, as well as a brief background of commonly used methods. Section 3 presents advances to describe BET aimed to develop selective and potent inhibitors. In Section 4 we present a case study, using the coumarin scaffold as starting point for virtual screening. Finally, Section 5 summarizes the current landscape of the field and some future directions for BET drug discovery.

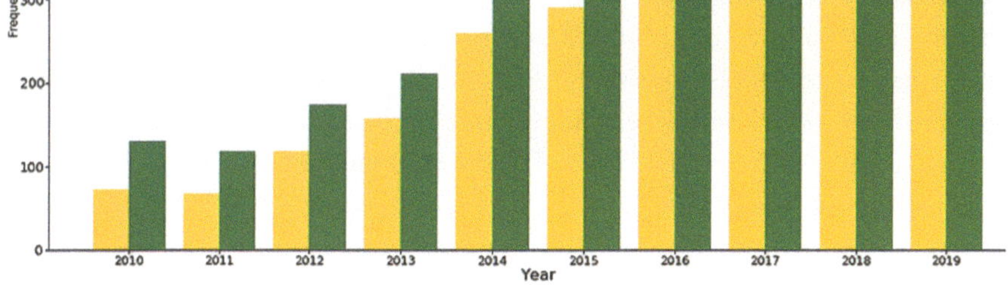


**Fig. 2** Frequency of "bromodomain" keyword, as obtained from two mainstream search engines.

## 2. *In silico* methods and computer-aided drug design (CADD)

Through the past 50 years the use of computational resources to tackle complex problems has been done on several disciplines. In chemistry, a first approach was done via *ab initio* calculations due to their exhaustiveness. Further progress was made with the rise of cheminformatics, which consists in the description, organization and management of 2D and 3D structures to mine and/or obtain new data (Willett, 2008). With novel research tools, this field grew at a steady pace during the 80s and 90s and continue growing until today in several areas including drug discovery, organic synthesis, food, and material science, to name a few (Ganesan, 2004; Larsson, Backlund, & Bohlin, 2008; Naveja, Pilón-Jiménez, Bajorath, & Medina-Franco, 2019; Naveja, Rico-Hidalgo, & Medina-Franco, 2018).

This has led to the natural extension of computer-aided design (CAD), now used by industry, academia, and government alike. Nowadays, it is easy to think in several *in silico* methodologies for chemical problem solving. Hence, in the next sections we discuss examples of useful methods for computer-aided drug design (CADD), with special attention to those used in BET inhibitor research. We begin our discussion focusing on structure-based drug design (SBDD), as these methods have been used quite extensively not only on BET research, but on a wide array of drug targets.

### 2.1 Structure-based drug design (SBDD)

In the early days drug design was often a mixture of careful thought and a tad of luck. It relied on organic chemistry knowledge to optimize lead compounds through common replacements and long synthesis-evaluation cycles. With the advent of combinatorial chemistry and high-throughput Screening (HTS), this seemed due to change. Alas, this was not the case, mainly due to low structural diversity found on combinatorial libraries and a high rate of false positives from HTS. A common criticism of these approaches is their complete reliance on chance to solve a problem that demands rational and creative input (Gane & Dean, 2000).

During the 90s, several success cases brought attention to the so-called structure-based drug design (SBDD). The knowledge gained from target identification and the ability to get crystal structures of proteins paved the shift the paradigm of most drug discovery endeavors. These methods take advantage of experimental evidence to improve description of structure-activity relationships (SAR). Still, in most cases drug optimization through SBDD may become more art than science, due to the iterative cycles and prior knowledge still needed (Salemme, Spurlino, & Bone, 1997). Traditional SBDD often relies on techniques like X-ray structures or NMR ensembles to identify and characterize binding modes, followed by exhaustive selection of promising scaffolds (van Dongen, Weigelt, Uppenberg, Schultz, & Wikström, 2002).

Almost simultaneously, many pharmaceutical companies turned their attention to software tools for drug development. This was mainly focused toward *in silico* screening of thousands of compounds. A first approach to this were 3D superpositions for pharmacophore elucidation; e.g., ALADDIN (developed in Abbot) or CONCORD (commercialized by Tripos) (Van Drie, 2010). In a similar manner, other groups took advantage of the growing computational resources and protein crystals to develop the first protocols for molecular docking.

By the turn of the century, several methods and algorithms were disclosed giving rise to a broad use of SBDD in the pharmaceutical industry and academia. In this sense, computational methods were highly regarded as the state-of-the-art, as it was now possible to evaluate protein-ligand interaction and even target construction by means of homology modeling (Anderson, 2003).

Recently, the SBDD toolbox has grown significantly, making it a well-rounded approach for virtual screening. Even so, there are some persisting challenges for computational methods such as structure validation, protein flexibility, low identity templates, pKa prediction/tautomerization, and solvent contribution (Kalyaanamoorthy & Chen, 2011).

In the following subsections we present a brief discussion of two main techniques for SBDD: molecular docking and dynamics, as well as free energy methods, which have gained notable attention due to their successful developments in recent years.

### 2.1.1 Molecular docking

As stated before, molecular docking has become the method of choice to start most SBDD and CADD campaigns, when the molecular target is known (Prieto-Martínez, Arciniega, & Medina-Franco, 2018). The fast calculations and (relative) ease of use, pose as significant advantages when compared to other computational methods. A molecular docking protocol can be divided on three main stages:

- Protein selection and preparation
- Conformational search
- Posing and scoring

The preparation step is often quick and simple, which sometimes is trivialized mostly by first time users. Nonetheless it is crucial for reliable results, as wrong decisions may lead to incompatibilities, numerical errors (Madhavi Sastry, Adzhigirey, Day, Annabhimoju, & Sherman, 2013), and misleading interpretation of the results.

For ligands, preparation includes atom type, charges and tautomeric state assignment; protein on the other hand requires similar checks, with the addition of careful revision of missing side chains or even residues on the structure. Thus, a thoughtful selection of the protein structure to be used is paramount, as it can reduce the steps needed for preparation and improve the significance of results. Such practice cannot be overstated, care must be taken when selecting a protein for docking, as the obtained poses may not be stable or reliable due to binding site influence (Huggins, Sherman, & Tidor, 2012). In addition, it is advisable to conduct a general energy minimization of both protein and ligand before carrying out docking.

After preparation, the site on which docking will be attempted must be defined. A common practice is to use a reference compound to define centroid coordinates, then a "topology" of the binding site is constructed. To this end, GRID is one of the most common methods, which maps interaction contours are based on electrostatic and steric features (Goodford, 1985). However, there may be cases where this is not possible. Due to this, several tools have been developed to predict putative binding pockets using geometric, template or fragment-based search (Rathi et al., 2017; Wu, Peng, Zhang, & Yang, 2018).

The next step is a conformational search, which must be exhaustive enough to provide enough sampling of the free energy landscape of the ligands. This poses a problem, as a truly exhaustive search is neither feasible or practical. Hence, several methodologies for conformer generation are on constant development and benchmarking (Friedrich et al., 2017; Gürsoy & Smieško, 2017; Hawkins, 2017). Common examples of conformer search in docking protocols are presented on Table 3.

Ligand conformations are then submitted to an evaluation process, which attempts to effectively identify the optimal arrangement which leads to biological activity. A naive docking practitioner may think that such evaluation can be approached by searching the global minima. However, this is not the case as in most instances the protein-ligand recognition is an induced-fit phenomenon, which may not be favorable at all, for ligand or protein or both. Scoring functions seek to provide a rational description of interaction terms or at least those more crucial for effective binding.

First generation scoring functions are "physics-based" using simplified forms of molecular force fields (*vide infra*). While reasonable, the dependence on forcefield description presented notable problems. Such as entropy description, reparametrization and the need of forcefield modification for

**Table 3** Algorithms commonly used for conformer search in molecular docking.

| **Algorithm** | **Description** | **Reference** |
|---|---|---|
| Anchor-and-grow | Ligands are fragmented based on rigid segments. Each fragment is then docked and scored individually until the main "anchor" is identified. From the anchor the ligand is reconstructed using a similar process. | Allen et al. (2015) |
| Simmulated annealing | Based on Monte Carlo methods, also known as the Metropolis criterion. It attempts to sample conformations outside local minima by the following evaluation: $e^{-\frac{\Delta E}{k_B T}}$ | Chéron, Jasty and Shakhnovich (2016) |
| Metaheuristics | "Nature-inspired" algorithms, which use a population of "individuals" to explore search space. Docking algorithms usually employ evolutionary metaphors. | García-Godoy, López-Camacho, García-Nieto, Nebro and Aldana-Montes (2015) |
| Numerical optimization | Implementations of "hill-climbing" methods, that search for local minima by means of iterative cycles. Examples include quasi-Newton methods, like the Broyden-Fletcher-Goldfarb-Shanno algorithm. | Ng, Fong and Siu (2015) |

special cases (e.g., metallo-proteins). Nevertheless, some advantages are the addition of solvation terms from implicit models, either Poisson-Boltzmann (PB) or Generalized Born (GB) surface area methods (*vide infra*).

With the development of protein-crystallography, access to firsthand information to binding modes prompted the development of new scoring functions. This gave rise to "empirical" scoring functions, as these were based on experimental information used to construct regression models for the description of interactions:

$$\Delta G = \sum_i W_i \cdot \Delta G_i \tag{1}$$

This general expression presents how the free energy is decomposed to different contributions ($\Delta G_i$), which are multiplied by weighted coefficients from the regression model ($W_i$). The main difference with forcefield approaches is their evaluation, as it is simpler which results on quicker calculations. Moreover, these functions do not evaluate energy directly, as these try to reproduce "optimal" interaction values.

Empirical functions significantly outperformed force field-based scoring. Still, these presented some notable problems, such as transferability, overfitting (Kitchen, Decornez, Furr, & Bajorath, 2004) and bias for "big" ligands (such as polyphenols, for example). Plus, it has been shown that binding energy cannot be decomposed on discrete contributions, easily or effectively (Böhm, 1992).

Considering these limitations, a third type of function was developed based on statistical mechanics description. Basically, protein-ligand interactions can be calculated as a summation of distance-dependent potentials of a pair of atoms $i$ and $j$:

$$\Delta G = \sum_i^{\text{lig}} \sum_j^{\text{prot}} \omega_{ij}(r) \tag{2}$$

Said potential comes from the inverse Boltzmann relation, which states:

$$\omega_{ij}(r) = -k_B T \ln \left[ \frac{\rho_{ij}(r)}{\rho_{ij}^*} \right] \tag{3}$$

While theoretically sound, the previous strategy has its problems, the main one being the so-called reference state (Muegge, 2000). Plus, the inverse Boltzmann relation was initially derived from gases and it is not directly homologous to a biological context (Liu & Wang, 2015).

**Table 4** Examples of marketed drugs developed with the aid of SBDD.

| Compound name | Biological activity | Reference |
|---|---|---|
| Dorzolamide | Anhidrase inhibitor | Ghosh and Gemma (2015) |
| Oseltamivir | Neuraminidase inhibitor | Lew, Chen and Kim (2012) |
| Indinavir | HIV protease inhibitor | Cohen (1996) |
| Boceprivir | HCV NS3/4A protease inhibitor | Howe and Venkatraman (2013) |
| Aliskiren | Renin inhibitor | Allikmets (2007) |
| Captopril | ACE inhibitor | Andrews, Carson, Caselli, Spark and Woods (1985) |

This goes to show the weaknesses of scoring functions, as there is not a universal solution for binding energy description based on discrete contributions. Additionally, current challenges in the field include a better sampling of the entropic terms involved, i.e., side-chain flexibility, solvent contributions, etc. Such drawback has led to the study and development of additional or improved methods (see Section 2.1.3.). Even though the relatively "limited" approach, SBDD has proven its worth in drug development. Notable examples are presented on Table 4.

In summary, molecular docking has proven to be a useful approach for drug discovery and development. Still, there is no denying that it has serious limitations and can be prone to errors. As there exist diverse problems and cases where docking is no better (or even worse) than an educated guess. For further reading on the subject, we suggest the following sources (Brás, Cerqueira, Sousa, Fernandes, & Ramos, 2014; Carlson & McCammon, 2000; Morris & Lim-Wilby, 2008; Prieto-Martínez et al., 2018). Recently, molecular dynamics simulations (MD) have been adopted as the novel standard for a more accurate description of protein-ligand interaction, at the cost of more computational time and resources. In the following subsection we present the basic theory on classical implementations of the method.

### *2.1.2 Molecular dynamics (MD)*

Since the 1950s there has been a growing interest in the description of biological phenomena with reasonable simulations. This gave rise to molecular mechanics, which uses the laws of physics to describe the movement and interaction of molecules. A common approach is to the electrostatic potential (Lennard-Jones and Coulomb), bond angles, dihedrals, and bond lengths. Thus, force fields emerged: Mathematical expressions which relate the energy of a system as function of the coordinates of its components (González, 2011). A general expression for such function may be as follows:

$$E_{\text{total}} = E_{\text{bonded}} + E_{\text{non-bonded}} \tag{4}$$

where the bonded contribution is calculated with Eq. (5):

$$U(r) = \sum_{\text{bonds}} k_b(b - b_0)^2 + \sum_{\text{angles}} k_\theta(\theta - \theta_0)^2 + \sum_{\text{torsions}} k_\varphi[\cos(n\varphi + \delta) + 1] + \sum_{\text{unbonded pairs}} \left[\frac{q_i q_j}{r_{ij}} + \frac{A_{ij}}{r_{ij}^{12}} - \frac{C_{ij}}{r_{ij}^{6}}\right] \tag{5}$$

and the non-bonded contribution is calculated with Eq. (6):

$$U(r) = \sum_{\text{nonbonded pairs}} \left(\varepsilon_{ij}\left[\left(\frac{R_{\min,ij}}{r_{ij}}\right)^{12} - 2 \times \left(\frac{R_{\min,ij}}{r_{ij}}\right)^{6}\right] + \frac{q_i q_j}{r_{ij}}\right) \tag{6}$$

where $b$ refers to interatomic distance, $k_b$ and $b_0$ are bond rigidity and length, $\theta$ describes the angle formed by two vectors, whereas $\theta_0$ and $k_\theta$ are angle rigidity and its constant. The torsional terms come from the cosine function of $\varphi$ (which is the torsion angle), $\delta$ (which is the phase) and $n$ represents the dihedral potential. The non-bonded contribution includes the electrostatic interaction $\varepsilon_{ij}$ of atoms $i$ and $j$, $q_i$ and $q_j$ denote the charges of each atom and finally, $R_{\min}$ represents the distance in which the Lennard-Jones potential is at its minimum (Vlachakis, Bencurova, Papangelopoulos, & Kossida, 2014).

Based on this, it is possible to extend this approach to Newtonian dynamics, as the potential energy can be related to force:

$$F = -\nabla U_i \tag{7}$$

Also, if we consider the classical expression of force and equal both, it yields:

$$-\frac{dU}{dr_i} = m_i \frac{d^2 r_i}{dt^2} \tag{8}$$

Therefore, all that is needed to obtain the trajectory of the system is solve the equations of motion. However, due to 3 N coordinates and velocities of each component, an analytical solution is out of reach. Fortunately, it is possible to give a numerical approximation to the problem, by means of Taylor expansion we arrive to the well-known kinetic equation:

$$r(t) \simeq r_0 + v_0 + \frac{1}{2}at^2 \tag{9}$$

In principle, MD simulations are an iterative process, best summarized in the following main steps:

- Based on atom positions, the initial energy of the system is calculated by force field means.
- Using these forces, the equations of motion are solved, yielding the acceleration and velocity of particles.
- Finally, with these values the positions are updated, and the cycle continues.

As simple as it may seem, there is an additional consideration that must be made. Due to the numerical nature of the integration it could lead to instability, hence the movement is integrated on discrete time intervals ($\delta t$) or "timesteps" as usually called (Hospital, Goñi, Orozco, & Gelpi, 2015). Furthermore, this integration interval must be sufficiently large to achieve reasonable time scales in the simulation, but small enough to be stable and yield reliable forces. In most MD implementations a timestep of 1.2 or 2 fs is used to balance computational effort and accuracy (De Vivo, Masetti, Bottegoni, & Cavalli, 2016).

Thus, using a force field (potential energy function) and the kinetic equation it is possible to describe the dynamic behavior of molecular systems. In fact, one of the very first protein simulations were reported using such an approach in 1977 (McCammon, Gelin, & Karplus, 1977). From this point forward, there were numerous developments to improve the description of these phenomena. One of the most significant is the use of statistical mechanics to achieve thermodynamic description of the simulated system. A first approach to this is to consider an isolated system, in which particle number (N), volume (V), and energy (E) remain constant. A more common denomination to this system is *microcanonical ensemble*. If we consider that such system can be in one of $\Omega$ microstates, the principle of equal probability states:

$$p_i = \frac{1}{\Omega} \tag{10}$$

From this expression, Ludwig Boltzmann successfully derived the entropy as a weighted sum of the logarithms of each possible state, arriving to:

$$S = -k_B \sum p_i \ln p_i = k_B \ln \Omega \tag{11}$$

which has some notable implications as it states that the entropy (a state variable) depends directly on the number of accessible microstates. Similarly, it is possible to describe other ensembles, where the system is in contact with a heat bath (NVT or canonical ensemble) and a pressure reservoir (NPT or isobaric-isothermic ensemble), to achieve more "natural" conditions. A detailed discussion on thermostats and barostats, as well as their physics is beyond the scope of this chapter, but the interested reader is referred to the following sources (Bou-Rabee, 2014; Hernández et al., 2008; Schlick, 2010; Tuckerman, 2010; Wereszczynski & McCammon, 2012).

Development on protein force fields began in the early 80s, most of these efforts resulted in the mainstream options used today, such as AMBER, CHARMM, and OPLS. At first, these functions were based on different base sets and experimental data for parametrization and relied on "united-atom" representations due to computational power (Ponder & Case, 2003). With the maturing of these studies came the identification of "cross-terms" which allowed a better description of intramolecular behavior (González, 2011). This marked the beginning of second-generation functions, a notable example being MMFF94, whose description of hydrogen-bonding was near top-class at the time (Halgren, 1996).

Another difference of MMFF94 was its focus to describe "complex" systems, i.e., protein-ligand, due to the growing interest in the modeling and simulation of drug-like molecules in biological systems. This led to the development of "generalized force fields", which have seen a similar evolution to their protein counterparts. Recently these tools have reached a significant maturity allowing robust description of small molecules (Zhu, 2019).

Yet, one of the most prominent problems in force field construction is parameter selection and optimization, as this determines the accuracy of calculated properties. In this regard initiatives like the open force field and "chemical perception" intend to reduce the number of parameters and avoid the "curse" of atom typing (Mobley et al., 2018; Zanette et al., 2019). Plus, the role of proper sampling is often understated during accuracy assessment of any given force field (Procacci, 2019).

With this, we would like to discuss the main goal for SBDD methods, the prediction of binding energies for protein-ligand complexes. We present general theory and developments on the field.

### 2.1.3 Free energy calculations

One of the main goals of computational chemistry is the prediction of properties or activities for given compounds. This includes the evaluation of protein-ligand affinity. However, this poses several technical difficulties which may not be apparent. If we begin from a chemistry perspective, the binding process is an equilibrium:

$$[EI_{aq}] \leftrightarrow [E_{aq}] + [I_{aq}] \tag{12}$$

To achieve a description of binding energy ($\Delta G$), we can relate this equilibrium to the dissociation constant ($K_i$):

$$\frac{[EI]}{[E][I]} = \frac{1}{K_i} = K_A \tag{13}$$

$$\Delta G = -RT \ln K_A \tag{14}$$

If we consider the phenomena involved, we can breakdown the contributions to $\Delta G$, which is no different than:

$$\Delta G = \Delta H - T\Delta S \tag{15}$$

Now here is where intricacies begin, as the enthalpic component comprises the solute-solvent interactions, while entropy accounts for conformational strain and solvent entropy (Fig. 3). Furthermore, both components must be determined for protein and ligand, making it a rather complex task. This goes to show that subtle changes have great impact on binding energies, moreover a ligand may have more than one binding mode often

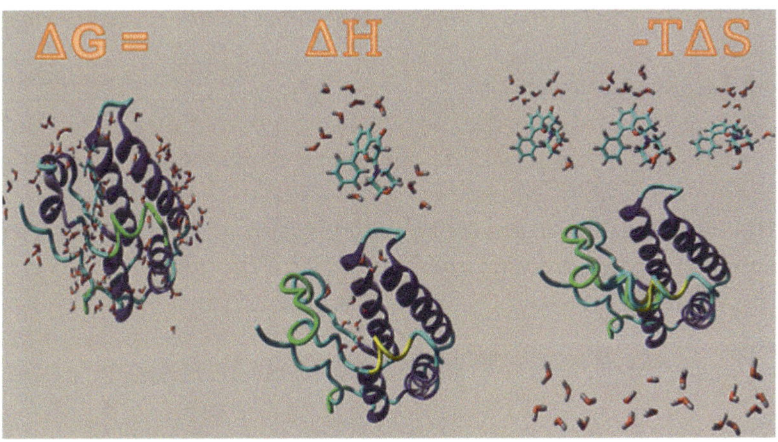


**Fig. 3** Representation of binding energy contributions for protein-ligand complexes.

separated by significant energy barriers, prompting a paradigm shift from "modes" to landscapes (Mobley & Dill, 2009).

From here, we will develop on the main methods to yield free energy predictions, with a special focus toward absolute binding free-energy (ABFE) methods.

#### 2.1.3.1 Scoring functions

As mentioned, in Section 2.1.1. The first attempts to approximate the binding energy of protein-ligand complexes were made with the use of force fields. While useful, the standalone implementation of force fields makes a truncated description (at best) of the phenomena involved. To overcome this problem, most physics-based scoring functions utilize solvation terms. For most cases, these come from implicit solvent approximations, such as Poisson-Boltzmann (PB) or Generalized Born (GB) models. Thus, a general expression for these scoring functions is as follows (Liu & Wang, 2015):

$$\Delta G = \Delta E_{vdW} + \Delta E_{\text{electrostatic}} + [\Delta E_{\text{Hbond}}] + \Delta G_{\text{solvation}} \qquad (16)$$

Empirical scoring functions have a better all-around performance, nonetheless these also come with their inherent problems. Their main disadvantage is their reliance on the training data, making their modification or combination rather difficult (Kitchen et al., 2004). Another limitation with these approximations is the technical impossibility of decomposing $\Delta G$ to a set of additive components. It has been shown several times that some interactions make major contributions to binding, which explains why fragment-based design and pharmacophore modeling provide significant results (van Drie, 2012).

#### 2.1.3.2 End-point calculations

These are an extension of the physical approach used in scoring functions. Such methods use force field description in conjunction with continuum solvent models, commonly known as MM/PBSA or MM/GBSA methods. End-point calculations compute a thermodynamic cycle to assess the energy difference between unbound and bond states. This can be expressed as follows (Decherchi, Masetti, Vyalov, & Rocchia, 2015):

$$\begin{aligned}\Delta G^{\text{MM-PB/GBSA}} = {} & \Delta G^{pl}_{\text{solv}} - \Delta G^{p}_{\text{solv}} - \Delta G^{l}_{\text{solv}} \\ & + \left(\Delta U_{\text{bind, gas}} - T\Delta S^{0}_{\text{bind, gas}}\right)\end{aligned} \qquad (17)$$

The main difference with scoring functions is their reliance on MD, using the obtained ensembles and solvent orientation to provide a better description of the phenomena (Mikulskis, Genheden, & Ryde, 2014). Still, the calculations may result quite expensive, hence the usual approach is to select a discrete set of snapshots to make the computations. Thus, it may be argued that the dividing line with scoring functions is quite faint. In addition, some limitations arise, the main one being entropy description. Normally, entropy values for such calculations come from normal mode analysis. However, it has been suggested that significant convergence problems are common for entropy. Therefore, the performance of the method is significantly affected based on simulation time and sampling (Sun et al., 2018). Furthermore, the electrostatic term is highly dependent on the dielectric constant value ($\varepsilon$). In the original implementation of the MM/PBSA method this was set to 1, recently it has been suggested that higher values ($\varepsilon = 2-4$) provide better results for highly polar or charged sites (Genheden & Ryde, 2015).

In summary, end point methods have shown a good potential for free energy approximation. With notable success cases in the literature, there is no denying the predictive power at a relatively "low" computational cost (Kuhn, Gerber, Schulz-Gasch, & Stahl, 2005). These approaches may be appropriate for series of related ligands, where the entropy term can be regarded as constant (Decherchi et al., 2015). Nevertheless, based on the highly variable results and meticulous attention to detail needed, these remain no better than an educated guess in many cases.

#### 2.1.3.3 Free energy perturbations (alchemical methods)

Also known as pathway methods, which consist of rigorous calculations, based on thermodynamic cycles. These consist of non-physical transformations of ligands during independent simulations, either to a different chemical entity or a non-interacting "dummy" molecule (Fig. 4). Whichever the case, the free energy change between ligands A and B comes from (Du et al., 2016):

$$\begin{aligned}\Delta G_{\text{bound}}(A \rightarrow B) - \Delta G_{\text{free}}(A \rightarrow B) &= \Delta G_{\text{bind}}(B) - \Delta G_{\text{bind}}(A) \\ &= \Delta\Delta G_{\text{bind}} \end{aligned} \quad (18)$$

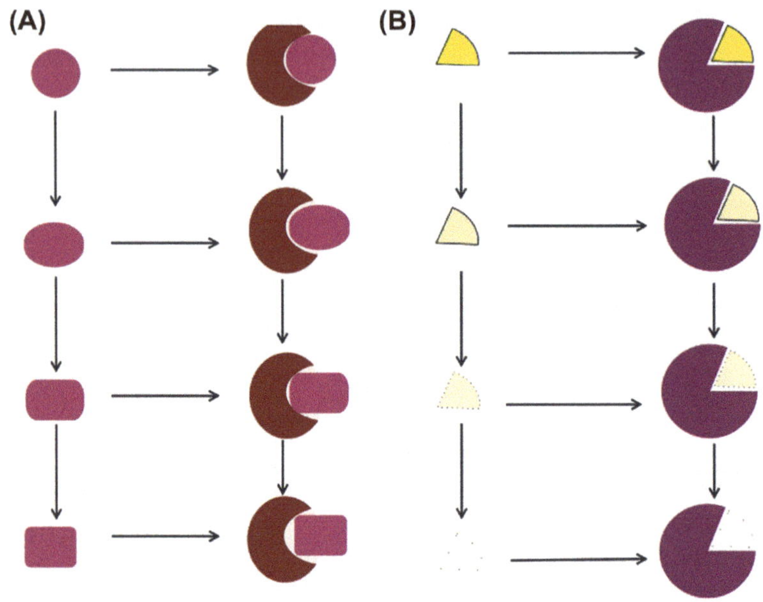


**Fig. 4** Example of a thermodynamic cycle for free energy calculations.

For a thermodynamic ensemble, the free energy difference between two states is related to the ratio of probabilities of said states, which in turn may be derived from partition functions (Q):

$$\Delta A_{ij} = -k_B T \ln \frac{Q_j}{Q_i} \tag{19}$$

$$Q = \int_{\Gamma} e^{-u(\vec{q})} d\vec{q} \tag{20}$$

where $\Gamma$ is the phase space volume, i.e., the traditional volume times the momentum space; $u$ represents a "reduced potential" (*vide infra*) and $q$ is a vector storing all positions and momentum. Taking this expression and applying it to an NVT ensemble, we first substitute the values in $u$:

$$u = \beta\left(U + PV - \sum \mu N_i\right) \tag{21}$$

which simplifies:

$$Q = \int_{\Gamma} e^{-\beta U(\vec{q})} d\vec{q} \tag{22}$$

Then we substitute for $\Delta A$, which now is the Helmholtz free energy:

$$\Delta A_{ij} = -k_B T \ln \frac{\int \Gamma_j e^{-H_j(\vec{q})\beta} d\vec{q}}{\int \Gamma_i e^{-H_i(\vec{q})\beta} d\vec{q}} \tag{23}$$

where $H$ is the system's Hamiltonian and $\beta$ is the inverse Boltzmann constant times $T$.

Alchemical methods have shown a notable predicting power, with several success cases where the root-mean squared error is around 2.0–2.5 kcal/mol when compared to experimental determinations (Aldeghi, Heifetz, Bodkin, Knapp, & Biggin, 2016). However, their performance varies significantly due to high dependence on several aspects such as: initial conformation (of protein and ligand), solvation (implicit or explicit), the number of independent simulations, sampling, etc. (Manzoni & Ryde, 2018). This may prove time-consuming for every case, prompting for the development of different solutions which could result in further confusion of first-time users.

To address some of these problems, several methodologies have been proposed, e.g., replica-exchange (RE), umbrella sampling (US), Bennett Acceptance Ratio (BAR/MBAR) (De Ruiter & Oostenbrink, 2011). The latter has been combined successfully with extended sampling, achieving notable performance increase (Kim & Allen, 2012). Also, its "multistate" variant (MBAR) provides an optimal way to use all simulation data with good estimates for error (*vide infra*). This with the added benefit of more control on simulation length when compared to other ABFE calculations, such as thermodynamic integration (Chodera et al., 2011).

The main reason for this is that the BAR method only requires configurational information of two states. For a two-sided estimation of free energy, performance depends on the ratio:

$$r = \frac{n_R}{n_F} \tag{24}$$

In other words, if we are to obtain free energy difference it is possible to use both: forward and reverse work values. This leads to an important question, which is the optimal value for the number of forward and reverse work values to analyze? Such estimator is intended as: $N = n_R + n_F$; hence we need to determine an optimal value for $r$ when $N$ remains constant (Hahn & Then, 2009). Bennet's contribution was the derivation which yields the optimal values for $n_R$ and $n_F$. Now we present a similar reasoning starting from the expression for free energy in Eq. (19). Previously, Zwanzig demonstrated that the exponential average of equilibrium energy differences among two states equals the free energy between said states (Zwanzig, 1954), thus:

$$\Delta A_{ij} = \frac{\langle \alpha(\overrightarrow{q}) e^{-\beta H_i(\vec{q})}{}_i \rangle}{\langle \alpha(\overrightarrow{q}) e^{-\beta H_j(\vec{q})}{}_j \rangle} \tag{25}$$

Such expression is true for every $\alpha(\overrightarrow{q}) > 0$.

Yet, it can be argued that such a set of data is both noisy and biased; as the results of exponential average are strongly dependent on the distribution tails, these however tend to be poorly sampled (Shirts, Bair, Hooker, & Pande, 2003). Therefore, we make use the Jarzynski relation for non-equilibrium work, which states:

$$e^{-\beta \Delta F} = \frac{\langle f(W) \rangle_F}{\langle f(-W) e^{-\beta W} \rangle_R} \tag{26}$$

Here $\Delta F$ is the change in free energy; This is true for any function, which defines the measures from an initial to a final state. Hence, we denote this as a "forward" direction (subscript $F$); whereas the measures from the final to initial state follow a "reverse" direction (subscript $R$) (Crooks, 2000). From Eq. (26), an alternative derivation of BAR is possible using the Maximum likelihood estimator (Shirts et al., 2003), thus if we start from the probability distributions for work processes in opposite directions:

$$\ln\left[\frac{P_F(W)}{P_R(-W)}\right] = \beta(W - \Delta F) \tag{27}$$

Eq. (27) may result familiar to the reader; as it is a logistic regression commonly used in the field of epidemiology. Therefore, we may rewrite the expression as a ratio of conditional probabilities: $P(W|F)/P(W|R)$; now from the rules of such probabilities we know that: $P(W|F) + P(W|R) = 1$; this leads to:

$$\frac{P(W|F)}{P(W|R)} = \frac{P(F|W)P(R)}{P(R|W)P(F)} = \frac{P(F|W)}{1 - P(F|W)} \frac{P(R)}{P(F)} \tag{28}$$

The reader may notice that $P(R)/P(F) = n_R/n_F$, now we define a constant to rewrite Eq. (27) (Crooks, 2000);

$$M = kT \ln \frac{n_F}{n_R} \tag{29}$$

$$\ln \frac{P(F|W)}{1 - P(F|W)} = \beta(M + W - \Delta F) \tag{30}$$

with this we may rewrite the probability of a single measurement or $P(F \mid W_i)$; $P(R \mid W_i)$

$$\frac{1}{1 + e^{-\beta(M + W_i - \Delta F)}} = P(F|W_i) \tag{31}$$

$$\frac{1}{1 + e^{\beta(M + W_i - \Delta F)}} = P(R|W_i) \tag{32}$$

This implies that we can estimate the value of $\Delta F$ that maximizes the likelihood of obtaining the specified values for forward/reverse values. Here we would like to compare BAR with other techniques, to better appreciate its power. If we return to the general approach of FEP methods, multi-state sampling is often used for accuracy improvement. Said sampling is made along the "alchemical" coordinate (often denoted as $\lambda$) for $n$ intermediate states, i.e., for $\lambda$ between 0 and 1; which is:

$$e^{-\beta\Delta F} = \prod_{i=0}^{n-1} \langle e^{-\beta(U(\lambda_{i+1}) - U(\lambda_i))} \rangle_{\lambda i} \tag{33}$$

The main drawback of other techniques such as exponential averaging can be avoided if the values of $U(\lambda_i)$ are stored for each stratum during simulation and post-processing (Procacci, 2019). With this in mind, we introduce the maximum likelihood $L$; expressed as a product the joint probabilities of obtaining the measurements of work (forwards and backwards) at the specified values (Shirts et al., 2003):

$$L(\Delta F) = \prod_{i=1}^{n_F} P(F|W_i) \prod_{j=1}^{n_R} P(R|W_i) \tag{34}$$

We can further simplify the previous expression solving for the logarithm of the likelihood and then deriving it with respect of F; finally, we equal to 0, arriving to:

$$\frac{\partial \ln L(\Delta F)}{\partial \Delta F} = \sum_{i=1}^{n_F} \frac{1}{1 + e^{\beta(M+W_i-\Delta F)}} - \sum_{j=1}^{n_R} \frac{1}{1 + e^{-\beta(M+W_j-\Delta F)}} = 0 \tag{35}$$

This implies that $\Delta F$ has its limits on $n_F$ and $-n_R$, which in turn assures it has a unique root. As stated, this reasoning is analogous to Bennett's original proposal (Bennett, 1976); showing that *r* is the minimum variance asymptotically unbiased estimate. Thus, BAR always yields better estimates when compared to exponential averaging (Shirts et al., 2003). Plus, it has been shown that BAR implementation is possible in rather unusual free energy differences, where thermodynamic integration or exponential averages are not practical (König, Bruckner, & Boresch, 2009).

Even with these developments and novel algorithms, the field of free energy calculations still has significant problems. However, it can be stated that most of these are mostly limited to the practical aspects, such as definitions for good practices or benchmarks (Chodera et al., 2011).

With this we end our focus in SBDD methods and introduce ligand-based drug design (LBDD). In this field the most prominent techniques come from cheminformatics, which in turn has its cornerstone on molecular description and similarity. Hence, the following subsection encompasses the fundamentals of these.

### *2.1.4 Similarity methods*

As an alternative to structure-based methods, where the coordinates of the biomolecular target should be known, are the ligand-based approaches. In general, in ligand based-methods the chemical structures of known active compounds and, if available, inactive molecules, are used as reference and "learn" about the biological system, in an indirect manner. Indeed, in many practical instances, the coordinates of the molecular targets are not known, and it is not possible to obtain reliable homology models. In other instances, the actual molecular targets have not been identified. The latter is a typical case of, for instance, phenotypic screening. On other cases that are frequent, the biological effect may be due to the interaction of the ligand with several target compounds, so it becomes in practical determining the exact interaction binding energies for real-case drug discovery efforts. This is also typical in constantly increasing multi-target drug discovery (Medina-Franco, Giulianotti, Welmaker, & Houghten, 2013). A workaround in the cases is then use ligand-based methods.

In addition to QSAR studies, one of the most successful ligand-based approaches is similarity searching. This crucial topic in drug discovery has been extensively reviewed in recent publications (Maggiora & Shanmugasundaram, 2011; Medina-Franco & Maggiora, 2013). Such reviews discuss the concept of similarity considering medicinal chemistry projects (Maggiora, Vogt, Stumpfe, & Bajorath, 2013).

The basic idea of similarity searching relies on the similarity principle where one expects that compounds with similar chemical structure will have similar properties, for instance, similar biological activity. Of course, this is a "principle" and there are very well known and notable exceptions of compounds that have similar chemical structure but very different biological activity. These are called activity cliffs (Maggiora, 2006) (or property cliffs) and have been the subject of intense studies and characterization over the past 15 years (Medina-Franco, 2013; Stumpfe, Hu, Dimova, & Bajorath, 2014). Putting aside the activity cliffs that, overall, are not that common in large screening databases (Iyer, Stumpfe, Vogt, Bajorath, & Maggiora, 2013), similarity searching has been quite useful to compare large chemical databases to the structures active compounds and the most similar compounds are also active. This process is also called similarity-based virtual screening (Cereto-Massagué et al., 2015).

As clearly discussed in the review papers, similarity searching involves two major components: (1) Structure representation of the compounds in which the similarity will be measured, and (2) A similarity function. Structure representation is the cornerstone of chemoinformatics and is the basis not only of similarity searching but also of the concept of chemical space (Willett, 2014).

A very common structure representation used in similarity searching is based on molecular or structure fingerprints, also extensively discussed elsewhere (Cereto-Massagué et al., 2015). Structure fingerprints is a bit string of zeros and ones that account for the absence or presence of structural motifs. Such motifs are either pre-determined (dictionary-based fingerprints) or are computed directly from the chemical structures (such as the extended connectivity fingerprints that are based on the Morgan algorithm) (Walters, 2019).

Most of the similarity functions ref used to compute the similarity between two chemical structures $A$ and $B$, are basically a ratio of the number of structural motifs that share two compounds divided by the total number of motifs present in both molecules (Medina-Franco & Maggiora, 2013). Different similarity functions are scaled differently in order to capture, for instance, the different molecular sizes that compounds A and B can have. One of the most common similarity functions used is the Tanimoto coefficient, Eq. (36) (Jaccard, 1901):

$$S_{AB} = \frac{c}{a+b} \tag{36}$$

where $S_{AB}$ is the similarity between compounds $A$ and $B$; $c$ is the number of bits common to both molecules; $a$ is the total number of features (bits turned "on") in molecule $A$, and $b$ is the total number of features (bits turned "on") in molecule $B$.

Perhaps the most relevant or crucial points in similarity searching is precisely to choose the structural representation to be used. Since a single representation cannot capture all structural features relevant in drug discovery, a common practice is using several representations and then decide which one gives the best performance for the specific study system of combining the outputs in what is called data fusion (analogs to consensus scoring in docking) (Willett, 2013).

Another important point when planning similarity searching studies is to decide whether using two- or three-dimensional structure representations. In principle it can be expected that 3D similarity searching outperforms 2D-similarity methods because it would capture the 3D arrangement of atoms required in molecular recognition (Anighoro & Bajorath, 2016; Prieto-Martinez & Medina-Franco, 2018). A downside of 3D similarity searching is choosing the relevant and bioactive conformation. As discussed above, this is a very challenging point. In most practical applications a low energy conformer is used but this is not always accurate because of the known induced-fit effect that happens in molecular recognition. Without knowing the 3D structure of the receptor in ligand-based approaches, predicting the biological relevant information remains as a challenging task. Still, for rigid or relatively rigid compounds this might not be a major issue.

We have described the theoretical basis and, to some extent, an overview of *in silico* methods used in drug discovery. Now let us discuss their implementation, with BETs as a case study.

## 3. Discovery of BETs inhibitors

As it often happens, BET inhibitors were discovered by a series of fortuitous events; which have been extensively discussed elsewhere (Ferri, Petosa, & McKenna, 2016; Fujisawa & Filippakopoulos, 2017; Galdeano & Ciulli, 2016; Hewings et al., 2012; G. Zhang, Smith, & Zhou, 2015). Herein we present a bird's eye view on the development of key BET inhibitors. In 2001, scientists at Glaxo-Smith-Kline R&D, were looking for novel up regulators of Apolipoprotein A1 (ApoA1); as it is a promising strategy for the development atherosclerosis therapies (Mirguet et al., 2012). From these efforts a benzodiazepine compound was identified as a potent promoter of ApoA1, optimizations around this scaffold lead to Molibresib which was later confirmed as a pan-BET ligand (Mirguet et al., 2013). Similarly, Filippakopoulos et al., designed JQ1 with SBDD; based on the description of thienodiazepines as BRD4 ligands (Filippakopoulos et al., 2010). In both cases the similarity toward alprazolam, led to the study of the SAR for benzodiazepines and benzotriazepines, as well as the putative relation between BETs and the BDZ site of GABA receptor; no such relation was found as JQ1 and Molibresib exhibit poor binding for the BDZ site (Filippakopoulos et al., 2012).

Further studies and HTS campaigns in GSK identified the tetrahydroquinoline scaffold as an ApoA1 up regulator. During the profiling and SAR studies it became apparent that akin to the previously identified benzodiazepines, BET inhibition was greatly biased to one isomer (*S*) for benzodiazepines and *cis* for tetrahydroquinolines (Gosmini et al., 2014). Circa 2011, Hewing et al., observed that *N*-methyl pyrrolidone was a weak binder of several K—Ac pockets of bromodomains. From this, they decided to study other heterocycles bearing the methyl moiety, finding an unexpected potency for 3,5-dimethylisoxazole scaffold (Hewings et al., 2011). Almost in parallel, GSK scientists began fragment-based (FBDD) where 4-Phenyl 3,5-dimethylisoxazole showed promising activity toward BRD3 and BRD4. Just as JQ1, the optimization procedure was made with the aid of SBDD (Bamborough et al., 2012).

Another notable advancement was made by Picaud et al., whom described PFI-1 a novel inhibitor based on the 3-methyl-tetrahydroquinazolinone scaffold. The ligand was co-crystallized with BRD4, providing valuable insight for BET inhibition. Some notable observations were the confirmation of the water network contribution for ligand binding; as well as the differential binding between BD1 and BD2 confirmed by isothermal calorimetry (ITC) titration (Picaud, Da Costa et al., 2013; Picaud, Well et al., 2013).

Finally, Dittman et al., identified PI3-kinase inhibitors (flavopiridol derivatives) as BET ligands. This was a rather intriguing but promising observation, as it opened a novel venue for polypharmacology of cancer treatment (Dittmann et al., 2014). As seen, development of BET inhibitors was due to careful study of rather serendipitous observations. In all the inhibitors described thus far, the role of an KAc mimetic moiety has been a crucial starting point for successful inhibition. Also, it is notable that these first efforts were mostly aided by "classical" drug design, around SBDD methods; we now take our attention on several developments and research aided on *in silico* tools, with special focus on the search for BET selectivity.

## 3.1 Computational studies for the design of novel BET inhibitors

As stated, discovery of novel inhibitors has been pursued to better understand the role of BETs and their safety as therapeutic agents. A prominent approach for virtual screening is the use of public repositories due to the high order of records present. In 2014 the ZINC database (Irwin, Sterling, Mysinger, Bolstad, & Coleman, 2012) was screened for BRD4 ligands, using high through-put docking; the identified hits presented poor diversity when compared to current inhibitors (Zhao, Gartenmann, Dong, Spiliotopoulos, & Caflisch, 2014). Due to the significant success of SDBB, other groups have used supplementary *in silico* methodologies; FBDD and LBDD. Table 6 summarizes computational studies toward the identification of BET inhibitors using different VS strategies. The table emphasizes the main type of structure or molecular scaffolds explored in each VS campaign.

The focus of ligand-based methods is the discovery of similar ligands or bioisosteres, in a cheminformatic context this is known as scaffold hoping. Such approach is desirable in BET research, in order to find KAc mimetics. Thus, similarity methods are key for the advancement of the field. Several computational studies have focused toward the screening and filtering, but few have really committed to the search of diverse inhibitors. A notable example is the development of ALTA-VS (anchor-based library tailoring), which uses a fragment-based approach followed by docking screening; with the added benefit of using academic or open-source tools (Spiliotopoulos & Caflisch, 2016).

Of this minority, a couple of studies have used the diversity-oriented synthesis, which is a combination of known fragments to obtain a putative inhibitor. For example, Hoffer et al., proposed a workflow based on SBDD to identify hits, then compared to different reactions and fragments this "raw library" is then diversified using *de novo* design; by means of such approach xanthine derivatives with $IC_{50}$ values on the nM range were found (Hoffer et al., 2018). Likewise, Batiste et al., presented AutoCouple, a tool based on *de novo* design by means of fragment growing, this yielded several promising compounds still under study (Batiste et al., 2018).

Finally, we discuss some developments made to study not only BET proteins or their inhibitors, but general methods which may be of use for multiple targets. We will give a brief description of the role of BETs as case study in the computational field and some of the insight these have provided.

### 3.1.1 BETs as case study for development of novel methods and algorithms

#### 3.1.1.1 Drug repurposing

Repurposing is an attractive approach for hit identification, as it uses previously known compounds as a starting point. There are many examples in the literature reinforcing the power of drug repurposing (Brindha et al., 2017; Méndez-Lucio, Tran, Medina-Franco, Meurice, & Muller, 2014; Pérez-Nueno et al., 2014), with SARS-CoV2 inhibitors being a recent example (Zhou et al., 2020).

As with previous methods, the virtual screening aimed for repurposing must be properly delimited and constructed to yield good results. Some inherent problems such as target complexity or structural features are factors often overlooked on repurposing campaigns. Wakchaure et al., conducted a repurposing of FDA-approved drugs toward BRD4, prior to docking screening they conducted an enrichment search of crystal structures based on ROC curves. The results showed promise as top scoring compounds showed a consensus interaction with N140, W81, P82, F83, V87, L92 & 94, Y139, D144 & 145, I146 and the water network (Wakchaure, Velayutham, & Roy, 2019).

#### 3.1.1.2 Machine learning

One of the main problems in binding energy prediction is the complexity of the problem and inability to breakdown such phenomena to "universal" functions. Still, it is possible to construct target specific scoring functions; in order to avoid the drawbacks of linear regression found in empirical scoring functions (*vide supra*) one may use machine learning techniques for variable weighting and model constructions. Take for example the development of BRD4LGR, a scoring function trained with structural data which outperformed general methods by a 20%–30% in success margin (Xing et al., 2017).

#### 3.1.1.3 Atropisomerism

The role of stereochemistry for biological activity is undisputed fact; only recently we begin to assess the role of atropisomerism, this type of geometry only arises in compounds which have steric hindrance along rotatable bonds. This stereoisomerism tends to evoke images of biaryl and heterobiaryl compounds; such behavior has been also observed on natural products (macrocycles) and a variety of druglike substances, e.g., benzodiazepines (Glunz, 2018).

For bromodomain inhibition, other scaffolds described with this behavior are pyrimido-benzodiazipinones (a dual BRD4 and ERK5 inhibitor) (Wang et al., 2018) and amentoflavone (Prieto-Martínez & Medina-Franco, 2018).

#### 3.1.1.4 Activity-cliffs

The concept of activity-cliff is under continuous evaluation, as new evidences arises to broaden its definition and scope. Such is the case for "off-pocket cliffs", a rather unexpected trait which results in significant potency differences as a result of changes in non-interactive parts of the ligand (Abramyan, An, & Kireev, 2020). BRD4 has shown this phenomenon in some of its inhibitors, presenting an opportunity for the elucidation of governing factors of this event and its implications for drug discovery.

#### 3.1.1.5 QSAR models

One of the first implementations of computer science and resources in chemistry, it has come a long way since the Hansch-Fujita implementation (Kubinyi, 1993). As for any method, QSAR performance is highly dependent on input data, thus the selection of descriptors is paramount for successful predictions.

Proteochemometric models are an extension of classic QSAR, whereas in this case descriptors are chosen for ligand and target (either as whole or just the pocket site) are used for prediction (Kanakaveti et al., 2020). Recently, Giblin et al., developed a proteochemometric model for the bromodomain protein family. One of the highlights was the significant selectivity showed by screened compounds for BRD4, considering that researchers made use of histone-peptides as ligands and its descriptors translated correctly for small molecules (Giblin, Hughes, Boyd, Hansson, & Bender, 2018).

#### 3.1.1.6 Free energy methods

Herein we have discussed the main advantages and challenges present in the field of binding prediction. As this field reaches maturity there are questions that have been solved while others remain open. One such question could be if it ever will be a general and robust method that performs well in multiple targets. To address this, it has been suggested that ABFE methods are robust enough for small binding sites such as those in BRDs, with the error being around 1.5 kcal/mol (Aldeghi et al., 2016). Moreover, it has been shown that end-point methods have a reasonable performance when compared to ABFE (at least for BET bromodomains that is) and may be of use in cases were computational resources are scarce (Aldeghi, Bodkin, Knapp, & Biggin, 2017).

#### 3.1.1.7 Binding site hydration and solvent contribution to binding

The role of water for ligand binding was previously seen as null; with the advent of protein crystals it soon became clear that there was more to it than mere desolvation of pockets. With this paradigm shift several methods have struggled to include water molecules as part of the simulation (Ladbury, 1996). Hence, there is a major interest on the accurate and reasonable description of both conserved waters and their role for ligand binding.

There have been several tools which have attempted to do this, however only recently there has been significant progress. On this regard, BRD4 offers a good benchmarking system due to its occluded pocket and highly conserved water network. As matter of fact, two recent studies have shown notable performance on this protein with a deviation of 1 Å for water placement (Masters, Mahmoud, Yang, & Lill, 2018; Nittinger et al., 2018). These approaches can further benefit from physical methods such as cosolvent dynamics (Huang, Rossini, Steiner, & Caflisch, 2014) and Monte Carlo simulation (Aldeghi et al., 2018), for thermodynamic description of its stability.

## 3.2 Probing for BET selectivity

One of the first computational studies on bromodomains was made by Vidler et al., which consisted on the characterization of several bromodomains. The main goal of this study was to assess the overall "druggability", i.e., proteins that show or are presumed to bind (small) drug-like entities.

From the gathered data, BETs were clearly druggable, mainly due to their pocket size and the role of the ZA channel which serves as a better contact surface for putative ligands (Vidler, Brown, Knapp, & Hoelder, 2012). Experimental confirmation for this came from later studies were a >10-fold ratio of affinity toward BRD4-BD1 was observed; based on ZA-loop behavior and ligand orientation (Raux et al., 2016).

As previously stated, one of the main problems in BET inhibition is the lack of selective ligands between isoforms. Sequence alignment of BET isoforms (Fig. 5) shows the extent of this problem, as not only complete sequences are significantly similar; also, the active site is highly conserved. Due to this, bromodomain selectivity is limited to subtle differences which are mostly seen between BD1 & BD2; these are limited to five residues commonly known as the "V-positions", due to their vertical orientation on the pocket (Table 5) (Shadrick et al., 2018).

Moreover, the pharmacological selectivity between BD1 and BD2 is still an open debate; current evidence clearly suggests that BD2 (of BRD4) has a strong interaction with the diacetylated peptide H3K4AcK9Ac, this preference extends to non-histone proteins (e.g., TWIST) which results on lesser transcriptional roles (Liu et al., 2017). With the aid of NMR techniques, it was confirmed that this preference is due to critical residues on the pocket of

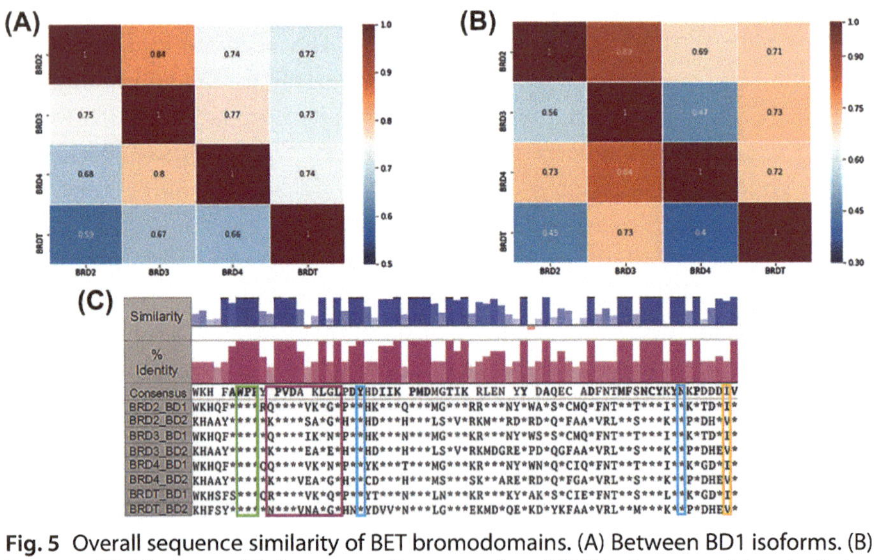


Fig. 5 Overall sequence similarity of BET bromodomains. (A) Between BD1 isoforms. (B) Between BD2 isoforms. (C) Sequence alignments for all BET isoforms; residue similarity, identity and consensus are shown. Conserved residues are marked by *, binding site hotspots follow the same color code as in Fig. 1.

**Table 5** Five "vertically positioned" residues; which differentiate BD1 and BD2 in BET proteins.

| Position | BRD2 | BRD3 | BRD4 | BRDT |
|---|---|---|---|---|
| V1 | I162//V397 | I122//V397 | I146//V439 | I115//V357 |
| V2 | D160//H433 | D120//H395 | D144//H437 | D113//H355 |
| V3 | K107//A380 | K67//A342 | K91//A384 | K60//A302 |
| V4 | Q101//K374 | Q61//K336 | Q85//K378 | R54//N296 |
| V5 | R100//Y373 | Y60//Y335 | Q84//Y377 | Q23//Y295 |

BD2, it has been suggested that this activity may explain the effect of BET inhibitors in xenograft models of cancer (Shi & Vakoc, 2014). Hence to further understand the role and partners of BD2, several attempts have been made such as the development of RVX-208 and RVX-297 (Kharenko et al., 2016; Picaud, Da Costa et al., 2013; Picaud, Well et al., 2013); bivalent inhibitors (Tanaka et al., 2016) and biological tests (Kougnassoukou Tchara, Filippakopoulos, & Lambert, 2019).

It has been suggested that the ZA loop diversity and dynamics provide a venue for selectivity, due to its plasticity and length difference between bromodomains (Steiner, Magno, Huang, & Caflisch, 2013). Therefore, a series of recent studies have focused on the description of BRD4 BD1 and BD2 by means of MD simulation. Cheng et al., carried out a combination of classic and hybrid simulations (QM/MM), these revealed that the ZA loop undergoes an "in/out" transition which results in the differential binding of RVX-208 & 297. Plus, it was shown that the dynamic behavior of the loop is greatly influenced by D144, as it forms a dyad with Y139 (Cheng et al., 2017). On the same year, an extensive ABFE study of several bromodomains was carried out. Again, with the use of RVX-208 & 297 it was observed that these methods are robust enough to calculate the binding profile and energy over multiple proteins while remaining sensible to differentiate BD1/BD2 preference (Aldeghi, Heifetz, Bodkin, Knapp, & Biggin, 2017).

Further profiling was done with *in silico* alanine scanning, in this case the computed contributions by residue supported the major role of the ZA channel, highlighting the hydrophobic interactions made by L92, W81 and F83 (Su et al., 2018). In the light of these findings, other groups turned their attention to the water network present in the channel. Using a combination of MD and binding energy predictions (LIA method), the role of the network and the V4 position on the binding of tetrahydroquinazolines for both domains. Results showed that the movement of the V4 position leads to the reorganization of the solvent network. This effect was significant for pyrrol-substituents and not shown by furan R-groups suggesting preference for weak H-bond donors (Bharatham et al., 2018). These observations were confirmed by ITC experiments, which showed that the water network can increase enthalpy contributions of weak donors/acceptors and be crucial for selectivity (Shadrick et al., 2018).

Finally, with the aid of ABFE paths of binding and unbinding it was shown that the major contributors for strong interactions are residues Q85, P86, N140 and I146; with the last two being the stronger of them (Wang et al., 2019).

## 4. Case study: optimizing the coumarin scaffold

It came to our attention the attempts of Zhang et al., whom chose the coumarin scaffold based on PFI-1. Based on this and previously reported activity of flavonoids as BET inhibitors (Table 6), we decided to further explore the coumarin scaffold and its putative SAR using *in silico* modeling

**Table 6** *In silico studies* focused on BET inhibition.

| Main scaffold[a],[b] | VS protocol※ | Reference |
|---|---|---|
| Dimethyl-isoxazole | SBDD | Hewings et al. (2011) |
| Phenyl isoxazole | FBDD | Bamborough et al. (2012) |
| Dihydro-methyl-quinazolinone | FBDD | Fish et al. (2012) |
| Triazolopyrimidines | Similarity searches | Vidler et al. (2013) |
| 2-thiazolidinones | Consensus workflow | Zhao et al. (2013) |
| Diazobenzene | SBDD | Zhang et al. (2013) |
| Imidazopyrimidines | FBDD | McKeown et al. (2014) |
| Benzodiazepine | SBDD | Zhao et al. (2014) |
| γ-Carbolines | SBDD | Ran et al. (2015) |
| Dihydroquinoxalinone | Consensus workflow | Duffy et al., (2015) |
| Flavonoids | SBDD | Raj, Kumar and Varadwaj (2016) |
| Benzoindolones | SBDD | Xue et al. (2016) |
| 2-thiazolidinone/ Benzoxazinone | Consensus workflow | Marchand, Dalle Vedove, Lolli and Caflisch (2017) |
| – | Machine-learning | Xing et al. (2017) |
| 3,5-dimethyl-isoxazole | SBDD | Zhao et al. (2017) |
| Pyridone | FBDD | Le Wang et al. (2017) |
| Dimethylisoxazole | SBDD | Zhao et al. (2017) |
| Pyrido-pyrimidintrione | SBDD | Ayoub et al. (2017) |

**Table 6** *In silico studies* focused on BET inhibition.—cont'd

| Main scaffold[a],[b] | VS protocol※ | Reference |
|---|---|---|
| Pheynyl-pyrrolidone | SBDD | Allen et al. (2017) |
| Dihydroquinazolione | SBDD | Shao et al. (2017) |
| 4-chloro-5-methy-tetrahydro naphthyridin-amine | QSAR | Speck-Planche and Scotti (2019) |
| Xanthine | Consensus workflow | Hoffer et al. (2018) |
| Dimethylbenzimidazolone/5-aminoindole | FBDD | Borysko et al. (2018) |
| Flavonoids/Steroids | Consensus workflow | Prieto-Martinez and Medina-Franco (2018) |
| Dimethylisoxazole | SBDD | Ma et al. (2019) |
| Dihydroquinoxalinone | SBDD | Hu et al. (2019) |
| Coumarin | Similarity methods | Zhang et al. (2019) |
| Dimethylisoxazole | SBDD | Sperandio et al. (2019) |
| N-methylpyrrolidone | FBDD | Hilton-Proctor et al. (2019) |
| Amino-methylcyclopentenone | SBDD | Wang et al. (2019) |
| — | SBDD | Wakchaure et al. (2019) |
| Phenanthridinone | SBDD | Zhi et al. (2019) |
| methylimidazopyrazinone | FBDD | Yang et al. (2019) |
| — | SBDD | Lifei Wang, Wang, Sun, Zhao and Wang (2019) |
| Dimethylpyridone | Consensus workflow | Wellaway et al. (2020) |

※ Main method(s) used for VS; Consensus workflow describes the use of several methods for filtering or screening.
[a]Either identified or studied.
[b]Refers to the KAc-mimic or main anchor.

and screening. The coumarin scaffold is a privileged structure in medicinal chemistry, as it has shown a broad spectum of bioactivities, such as anti-inflammatory, antibacterial, anticoagulant, ROS scavenger, antithrombotic, and CNS stimulant to name a few (Sandhu, Bansal, Silakari, & Bansal, 2014). Furthermore, coumarins have been studied extensively as putative anticancer agents, showing multitarget activity and rich SAR (Thakur, Singla, & Jaitak, 2015; Zhang & Xu, 2019).

As a first step, all compounds with a coumarin scaffold present in ZINC15 (https://zinc.docking.org/) (Sterling & Irwin, 2015) were downloaded. The dataset was curated using MOE (Molecular Operating Environment MOE, 2019.01, 2019), removing duplicates, neutralizing and optimizing the structures with energy minimization using MMFF94x forcefield. This resulted in a total of 229,000 compounds used for VS.

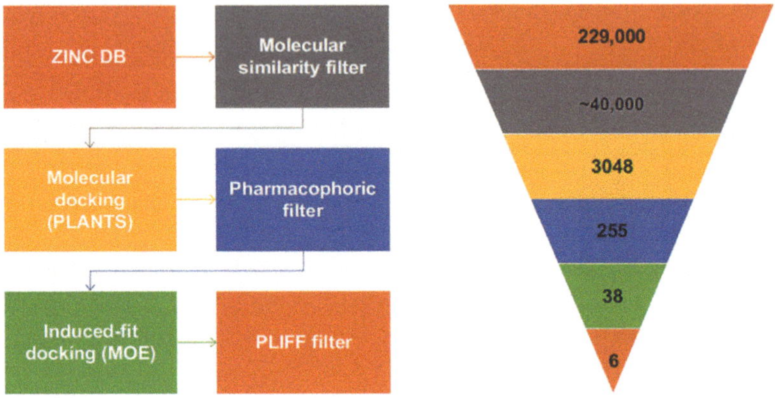


**Fig. 6** General workflow and results during the proposed stages for coumarin-based virtual screening.

A previously curated database of BET inhibitors (Prieto-Martínez et al., 2016) was used for similarity comparison and compound filtering. The VS was carried out in five steps, see Fig. 6 for a general breakdown.

Fingerprint models of the coumarin-compounds were constructed with MOE, using ECFP4, GpiDPH3 and 3D shape fingerprint were used to assess potential similarity to reference compounds. Almost 40,000 compounds showed similarity to BET inhibitors. These were docked to BRD4 (PDB ID: 4MR4) with PLANTS (Korb, Stützle, & Exner, 2009). The binding site was defined within a 10 Å sphere with a centroid on N140. Additional parameters were kept in default values. The top five poses of the resulting candidates were then compared to the pharmacophoric query (*vide infra*) resulting in 255 potential hits. The pharmacophoric model was constructed as follows: first, a diverse subset of top active compounds from the BET dataset, was selected for molecular superposition. In addition to reference compounds, amentoflavone was included in the pharmacophoric alignments. The pharmacophoric model was obtained using pharmACOphore (for alignment and scoring) (Hessler, Korb, Monecke, Stützle, & Exner, 2010) and a manual query selection with MOE.

As a final step of filtering, an induced-fit docking was carried out with MOE, yielding 38 compounds. Protein-ligand Interaction fingerprints (PLIFs) were used to identify compounds with significant contacts. Only six coumarin derivatives (Chart 2) were found with an acceptable interaction profile and thus were used for scaffold optimization.

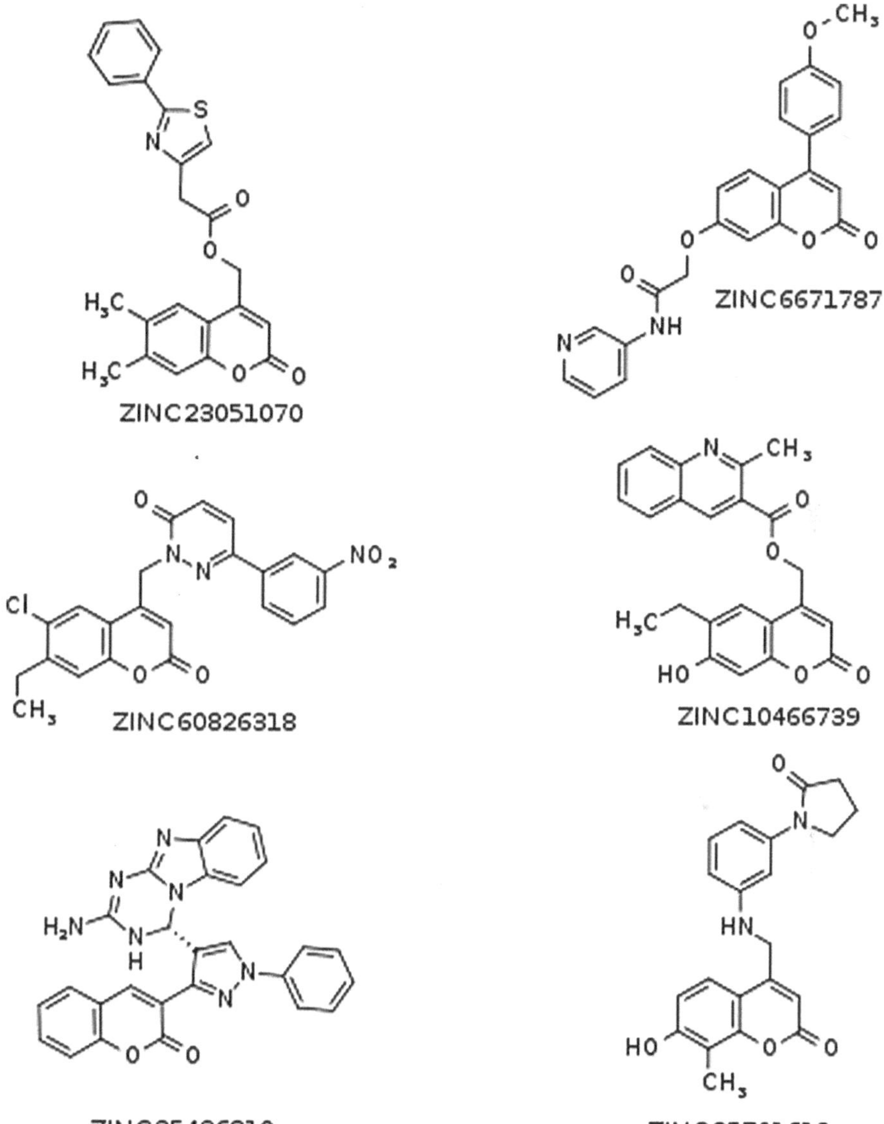


**Chart 2** Structures of the computational hits identified by virtual screening.

Based on the binding poses of the putative hits we confirmed the coumarin carbonyl is the main anchor toward N140. Considering this, we focused on positions 4 and 7 of the coumarin ring as putative optimization points. Hence, it was decided to add a hydrophobic substituent with a

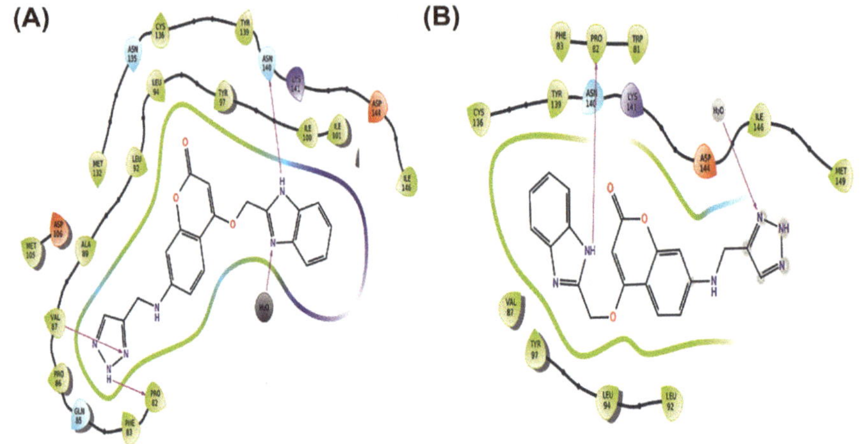


**Fig. 7** Docking poses identified for the designed coumarin.

flexible spacer on position 7, to improve contact with the WPF shelf. For position 4 we opted for a bioisoster of carboxylic groups present on JQ1 and Molibresib. This was done to improve interaction with positions D144 & 145 which may provide BRD4-BD1 selectivity (Jung et al., 2014).

Following optimization, the designed coumarin was docked to BRD4 with MOE using an induced fit approach. After a visual inspection of the binding poses, two different arrangements were identified (Fig. 7) and submitted to MD simulations. These were run with YASARA (v.19.12.14) (Krieger & Vriend, 2014). After inspection of the solute RMSD as a function of simulation time, the first 50ns were considered equilibration time and excluded from further analysis. Binding energy calculations were done with MM/PBSA method, using the snapshots from production runs (150 ns). Calculations were made with the Adaptative Poisson-Boltzmann Solver (APBS) software (Baker, Sept, Joseph, Holst, & McCammon, 2001; Jurrus et al., 2018).

The interaction profiles from MD simulations (Fig. 8) showed that pose A made better contacts with crucial positions of the pocket (P82, V87, L92, N140 and I146); whereas for pose B the most conserved contacts were with Q85 (position V4) and D105. Common contacts between poses were found in the WPF shelf, V87, L92&94 and I146. Of these, interactions with P82 may prove significant as this residue may be related to the P-binding, commonly observed on other bromodomain families (Magno, Steiner, & Caflisch, 2013) and has been related to binding stability (Wang et al.,

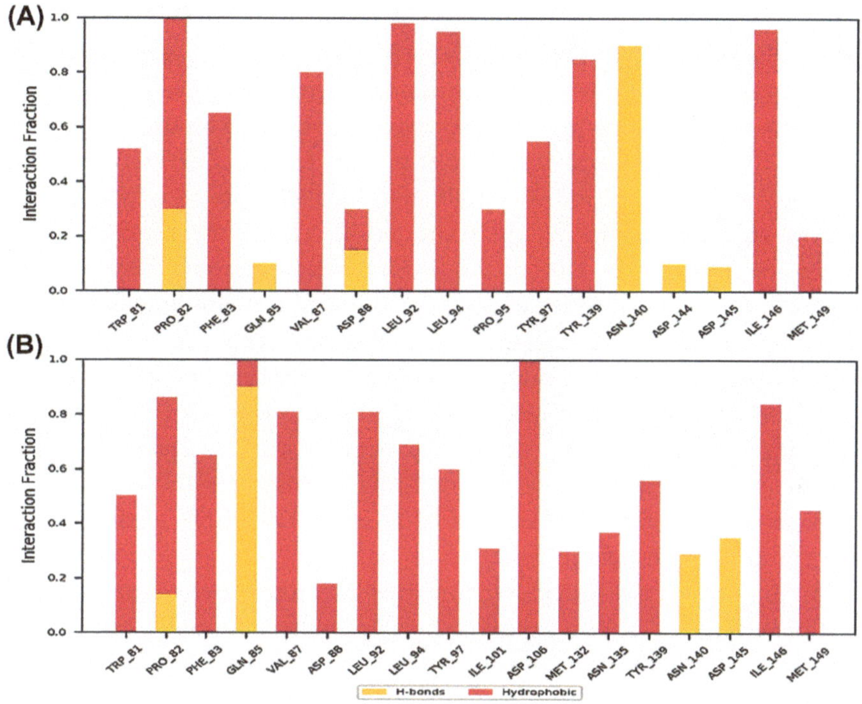


**Fig. 8** Protein-ligand interactions of the designed coumarin and BRD4, H-bonds (yellow) and hydrophobic contacts (magenta) for poses A & B, obtained with molecular dynamics.

2019). In addition, the contact with V87 suggests that this compound may occupy the subpocket recently described for phenanthridine derivatives (Zhi et al., 2019). Just as expected, the free energy values were quite different between poses: 140 ± 5 kcal/mol and -90 ± 2.5 kcal/mol, respectively. These observations are supported by the dynamics of the ZA channel, as suggested by other studies (Cheng et al., 2017; Kuang et al., 2015; Wang et al., 2019).

Hence, optimization around the coumarin scaffold shows promise, considering it may provide putative selectivity for BRD4-BD1. Nevertheless, further studies are required to evaluate the potential and putative potency of such compounds. A first step toward this may be with the use of ABFE and MBAR methods, as well as extended sampling for proper description and comparison between the observed binding modes. We are working on such approaches but believe that these preliminary results may serve to others looking for novel scaffolds or fragments for use in BET inhibition.

## 5. Closing remarks

In this chapter, we discussed fundamental aspects of computational methods and their role in drug discovery efforts, focusing on the development of BET inhibitors and their structural features. BETs continue to show promise as novel and effective targets for a plethora of chronic diseases. However, this potential may be truncated if a clear understanding of their physiology is not met. Due to this, several research groups have proposed methods for probe identification and binding site requirements for selectivity. This has led to rational improvements and critical assessments of *in silico* methodologies, as bromodomains pose an interesting case study. Ranging from the development of novel algorithms, computational workflows, to lead development and free energy calculations; most of these efforts have taken advantage of in-house data or tools, highlighting the pressing need for open-science frameworks. The latter applies not only to this case but the computational field in general.

In this chapter, we also present a virtual screening based on the coumarin scaffold as a proof-of-concept. Based on its results, positions 4 and 7 on the ring arose as promising substituents for the interaction with WPF and the ZA channel. The proposed compound is an attempt to improve selectivity around the hydrophobic sub-pocket in the ZA channel (V87 and Leu92&94) and semi-polar region on the BC loop (Y139 and Asp 144&145). MD simulations and free energy calculations showed high contacts with these regions, prompting further study on coumarins with SBDD.

We hope that the information presented herein serves as a starting point for further advances toward the understanding of bromodomains, as well as an introduction to the scope and capabilities of current computational methodologies.

## Acknowledgments

F.D.P.-M., thanks the PhD scholarship grant from CONACyT No. 660465/576637. We thank NUATEI Program (IIB—UNAM) for providing funding to purchase YASARA software. Also, we are thankful for the computational resources granted by the Dirección General de Cómputo y de Tecnologías de Información y Comunicación (DGTIC), project grant LANCAD-UNAM-DGTIC-335. Which allowed the use of the Miztli supercomputer at UNAM, for the high throughput docking presented herein. Finally, the authors are grateful to Dr. Gabriel Navarrete-Vázquez, Dr. Diego Prada-Gracia and M. Sc. Francisco Barrera-Téllez for helpful discussions.

## References

Abramyan, T. M., An, Y., & Kireev, D. (2020). Off-pocket activity cliffs: A puzzling facet of molecular recognition. *Journal of Chemical Information and Modeling, 60*(1), 152–161. http://dx.doi.org/10.1021/acs.jcim.9b00731.

Aldeghi, M., Bodkin, M. J., Knapp, S., & Biggin, P. C. (2017). Statistical analysis on the performance of molecular mechanics Poisson-Boltzmann surface area versus absolute binding free energy calculations: Bromodomains as a case study. *Journal of Chemical Information and Modeling, 57*(9), 2203–2221. http://dx.doi.org/10.1021/acs.jcim.7b00347.

Aldeghi, M., Heifetz, A., Bodkin, M. J., Knapp, S., & Biggin, P. C. (2016). Accurate calculation of the absolute free energy of binding for drug molecules. *Chemical Science, 7*(1), 207–218. http://dx.doi.org/10.1039/C5SC02678D.

Aldeghi, M., Heifetz, A., Bodkin, M. J., Knapp, S., & Biggin, P. C. (2017). Predictions of ligand selectivity from absolute binding free energy calculations. *Journal of the American Chemical Society, 139*(2), 946–957. http://dx.doi.org/10.1021/jacs.6b11467.

Aldeghi, M., Ross, G. A., Bodkin, M. J., Essex, J. W., Knapp, S., & Biggin, P. C. (2018). Large-scale analysis of water stability in bromodomain binding pockets with grand canonical Monte Carlo. *Communications Chemistry, 1*(1), 19. http://dx.doi.org/10.1038/s42004-018-0019-x.

Allen, W. J., Balius, T. E., Mukherjee, S., Brozell, S. R., Moustakas, D. T., Lang, P. T., et al. (2015). Dock 6: Impact of new features and current docking performance. *Journal of Computational Chemistry, 36*(15), 1132–1156. http://dx.doi.org/10.1002/jcc.23905.

Allen, B. K., Mehta, S., Ember, S. W. J., Zhu, J.-Y., Schönbrunn, E., Ayad, N. G., et al. (2017). Identification of a novel class of BRD4 inhibitors by computational screening and binding simulations. *ACS Omega, 2*(8), 4760–4771. http://dx.doi.org/10.1021/acsomega.7b00553.

Allikmets, K. (2007). Aliskiren - an orally active renin inhibitor. Review of pharmacology, pharmacodynamics, kinetics, and clinical potential in the treatment of hypertension. *Vascular Health and Risk Management, 3*(6), 809–815.

Anderson, A. C. (2003). The process of structure-based drug design. *Chemistry & Biology, 10*(9), 787–797. http://dx.doi.org/10.1016/j.chembiol.2003.09.002.

Andrews, P. R., Carson, J. M., Caselli, A., Spark, M. J., & Woods, R. (1985). Conformational analysis and active site modelling of angiotensin-converting enzyme inhibitors. *Journal of Medicinal Chemistry, 28*(3), 393–399. http://dx.doi.org/10.1021/jm00381a021.

Andrieu, G., Belkina, A. C., & Denis, G. V. (2016). Clinical trials for BET inhibitors run ahead of the science. *Drug Discovery Today: Technologies, 19*, 45–50. http://dx.doi.org/10.1016/j.ddtec.2016.06.004.

Anighoro, A., & Bajorath, J. (2016). Three-dimensional similarity in molecular docking: Prioritizing ligand poses on the basis of experimental binding modes. *Journal of Chemical Information and Modeling, 56*(3), 580–587. http://dx.doi.org/10.1021/acs.jcim.5b00745.

Ayoub, A. M., Hawk, L. M. L., Herzig, R. J., Jiang, J., Wisniewski, A. J., Gee, C. T., et al. (2017). BET bromodomain inhibitors with one-step synthesis discovered from virtual screen. *Journal of Medicinal Chemistry, 60*(12), 4805–4817. http://dx.doi.org/10.1021/acs.jmedchem.6b01336.

Baker, N. A., Sept, D., Joseph, S., Holst, M. J., & McCammon, J. A. (2001). Electrostatics of nanosystems: Application to microtubules and the ribosome. *Proceedings of the National Academy of Sciences of the United States of America, 98*(18), 10037–10041. http://dx.doi.org/10.1073/pnas.181342398.

Bamborough, P., Diallo, H., Goodacre, J. D., Gordon, L., Lewis, A., Seal, J. T., et al. (2012). Fragment-based discovery of bromodomain inhibitors part 2: Optimization of

phenylisoxazole sulfonamides. *Journal of Medicinal Chemistry, 55*(2), 587–596. http://dx.doi.org/10.1021/jm201283q.
Basheer, F., & Huntly, B. J. P. (2015). BET bromodomain inhibitors in leukemia. *Experimental Hematology, 43*(8), 718–731. http://dx.doi.org/10.1016/j.exphem.2015.06.004.
Batiste, L., Unzue, A., Dolbois, A., Hassler, F., Wang, X., Deerain, N., et al. (2018). Chemical space expansion of bromodomain ligands guided by in silico virtual couplings (AutoCouple). *ACS Central Science, 4*(2), 180–188. http://dx.doi.org/10.1021/acscentsci.7b00401.
Bechter, O., & Schöffski, P. (2020). Make your best BET the emerging role of BET inhibitor treatment in malignant tumors. *Pharmacology & Therapeutics, 208*, 2–14. http://dx.doi.org/10.1016/j.pharmthera.2020.107479. Elsevier Inc.
Bennett, C. H. (1976). Efficient estimation of free energy differences from Monte Carlo data. *Journal of Computational Physics, 22*(2), 245–268. http://dx.doi.org/10.1016/0021-9991(76)90078-4.
Bharatham, N., Slavish, P. J., Young, B. M., & Shelat, A. A. (2018). The role of ZA channel water-mediated interactions in the design of bromodomain-selective BET inhibitors. *Journal of Molecular Graphics and Modelling, 81*, 197–210. http://dx.doi.org/10.1016/j.jmgm.2018.03.005.
Böhm, H. J. (1992). The computer program ludi: A new method for the de novo design of enzyme inhibitors. *Journal of Computer-Aided Molecular Design, 6*(1), 61–78. http://dx.doi.org/10.1007/BF00124387.
Borysko, P., Moroz, Y. S., Vasylchenko, O. V., Hurmach, V. V., Starodubtseva, A., Stefanishena, N., et al. (2018). Straightforward hit identification approach in fragment-based discovery of bromodomain-containing protein 4 (BRD4) inhibitors. *Bioorganic & Medicinal Chemistry, 26*(12), 3399–3405. http://dx.doi.org/10.1016/j.bmc.2018.05.010.
Bou-Rabee, N. (2014). Time integrators for molecular dynamics. *Entropy, 16*(1), 138–162. http://dx.doi.org/10.3390/e16010138.
Brás, N. F., Cerqueira, N. M. F. S. A., Sousa, S. F., Fernandes, P. A., & Ramos, M. J. (2014). Protein ligand docking in drug discovery. *Protein Modelling*. http://dx.doi.org/10.1007/978-3-319-09976-7_11.
Bredfeldt, T. G., & Walker, C. L. (2010). Epigenetics. In *Comprehensive toxicology* (pp. 335–358). Elsevier. http://dx.doi.org/10.1016/B978-0-08-046884-6.00219-0.
Brindha, S., Vincent, S., Velmurugan, D., Ananthakrishnan, D., Sundaramurthi, J. C., & Gnanadoss, J. J. (2017). Bioinformatics approach to prioritize known drugs towards repurposing for tuberculosis. *Medical Hypotheses, 103*, 39–45. http://dx.doi.org/10.1016/J.MEHY.2017.04.005.
Carlson, H. A., & McCammon, J. A. (2000). Accommodating protein flexibility in computational drug design. *Molecular Pharmacology, 57*(2), 213–218.
Cereto-Massagué, A., Ojeda, M. J., Valls, C., Mulero, M., Pujadas, G., & Garcia-Vallve, S. (2015). Tools for in silico target fishing. *Methods, 71*, 98–103. http://dx.doi.org/10.1016/j.ymeth.2014.09.006.
Cheng, C., Diao, H., Zhang, F., Wang, Y., Wang, K., & Wu, R. (2017). Deciphering the mechanisms of selective inhibition for the tandem BD1/BD2 in the BET-bromodomain family. *Physical Chemistry Chemical Physics, 19*(35), 23934–23941. http://dx.doi.org/10.1039/c7cp04608a.
Chéron, N., Jasty, N., & Shakhnovich, E. I. (2016). OpenGrowth: An automated and rational algorithm for finding new protein ligands. *Journal of Medicinal Chemistry, 59*(9), 4171–4188. http://dx.doi.org/10.1021/acs.jmedchem.5b00886.
Chodera, J. D., Mobley, D. L., Shirts, M. R., Dixon, R. W., Branson, K., & Pande, V. S. (2011). Alchemical free energy methods for drug discovery: Progress and challenges.

*Current Opinion in Structural Biology, 21*(2), 150–160. http://dx.doi.org/10.1016/j.sbi.2011.01.011.

Cohen, J. (1996). Drug development: Protease inhibitors: A tale of two companies. *Science, 272*(5270), 1882–1883. http://dx.doi.org/10.1126/science.272.5270.1882.

Conway, S. J. (2012). Bromodomains: Are readers right for epigenetic therapy? *ACS Medicinal Chemistry Letters, 3*(9), 691–694. http://dx.doi.org/10.1021/ml300221t.

Crooks, G. E. (2000). Path-ensemble averages in systems driven far from equilibrium. *Physical Review E - Statistical Physics, Plasmas, Fluids, and Related Interdisciplinary Topics, 61*(3), 2361–2366. http://dx.doi.org/10.1103/PhysRevE.61.2361.

De Ruiter, A., & Oostenbrink, C. (2011). Free energy calculations of protein-ligand interactions. *Current Opinion in Chemical Biology, 15*(4), 547–552. http://dx.doi.org/10.1016/j.cbpa.2011.05.021.

De Vivo, M., Masetti, M., Bottegoni, G., & Cavalli, A. (2016). Role of molecular dynamics and related methods in drug discovery. *Journal of Medicinal Chemistry, 59*(9), 4035–4061. http://dx.doi.org/10.1021/acs.jmedchem.5b01684.

Decherchi, S., Masetti, M., Vyalov, I., & Rocchia, W. (2015). Implicit solvent methods for free energy estimation. *European Journal of Medicinal Chemistry, 91*, 27–42. http://dx.doi.org/10.1063/1.2177251.

Dhalluin, C., Carlson, J. E., Zeng, L., He, C., Aggarwal, A. K., & Zhou, M. M. (1999). Structure and ligand of a histone acetyltransferase bromodomain. *Nature, 399*(6735), 491–496. http://dx.doi.org/10.1038/20974.

Dittmann, A., Werner, T., Chung, C. W., Savitski, M. M., Fälth Savitski, M., Grandi, P., et al. (2014). The commonly used PI3-kinase probe LY294002 is an inhibitor of BET bromodomains. *ACS Chemical Biology, 9*(2), 495–502. http://dx.doi.org/10.1021/cb400789e.

van Dongen, M., Weigelt, J., Uppenberg, J., Schultz, J., & Wikström, M. (2002). Structure-based screening and design in drug discovery. *Drug Discovery Today, 7*(8), 471–478. http://dx.doi.org/10.1016/S1359-6446(02)02233-X.

Doroshow, D. B., Eder, J. P., & LoRusso, P. M. (2017). BET inhibitors: A novel epigenetic approach. *Annals of Oncology, 28*(8), 1776–1787. http://dx.doi.org/10.1093/annonc/mdx157.

van Drie, J. H. (2012). Monty Kier and the origin of the pharmacophore concept. *Internet Electronic Journal of Molecular Design, 11*(1), 271–279.

Duffy, B. C., Liu, S., Martin, G. S., Wang, R., Hsia, M. M., Zhao, H., et al. (2015). Discovery of a new chemical series of BRD4(1) inhibitors using protein-ligand docking and structure-guided design. *Bioorganic & Medicinal Chemistry Letters, 25*(14), 2818–2823. http://dx.doi.org/10.1016/j.bmcl.2015.04.107.

Du, X., Li, Y., Xia, Y. L., Ai, S. M., Liang, J., Sang, P., et al. (2016). Insights into protein–ligand interactions: Mechanisms, models, and methods. *International Journal of Molecular Sciences, 17*(2), 1–34. http://dx.doi.org/10.3390/ijms17020144.

Eberharter, A., & Becker, P. B. (2002). Histone acetylation: A switch between repressive and permissive chromatin. Second in review on chromatin dynamics. *EMBO Reports, 3*(3), 224–229. http://dx.doi.org/10.1093/embo-reports/kvf053.

Ferri, E., Petosa, C., & McKenna, C. E. (2016). Bromodomains: Structure, function and pharmacology of inhibition. *Biochemical Pharmacology, 106*, 1–18. http://dx.doi.org/10.1016/j.bcp.2015.12.005.

Figge, D. A., & Standaert, D. G. (2017). Dysregulation of BET proteins in levodopa-induced dyskinesia. *Neurobiology of Disease, 102*, 125–132. http://dx.doi.org/10.1016/j.nbd.2017.03.003.

Filippakopoulos, P., & Knapp, S. (2014). Targeting bromodomains: Epigenetic readers of lysine acetylation. *Nature Reviews Drug Discovery, 13*(5), 337–356. http://dx.doi.org/10.1038/nrd4286.

Filippakopoulos, P., Picaud, S., Fedorov, O., Keller, M., Wrobel, M., Morgenstern, O., et al. (2012). Benzodiazepines and benzotriazepines as protein interaction inhibitors targeting bromodomains of the BET family. *Bioorganic & Medicinal Chemistry, 20*(6), 1878–1886. http://dx.doi.org/10.1016/j.bmc.2011.10.080.

Filippakopoulos, P., Qi, J., Picaud, S., Shen, Y., Smith, W. B., Fedorov, O., et al. (2010). Selective inhibition of BET bromodomains. *Nature, 468*(7327), 1067–1073. http://dx.doi.org/10.1038/nature09504.

Fish, P. V., Filippakopoulos, P., Bish, G., Brennan, P. E., Bunnage, M. E., Cook, A. S., et al. (2012). Identification of a chemical probe for bromo and extra C-terminal bromodomain inhibition through optimization of a fragment-derived hit. *Journal of Medicinal Chemistry, 55*(22), 9831–9837. http://dx.doi.org/10.1021/jm3010515.

Friedrich, N. O., De Bruyn Kops, C., Flachsenberg, F., Sommer, K., Rarey, M., & Kirchmair, J. (2017). Benchmarking commercial conformer ensemble generators. *Journal of Chemical Information and Modeling, 57*(11), 2719–2728. http://dx.doi.org/10.1021/acs.jcim.7b00505.

Fujisawa, T., & Filippakopoulos, P. (2017). Functions of bromodomain-containing proteins and their roles in homeostasis and cancer. *Nature Reviews Molecular Cell Biology, 18*(4), 246–262. http://dx.doi.org/10.1038/nrm.2016.143.

Galdeano, C., & Ciulli, A. (2016). Selectivity on-target of bromodomain chemical probes by structure-guided medicinal chemistry and chemical biology. *Future Medicinal Chemistry, 8*(13), 1655–1680. http://dx.doi.org/10.4155/fmc-2016-0059.

Gane, P. J., & Dean, P. M. (2000). Recent advances in structure-based rational drug design. *Current Opinion in Structural Biology, 10*(4), 401–404. http://dx.doi.org/10.1016/S0959-440X(00)00105-6.

Ganesan, A. (2004). Natural products as a hunting ground for combinatorial chemistry. *Current Opinion in Biotechnology, 15*(6), 584–590. http://dx.doi.org/10.1016/j.copbio.2004.09.002.

García-Godoy, M. J., López-Camacho, E., García-Nieto, J., Nebro, A. J., & Aldana-Montes, J. F. (2015). Solving molecular docking problems with multi-objective metaheuristics. *Molecules, 20*(6), 10154–10183. http://dx.doi.org/10.3390/molecules200610154.

Gaucher, J., Boussouar, F., Montellier, E., Curtet, S., Buchou, T., Bertrand, S., et al. (2012). Bromodomain-dependent stage-specific male genome programming by Brdt. *The EMBO Journal, 31*(19), 3809–3820. http://dx.doi.org/10.1038/emboj.2012.233.

Genheden, S., & Ryde, U. (2015). The MM/PBSA and MM/GBSA methods to estimate ligand-binding affinities. *Expert Opinion on Drug Discovery, 10*(5), 449–461. http://dx.doi.org/10.1517/17460441.2015.1032936.

Ghosh, A. K., & Gemma, S. (2015). Carbonic anhydrase inhibitors for the treatment of glaucoma: Design and discovery of dorzolamide. In *Structure-based design of drugs and other bioactive molecules* (pp. 411–419). Weinheim, Germany: Wiley-VCH Verlag GmbH & Co. KGaA. http://dx.doi.org/10.1002/9783527665211.ch19.

Giblin, K. A., Hughes, S. J., Boyd, H., Hansson, P., & Bender, A. (2018). Prospectively validated proteochemometric models for the prediction of small-molecule binding to bromodomain proteins. *Journal of Chemical Information and Modeling, 58*(9), 1870–1888. http://dx.doi.org/10.1021/acs.jcim.8b00400. research-article.

Gil, J., Ramírez-Torres, A., & Encarnación-Guevara, S. (2017). Lysine acetylation and cancer: A proteomics perspective. *Journal of Proteomics, 150*, 297–309. http://dx.doi.org/10.1016/j.jprot.2016.10.003.

Glunz, P. W. (2018). Recent encounters with atropisomerism in drug discovery. *Bioorganic & Medicinal Chemistry Letters, 28*(2), 53–60. http://dx.doi.org/10.1016/j.bmcl.2017.11.050.

González, M. A. (2011). Force fields and molecular dynamics simulations. *Collection SFN, 12*, 169–200. http://dx.doi.org/10.1051/sfn/201112009.

Goodford, P. J. (1985). A computational procedure for determining energetically favorable binding sites on biologically important macromolecules. *Journal of Medicinal Chemistry, 28*(7), 849–857. http://dx.doi.org/10.1021/jm00145a002.

Gosmini, R., Nguyen, V. L., Toum, J., Simon, C., Brusq, J. M. G., Krysa, G., et al. (2014). The discovery of I-BET726 (GSK1324726A), a potent tetrahydroquinoline ApoA1 up-regulator and selective BET bromodomain inhibitor. *Journal of Medicinal Chemistry, 57*(19), 8111–8131. http://dx.doi.org/10.1021/jm5010539.

Guha, M. (2015). Erratum: HDAC inhibitors still need a home run, despite recent approval. *Nature Reviews Drug Discovery, 14*(5). http://dx.doi.org/10.1038/nrd4632, 365–365.

Gürsoy, O., & Smieško, M. (2017). Searching for bioactive conformations of drug-like ligands with current force fields: How good are we? *Journal of Cheminformatics, 9*(1), 29. http://dx.doi.org/10.1186/s13321-017-0216-0.

Hahn, Aljoscha, & Then, Holger (2009). Characteristic of Bennett's acceptance ratio method. *Physical Review E - Statistical, Nonlinear, and Soft Matter Physics, 80*(3), 1–10. http://dx.doi.org/10.1103/PhysRevE.80.031111.

Hajmirza, A., Emadali, A., Gauthier, A., Casasnovas, O., Gressin, R., & Callanan, M. B. (2018). BET family protein BRD4: An emerging actor in NFκB signaling in inflammation and cance. *Biomedicines, 6*(1), 1–9. http://dx.doi.org/10.3390/biomedicines6010016.

Halgren, T.a. (1996). Merck molecular force field. *Journal of Computational Chemistry, 17*(5–6), 490–519. http://dx.doi.org/10.1002/(SICI)1096-987X(199604)17:5/6<520::AID-JCC2>3.0.CO;2-W.

Hassler, M. R., & Egger, G. (2012). Epigenomics of cancer - emerging new concepts. *Biochimie, 94*(11), 2219–2230. http://dx.doi.org/10.1016/j.biochi.2012.05.007.

Hawkins, P. C. D. (2017). Conformation generation: The state of the art. *Journal of Chemical Information and Modeling, 57*(8), 1747–1756. http://dx.doi.org/10.1021/acs.jcim.7b00221.

Hernández, E. R., Zetina, L. M. M., Vega, G. T., Rocha, M. G., Ochoa, L. F. R., & Fernandez, R. L. (2008). Molecular dynamics: From basic techniques to applications (A molecular dynamics primer). *AIP Conference Proceedings, 95–123.* http://dx.doi.org/10.1063/1.3040265.

Hessler, G., Korb, O., Monecke, P., Stützle, T., & Exner, T. (2010). pharmACOphore: multiple flexible ligand alignment based on ant colony optimization. *Journal of Cheminformatics, 2*(S1), 2010. http://dx.doi.org/10.1186/1758-2946-2-s1-p17.

Hewings, D. S., Rooney, T. P. C., Jennings, L. E., Hay, D. A., Schofield, C. J., Brennan, P. E., et al. (2012). Progress in the development and application of small molecule inhibitors of bromodomain–acetyl-lysine interactions. *Journal of Medicinal Chemistry, 55*(22), 9393–9413. http://dx.doi.org/10.1021/jm300915b.

Hewings, D. S., Wang, M., Philpott, M., Fedorov, O., Uttarkar, S., Filippakopoulos, P., et al. (2011). 3,5-Dimethylisoxazoles act as acetyl-lysine-mimetic bromodomain ligands. *Journal of Medicinal Chemistry, 54*(19), 6761–6770. http://dx.doi.org/10.1021/jm200640v.

Hilton-Proctor, J. P., Ilyichova, O., Zheng, Z., Jennings, I. G., Johnstone, R. W., Shortt, J., et al. (2019). Synthesis and elaboration of N-methylpyrrolidone as an acetamide fragment substitute in bromodomain inhibition. *Bioorganic & Medicinal Chemistry, 27*(24), 115157. http://dx.doi.org/10.1016/j.bmc.2019.115157.

Hoffer, L., Voitovich, Y. V., Raux, B., Carrasco, K., Muller, C., Fedorov, A. Y., et al. (2018). Integrated strategy for lead optimization based on fragment growing: The diversity-oriented-target-focused-synthesis approach. *Journal of Medicinal Chemistry, 61*(13), 5719–5732. http://dx.doi.org/10.1021/acs.jmedchem.8b00653.

Hospital, A., Goñi, J. R., Orozco, M., & Gelpi, J. (2015). Molecular dynamics simulations: Advances and applications. *Advances and Applications in Bioinformatics and Chemistry, 8*, 37–47. http://dx.doi.org/10.2147/AABC.S70333.

Howe, A. Y. M., & Venkatraman, S. (2013). The discovery and development of boceprevir: A novel, first-generation inhibitor of the hepatitis C virus NS3/4A serine protease. *Journal of Clinical and Translational Hepatology, 1*(1), 22–32. http://dx.doi.org/10.14218/jcth.2013.002xx.

Huang, D., Rossini, E., Steiner, S., & Caflisch, A. (2014). Structured water molecules in the binding site of Bromodomains can be displaced by cosolvent. *ChemMedChem, 9*(3), 573–579. http://dx.doi.org/10.1002/cmdc.201300156.

Huggins, D. J., Sherman, W., & Tidor, B. (2012). Rational approaches to improving selectivity in drug design. *Journal of Medicinal Chemistry, 55*(4), 1424–1444. http://dx.doi.org/10.1021/jm2010332.

Hu, J., Tian, C. Q., Damaneh, M. S., Li, Y., Cao, D., Lv, K., et al. (2019). Structure-based discovery and development of a series of potent and selective bromodomain and extra-terminal protein inhibitors. *Journal of Medicinal Chemistry, 62*(18), 8642–8663. http://dx.doi.org/10.1021/acs.jmedchem.9b01094. research-article.

Irwin, J. J., Sterling, T., Mysinger, M. M., Bolstad, E. S., & Coleman, R. G. (2012). Zinc: A free tool to discover chemistry for biology. *Journal of Chemical Information and Modeling, 52*(7), 1757–1768. http://dx.doi.org/10.1021/ci3001277.

Iyer, P., Stumpfe, D., Vogt, M., Bajorath, J., & Maggiora, G. M. (2013). Activity landscapes, information theory, and structure - activity relationships. *Molecular Informatics, 32*(5–6), 421–430. http://dx.doi.org/10.1002/minf.201200120.

Jaccard, P. (1901). Étude comparative de la distribution florale dans une portion des Alpes et du Jura. *Bulletin de La Société Vaudoise Des Sciences Naturelles, 37*(142), 547. http://dx.doi.org/10.5169/seals-266450.

Jain, A. K., & Barton, M. C. (2017). Bromodomain histone readers and cancer. *Journal of Molecular Biology, 429*(13), 2003–2010. http://dx.doi.org/10.1016/j.jmb.2016.11.020.

Jung, M., Philpott, M., Müller, S., Schulze, J., Badock, V., Eberspächer, U., et al. (2014). Affinity map of bromodomain protein 4 (BRD4) interactions with the histone H4 tail and the small molecule inhibitor JQ1. *Journal of Biological Chemistry, 289*(13), 9304–9319. http://dx.doi.org/10.1074/jbc.M113.523019.

Jurrus, E., Engel, D., Star, K., Monson, K., Brandi, J., Felberg, L. E., et al. (2018). Improvements to the APBS biomolecular solvation software suite. *Protein Science, 27*(1), 112–128. http://dx.doi.org/10.1002/pro.3280.

Kalyaanamoorthy, S., & Chen, Y. P. P. (2011). Structure-based drug design to augment hit discovery. *Drug Discovery Today, 16*(17–18), 831–839. http://dx.doi.org/10.1016/j.drudis.2011.07.006.

Kanakaveti, V., Shanmugam, A., Ramakrishnan, C., Anoosha, P., Sakthivel, R., Rayala, S. K., et al. (2020). Computational approaches for identifying potential inhibitors on targeting protein interactions in drug discovery. In *Advances in protein chemistry and structural biology* (Vol. 121, pp. 1–23). Elsevier Ltd. http://dx.doi.org/10.1016/bs.apcsb.2019.11.013.

Kharenko, O. A., Gesner, E. M., Patel, R. G., Norek, K., White, A., Fontano, E., et al. (2016). RVX-297- a novel BD2 selective inhibitor of BET bromodomains. *Biochemical and Biophysical Research Communications, 477*(1), 62–67. http://dx.doi.org/10.1016/j.bbrc.2016.06.021.

Kim, I., & Allen, T. W. (2012). Bennetts acceptance ratio and histogram analysis methods enhanced by umbrella sampling along a reaction coordinate in configurational space. *The Journal of Chemical Physics, 136*(16). http://dx.doi.org/10.1063/1.3701766.

Kitchen, D. B., Decornez, H., Furr, J. R., & Bajorath, J. (2004). Docking and scoring in virtual screening for drug discovery: Methods and applications. *Nature Reviews Drug Discovery, 3*(11), 935–949. http://dx.doi.org/10.1038/nrd1549.

König, G., Bruckner, S., & Boresch, S. (2009). Unorthodox uses of Bennett's acceptance ratio method. *Journal of Computational Chemistry, 30*(11), 1712–1718. http://dx.doi.org/10.1002/jcc.21255.

Korb, O., Stützle, T., & Exner, T. E. (2009). Empirical scoring functions for advanced Protein-Ligand docking with PLANTS. *Journal of Chemical Information and Modeling, 49*(1), 84–96. http://dx.doi.org/10.1021/ci800298z.

Kougnassoukou Tchara, P.-E., Filippakopoulos, P., & Lambert, J.-P. (2019). Emerging tools to investigate bromodomain functions. *Methods*, 1–13. http://dx.doi.org/10.1016/j.ymeth.2019.11.003 (October).

Krieger, E., & Vriend, G. (2014). YASARA View—molecular graphics for all devices—from smartphones to workstations. *Bioinformatics, 30*(20), 2981–2982. http://dx.doi.org/10.1093/bioinformatics/btu426.

Kuang, M., Zhou, J., Wang, L., Liu, Z., Guo, J., & Wu, R. (2015). Binding kinetics versus affinities in BRD4 inhibition. *Journal of Chemical Information and Modeling, 55*(9), 1926–1935. http://dx.doi.org/10.1021/acs.jcim.5b00265.

Kubinyi, H. (1993). *Qsar: Hansch analysis and related approaches*. VCH. Retrieved from https://www.biblio.com/9783527300358.

Kuhn, B., Gerber, P., Schulz-Gasch, T., & Stahl, M. (2005). Validation and use of the MM-PBSA approach for drug discovery. *Journal of Medicinal Chemistry, 48*(12), 4040–4048. http://dx.doi.org/10.1021/jm049081q.

Ladbury, J. E. (1996). Just add water! the effect of water on the specificity of protein-ligand binding sites and its potential application to drug design. *Chemistry & Biology, 3*(12), 973–980. http://dx.doi.org/10.1016/S1074-5521(96)90164-7.

Larsson, S., Backlund, A., & Bohlin, L. (2008). Reappraising a decade old explanatory model for pharmacognosy. *Phytochemistry Letters, 1*(3), 131–134. http://dx.doi.org/10.1016/j.phytol.2008.09.001.

Lew, W., Chen, X., & Kim, C. (2012). Discovery and development of GS 4104 (oseltamivir) an orally active influenza neuraminidase inhibitor. *Current Medicinal Chemistry, 7*(6), 663–672. http://dx.doi.org/10.2174/0929867003374886.

Liu, J., & Wang, R. (2015). Classification of current scoring functions. *Journal of Chemical Information and Modeling, 55*(3), 475–482. http://dx.doi.org/10.1021/ci500731a.

Liu, Z., Wang, P., Chen, H., Wold, E. A., Tian, B., Brasier, A. R., et al. (2017). Drug discovery targeting bromodomain-containing protein 4. *Journal of Medicinal Chemistry, 60*(11), 4533–4558. http://dx.doi.org/10.1021/acs.jmedchem.6b01761.

Ma, J., Chen, H., Yang, J., Yu, Z., Huang, P., Yang, H., et al. (2019). Binding pocket-based design, synthesis and biological evaluation of novel selective BRD4-BD1 inhibitors. *Bioorganic & Medicinal Chemistry, 27*(9), 1871–1881. http://dx.doi.org/10.1016/j.bmc.2019.03.037.

Madhavi Sastry, G., Adzhigirey, M., Day, T., Annabhimoju, R., & Sherman, W. (2013). Protein and ligand preparation: Parameters, protocols, and influence on virtual screening enrichments. *Journal of Computer-Aided Molecular Design, 27*(3), 221–234. http://dx.doi.org/10.1007/s10822-013-9644-8.

Maggiora, G. M. (2006). On outliers and activity cliffs - why QSAR often disappoints. *Journal of Chemical Information and Modeling, 46*(4), 1535. http://dx.doi.org/10.1021/ci060117s.

Maggiora, G. M., & Shanmugasundaram, V. (2011). *Molecular similarity measures* (pp. 39–100). Totowa, NJ: Humana Press. http://dx.doi.org/10.1007/978-1-60761-839-3_2.

Maggiora, G. M., Vogt, M., Stumpfe, D., & Bajorath, J. J. (2013). Molecular similarity in medicinal chemistry. *Journal of Medicinal Chemistry, 57*(8), 3186–3204. http://dx.doi.org/10.1021/jm401411z.

Magno, A., Steiner, S., & Caflisch, A. (2013). Mechanism and kinetics of acetyl-lysine binding to bromodomains. *Journal of Chemical Theory and Computation, 9*(9), 4225–4232. http://dx.doi.org/10.1021/ct400361k.

Manzoni, F., & Ryde, U. (2018). Assessing the stability of free-energy perturbation calculations by performing variations in the method. *Journal of Computer-Aided Molecular Design, 32*(4), 529–536. http://dx.doi.org/10.1007/s10822-018-0110-5.

Marchand, J. R., Dalle Vedove, A., Lolli, G., & Caflisch, A. (2017). Discovery of inhibitors of four bromodomains by fragment-anchored ligand docking. *Journal of Chemical Information and Modeling, 57*(10), 2584–2597. http://dx.doi.org/10.1021/acs.jcim.7b00336.

Marmorstein, R., & Berger, S. L. (2001). Structure and function of bromodomains in chromatin-regulating complexes. *Gene, 272*(1–2), 1–9. http://dx.doi.org/10.1016/S0378-1119(01)00519-4.

Masters, M. R., Mahmoud, A. H., Yang, Y., & Lill, M. A. (2018). Efficient and accurate hydration site profiling for enclosed binding sites. *Journal of Chemical Information and Modeling, 58*(11), 2183–2188. http://dx.doi.org/10.1021/acs.jcim.8b00544. acs.jcim.8b00544.

Matzuk, M. M., McKeown, M. R., Filippakopoulos, P., Li, Q., Ma, L., Agno, J. E., et al. (2012). Small-molecule inhibition of BRDT for male contraception. *Cell, 150*(4), 673–684. http://dx.doi.org/10.1016/j.cell.2012.06.045.

McCammon, J. A., Gelin, B. R., & Karplus, M. (1977). Dynamics of folded proteins. *Nature, 267*(5612), 585–590. http://dx.doi.org/10.1038/267585a0.

McCullough, C. E., & Marmorstein, R. (2016). Molecular basis for histone acetyltransferase regulation by binding partners, associated domains, and autoacetylation. *ACS Chemical Biology, 11*(3), 632–642. http://dx.doi.org/10.1021/acschembio.5b00841.

McKeown, M. R., Shaw, D. L., Fu, H., Liu, S., Xu, X., Marineau, J. J., et al. (2014). Biased multicomponent reactions to develop novel bromodomain inhibitors. *Journal of Medicinal Chemistry, 57*(21), 9019–9027. http://dx.doi.org/10.1021/jm501120z.

Medina-Franco, J. L. (2013). Activity cliffs: Facts or artifacts? *Chemical Biology & Drug Design, 81*(5), 553–556. http://dx.doi.org/10.1111/cbdd.12115.

Medina-Franco, J. L., Giulianotti, M.a., Welmaker, G. S., & Houghten, R.a. (2013). Shifting from the single to the multitarget paradigm in drug discovery. *Drug Discovery Today, 18*(9–10), 495–501. http://dx.doi.org/10.1016/j.drudis.2013.01.008.

Medina-Franco, J. L., & Maggiora, G. M. (2013). Molecular similarity analysis. In *Chemoinformatics for drug discovery* (pp. 343–399). Hoboken, NJ: John Wiley & Sons, Inc. http://dx.doi.org/10.1002/9781118742785.ch15.

Méndez-Lucio, O., Tran, J., Medina-Franco, J. L., Meurice, N., & Muller, M. (2014). Toward drug repurposing in epigenetics: Olsalazine as a hypomethylating compound active in a cellular context. *ChemMedChem, 9*(3), 560–565. http://dx.doi.org/10.1002/cmdc.201300555.

Mikulskis, P., Genheden, S., & Ryde, U. (2014). Effect of explicit water molecules on ligand-binding affinities calculated with the MM/GBSA approach. *Journal of Molecular Modeling, 20*(6). http://dx.doi.org/10.1007/s00894-014-2273-x.

Mirguet, O., Gosmini, R., Toum, J., Clément, C. A., Barnathan, M., Brusq, J. M., et al. (2013). Discovery of epigenetic regulator i-bet762: Lead optimization to afford a clinical candidate inhibitor of the bet bromodomains. *Journal of Medicinal Chemistry, 56*(19), 7501–7515. http://dx.doi.org/10.1021/jm401088k.

Mirguet, O., Lamotte, Y., Donche, F., Toum, J., Gellibert, F., Bouillot, A., et al. (2012). From ApoA1 upregulation to BET family bromodomain inhibition: Discovery of I-BET151. *Bioorganic & Medicinal Chemistry Letters, 22*(8), 2963–2967. http://dx.doi.org/10.1016/j.bmcl.2012.01.125.

Mobley, D. L., Bannan, C. C., Rizzi, A., Bayly, C. I., Chodera, J. D., Lim, V. T., et al. (2018). Escaping atom types in force fields using direct chemical perception. *Journal of Chemical Theory and Computation, 14*(11), 6076–6092. http://dx.doi.org/10.1021/acs.jctc.8b00640.

Mobley, D. L., & Dill, K. A. (2009). Binding of small-molecule ligands to proteins: "What you see" is not always "what you get. *Structure, 17*(4), 489–498. http://dx.doi.org/10.1016/j.str.2009.02.010.

Molecular Operating Environment (MOE), 2019.01. (2019). 1010 *sherbooke st. West, suite #910, Montreal, QC, Canada, H3A 2R7*. Chemical Computing Group ULC. https://www.chemcomp.com/.

Morris, G. M., & Lim-Wilby, M. (2008). Molecular docking. In *Methods in molecular biology (Clifton, N.J.)* (Vol. 443, pp. 365–382). http://dx.doi.org/10.1007/978-1-59745-177-2_19.

Muegge, I. (2000). A knowledge-based scoring function for protein-ligand interactions: Probing the reference state. In *Perspectives in drug discovery and design* (Vol. 20, pp. 99–114). http://dx.doi.org/10.1023/A:1008729005958.

Mumby, S., Gambaryan, N., Meng, C., Perros, F., Humbert, M., Wort, S. J., et al. (2017). Bromodomain and extra-terminal protein mimic JQ1 decreases inflammation in human vascular endothelial cells: Implications for pulmonary arterial hypertension. *Respirology, 22*(1), 157–164. http://dx.doi.org/10.1111/resp.12872.

Nadeem, A., Al-Harbi, N. O., Al-Harbi, M. M., El-Sherbeeny, A. M., Ahmad, S. F., Siddiqui, N., et al. (2015). Imiquimod-induced psoriasis-like skin inflammation is suppressed by BET bromodomain inhibitor in mice through RORC/IL-17A pathway modulation. *Pharmacological Research, 99*, 248–257. http://dx.doi.org/10.1016/j.phrs.2015.06.001.

Naveja, J. J., Pilón-Jiménez, B. A., Bajorath, J., & Medina-Franco, J. L. (2019). A general approach for retrosynthetic molecular core analysis. *Journal of Cheminformatics, 11*(1), 1–9. http://dx.doi.org/10.1186/s13321-019-0380-5.

Naveja, J. J., Rico-Hidalgo, M. P., & Medina-Franco, J. L. (2018). Analysis of a large food chemical database: Chemical space, diversity, and complexity. *F1000Research*, 7, 993. http://dx.doi.org/10.12688/f1000research.15440.2.

Ng, M. C. K., Fong, S., & Siu, S. W. I. (2015). PSOVina: The hybrid particle swarm optimization algorithm for protein–ligand docking. *Journal of Bioinformatics and Computational Biology, 13*(03), 1541007. http://dx.doi.org/10.1142/S0219720015410073.

Nittinger, E., Flachsenberg, F., Bietz, S., Lange, G., Klein, R., & Rarey, M. (2018). Placement of water molecules in protein structures: From large-scale evaluations to single-case examples. *Journal of Chemical Information and Modeling, 58*(8), 1625–1637. http://dx.doi.org/10.1021/acs.jcim.8b00271. research-article.

Padmanabhan, B., Mathur, S., Manjula, R., & Tripathi, S. (2016). Bromodomain and extra-terminal (BET) family proteins: New therapeutic targets in major diseases. *Journal of Biosciences, 41*(2), 295–311. http://dx.doi.org/10.1007/s12038-016-9600-6.

Paneni, F., Costantino, S., Volpe, M., Lüscher, T. F., & Cosentino, F. (2013). Epigenetic signatures and vascular risk in type 2 diabetes: A clinical perspective. *Atherosclerosis, 230*(2), 191–197. http://dx.doi.org/10.1016/j.atherosclerosis.2013.07.003.

Pérez-Nueno, V. I., Karaboga, A. S., Souchet, M., Ritchie, D. W., Pharma, H., Transfert, E., et al. (2014). GES polypharmacology fingerprints: A novel approach for drug repositioning. *Journal of Chemical Information and Modeling, 54*(3), 720–734. http://dx.doi.org/10.1021/ci4006723.

Picaud, S., Da Costa, D., Thanasopoulou, A., Filippakopoulos, P., Fish, P. V., Philpott, M., et al. (2013). PFI-1, a highly selective protein interaction inhibitor, targeting BET bromodomains. *Cancer Research, 73*(11), 3336–3346. http://dx.doi.org/10.1158/0008-5472.CAN-12-3292.

Picaud, S., Wells, C., Felletar, I., Brotherton, D., Martin, S., Savitsky, P., et al. (2013). RVX-208, an inhibitor of BET transcriptional regulators with selectivity for the second bromodomain. *Proceedings of the National Academy of Sciences, 110*(49), 19754–19759. http://dx.doi.org/10.1073/pnas.1310658110.

Ponder, J. W., & Case, D. A. (2003). Force fields for protein simulations. *Advances in Protein Chemistry, 66*, 27–85. http://dx.doi.org/10.1016/S0065-3233(03)66002-X.

Prieto-Martínez, F. D., Arciniega, M., & Medina-Franco, J. L. (2018). Molecular docking: Current advances and challenges. *Tip. Revista Especializada en Ciencias Químico-Biológicas, 21*. http://dx.doi.org/10.22201/fesz.23958723e.2018.0.143.

Prieto-Martínez, F. D., Fernandez-de Gortari, E., Méndez-Lucio, O., & Medina-Franco, J. L. (2016). A chemical space odyssey of inhibitors of histone deacetylases and bromodomains. *RSC Advances, 6*(61), 56225–56239. http://dx.doi.org/10.1039/C6RA07224K.

Prieto-Martinez, F. D., & Medina-Franco, J. L. (2018). Charting the bromodomain BRD4: Towards the identification of novel inhibitors with molecular similarity and receptor mapping. *Letters in Drug Design and Discovery, 15*(9), 1002–1011. http://dx.doi.org/10.2174/1570180814666171121145731.

Prieto-Martínez, F. D., & Medina-Franco, J. L. (2018). Flavonoids as putative epi-modulators: Insight into their binding mode with BRD4 bromodomains using molecular docking and dynamics. *Biomolecules, 8*(3), 61. http://dx.doi.org/10.3390/biom8030061.

Prinjha, R. K., Witherington, J., & Lee, K. (2012). Place your BETs: The therapeutic potential of bromodomains. *Trends in Pharmacological Sciences, 33*(3), 146–153. http://dx.doi.org/10.1016/j.tips.2011.12.002.

Procacci, P. (2019). Solvation free energies via alchemical simulations: Let's get honest about sampling, once more. *Physical Chemistry Chemical Physics, 21*(25), 13826–13834. http://dx.doi.org/10.1039/c9cp02808k.

Raj, U., Kumar, H., & Varadwaj, P. K. (2016). Molecular docking and dynamics simulation study of flavonoids as BET bromodomain inhibitors. *Journal of Biomolecular Structure and Dynamics, 1102*(August), 1–12. http://dx.doi.org/10.1080/07391102.2016.1217276.

Ran, X., Zhao, Y., Liu, L., Bai, L., Yang, C. Y., Zhou, B., et al. (2015). Structure-based design of γ-carboline analogues as potent and specific BET bromodomain inhibitors. *Journal of Medicinal Chemistry, 58*(12), 4927–4939. http://dx.doi.org/10.1021/acs.jmedchem.5b00613.

Rathi, P. C., Ludlow, R. F., Hall, R. J., Murray, C. W., Mortenson, P. N., & Verdonk, M. L. (2017). Predicting "hot" and "warm" spots for fragment binding. *Journal of Medicinal Chemistry, 60*(9), 4036–4046. http://dx.doi.org/10.1021/acs.jmedchem.7b00366.

Raux, B., Voitovich, Y., Derviaux, C., Lugari, A., Rebuffet, E., Milhas, S., et al. (2016). Exploring selective inhibition of the first bromodomain of the human bromodomain and extra-terminal domain (BET) proteins. *Journal of Medicinal Chemistry, 59*(4), 1634–1641. http://dx.doi.org/10.1021/acs.jmedchem.5b01708.

Robert, C., & Rassool, F. V. (2012). HDAC inhibitors. In*Histone deacetylase inhibitors as cancer therapeutics* (1st ed., Vol. 116, pp. 87–129). Elsevier Inc. http://dx.doi.org/10.1016/B978-0-12-394387-3.00003-3.

Salemme, F. R., Spurlino, J., & Bone, R. (1997). Serendipity meets precision: The integration of structure-based drug design and combinatorial chemistry for efficient drug discovery. *Structure, 5*(3), 319–324. http://dx.doi.org/10.1016/S0969-2126(97)00189-5.

Sandhu, S., Bansal, Y., Silakari, O., & Bansal, G. (2014). Coumarin hybrids as novel therapeutic agents. *Bioorganic & Medicinal Chemistry, 22*(15), 3806–3814. http://dx.doi.org/10.1016/j.bmc.2014.05.032.

Schiano, C., Benincasa, G., Franzese, M., Della Mura, N., Pane, K., Salvatore, M., et al. (2020). Epigenetic-sensitive pathways in personalized therapy of major cardiovascular diseases. *Pharmacology & Therapeutics*, 107514. http://dx.doi.org/10.1016/j.pharmthera.2020.107514.

Schlick, T. (2010). Molecular dynamics: Basics. In *Molecular modeling and simulation an interdisciplinary guide* (Vol. 21, pp. 425–461). http://dx.doi.org/10.1007/978-1-4419-6351-2_13.

Shadrick, W. R., Slavish, P. J., Chai, S. C., Waddell, B., Connelly, M., Low, J. A., et al. (2018). Exploiting a water network to achieve enthalpy-driven, bromodomain-selective BET inhibitors. *Bioorganic & Medicinal Chemistry, 26*(1), 25–36. http://dx.doi.org/10.1016/j.bmc.2017.10.042.

Shao, M., He, L., Zheng, L., Huang, L., Zhou, Y., Wang, T., et al. (2017). Structure-based design, synthesis and in vitro antiproliferative effects studies of novel dual BRD4/HDAC inhibitors. *Bioorganic & Medicinal Chemistry Letters, 27*(17), 4051–4055. http://dx.doi.org/10.1016/j.bmcl.2017.07.054.

Shi, J., & Vakoc, C. R. (2014). The mechanisms behind the therapeutic activity of BET bromodomain inhibition. *Molecular Cell, 54*(5), 728–736. http://dx.doi.org/10.1016/j.molcel.2014.05.016.

Shirts, M. R., Bair, E., Hooker, G., & Pande, V. S. (2003). Equilibrium free energies from nonequilibrium measurements using maximum-likelihood methods. *Physical Review Letters, 91*(14), 1–4. http://dx.doi.org/10.1103/PhysRevLett.91.140601.

Simon, R. P., Robaa, D., Alhalabi, Z., Sippl, W., & Jung, M. (2016). KATching-up on small molecule modulators of lysine acetyltransferases. *Journal of Medicinal Chemistry, 59*(4), 1249–1270. http://dx.doi.org/10.1021/acs.jmedchem.5b01502.

Smith, S. G., & Zhou, M.-M. (2015). The bromodomain as the acetyl-lysine binding domain in gene transcription. In *Histone recognition* (pp. 1–26). Cham: Springer International Publishing. http://dx.doi.org/10.1007/978-3-319-18102-8_1.

Speck-Planche, A., & Scotti, M. T. (2019). BET bromodomain inhibitors: Fragment-based in silico design using multi-target QSAR models. *Molecular Diversity, 23*(3), 555–572. http://dx.doi.org/10.1007/s11030-018-9890-8.

Sperandio, D., Aktoudianakis, V., Babaoglu, K., Chen, X., Elbel, K., Chin, G., et al. (2019). Structure-guided discovery of a novel, potent, and orally bioavailable 3,5-dimethylisoxazole aryl-benzimidazole BET bromodomain inhibitor. *Bioorganic & Medicinal Chemistry, 27*(3), 457–469. http://dx.doi.org/10.1016/j.bmc.2018.11.020.

Spiliotopoulos, D., & Caflisch, A. (2016). Fragment-based in silico screening of bromodomain ligands. *Drug Discovery Today: Technologies, 19*(March), 81–90. http://dx.doi.org/10.1016/j.ddtec.2016.06.003.

Steiner, S., Magno, A., Huang, D., & Caflisch, A. (2013). Does bromodomain flexibility influence histone recognition? *FEBS Letters, 587*(14), 2158–2163. http://dx.doi.org/10.1016/j.febslet.2013.05.032.

Sterling, T., & Irwin, J. J. (2015). Zinc 15 - ligand discovery for everyone. *Journal of Chemical Information and Modeling, 55*(11), 2324–2337. http://dx.doi.org/10.1021/acs.jcim.5b00559.

Stumpfe, D., Hu, Y., Dimova, D., & Bajorath, J. (2014). Recent progress in understanding activity cliffs and their utility in medicinal chemistry. *Journal of Medicinal Chemistry, 57*(1), 18–28. http://dx.doi.org/10.1021/jm401120g.

Su, J., Liu, X., Zhang, S., Yan, F., Zhang, Q., & Chen, J. (2018). A theoretical insight into selectivity of inhibitors toward two domains of bromodomain-containing protein 4 using molecular dynamics simulations. *Chemical Biology & Drug Design, 91*(3), 828–840. http://dx.doi.org/10.1111/cbdd.13148.

Sun, H., Duan, L., Chen, F., Liu, H., Wang, Z., Pan, P., et al. (2018). Assessing the performance of MM/PBSA and MM/GBSA methods. 7. Entropy effects on the performance of end-point binding free energy calculation approaches. *Physical Chemistry Chemical Physics, 20*(21), 14450–14460. http://dx.doi.org/10.1039/c7cp07623a.

Taniguchi, Y. (2016). The bromodomain and extra-terminal domain (BET) family: Functional anatomy of BET paralogous proteins. *International Journal of Molecular Sciences, 17*(11). http://dx.doi.org/10.3390/ijms17111849.

Thakur, A., Singla, R., & Jaitak, V. (2015). Coumarins as anticancer agents: A review on synthetic strategies, mechanism of action and SAR studies. *European Journal of Medicinal Chemistry, 101*, 476–495. http://dx.doi.org/10.1016/j.ejmech.2015.07.010.

Tuckerman, M. E. (2010). *Statistical mechanics: Theory and molecular simulation*. Oxford University Press.

Van Drie, J. H. (2010). History of 3D pharmacophore searching: Commercial, academic and open-source tools. *Drug Discovery Today: Technologies, 7*(4), e255–e262. http://dx.doi.org/10.1016/j.ddtec.2010.12.002.

VanderMolen, K. M., McCulloch, W., Pearce, C. J., & Oberlies, N. H. (2011). Romidepsin (istodax, NSC 630176, FR901228, FK228, depsipeptide): A natural product recently approved for cutaneous T-cell lymphoma. *Journal of Antibiotics, 64*(8), 525–531. http://dx.doi.org/10.1038/ja.2011.35.

Vidler, L. R., Brown, N., Knapp, S., & Hoelder, S. (2012). Druggability analysis and structural classi fi cation of bromodomain acetyl-lysine binding sites. *Journal of Medicinal Chemistry, 55*(Figure 2), 7346–7359. http://dx.doi.org/10.1021/jm300346w.

Vidler, L. R., Filippakopoulos, P., Fedorov, O., Picaud, S., Martin, S., Tomsett, M., et al. (2013). Discovery of novel small-molecule inhibitors of BRD4 using structure-based virtual screening. *Journal of Medicinal Chemistry, 56*(20), 8073–8088. http://dx.doi.org/10.1021/jm4011302.

Vita, M., & Henriksson, M. (August 2006). The Myc oncoprotein as a therapeutic target for human cancer. *Seminars in Cancer Biology, 16*(4), 318–330. http://dx.doi.org/10.1016/j.semcancer.2006.07.015.

Vlachakis, D., Bencurova, E., Papangelopoulos, N., & Kossida, S. (2014). Current state-of-the-art molecular dynamics methods and applications. In*Advances in protein chemistry and structural biology* (1st ed., Vol. 94). Elsevier Inc. http://dx.doi.org/10.1016/B978-0-12-800168-4.00007-X.

Wakchaure, P., Velayutham, R., & Roy, K. K. (2019). Structure investigation, enrichment analysis and structure-based repurposing of FDA-approved drugs as inhibitors of BET-BRD4. *Journal of Biomolecular Structure and Dynamics, 37*(12), 3048–3057. http://dx.doi.org/10.1080/07391102.2018.1507838.

Walters, W. P. (2019). Virtual chemical libraries. *Journal of Medicinal Chemistry, 62*(3), 1116–1124. http://dx.doi.org/10.1021/acs.jmedchem.8b01048. American Chemical Society.

Wang, J., Erazo, T., Ferguson, F. M., Buckley, D. L., Gomez, N., Muñoz-Guardiola, P., et al. (2018). Structural and atropisomeric factors governing the selectivity of pyrimido-benzodiazipinones as inhibitors of kinases and bromodomains. *ACS Chemical Biology, 13*(9), 2438–2448. http://dx.doi.org/10.1021/acschembio.7b00638.

Wang, C. Y., & Filippakopoulos, P. (2015). Beating the odds: BETs in disease. *Trends in Biochemical Sciences, 40*(8), 468–479. http://dx.doi.org/10.1016/j.tibs.2015.06.002.

Wang, Q., Li, Y., Xu, J., Wang, Y., Shi, D., Liu, L., et al. (2019). Computational study on the selective inhibition mechanism of MS402 to the first and second bromodomains of BRD4. *Proteins: Structure, Function and Bioinformatics, 87*(1), 3–11. http://dx.doi.org/10.1002/prot.25611.

Wang, Le, Pratt, J. K., Soltwedel, T., Sheppard, G. S., Fidanze, S. D., Liu, D., et al. (2017). Fragment-based, structure-enabled discovery of novel pyridones and pyridone macrocycles as potent bromodomain and extra-terminal domain (BET) family bromodomain inhibitors. *Journal of Medicinal Chemistry, 60*(9), 3828–3850. http://dx.doi.org/10.1021/acs.jmedchem.7b00017.

Wang, L., Wang, Y., Sun, H., Zhao, J., & Wang, Q. (2019). Theoretical insight into molecular mechanisms of inhibitor bindings to bromodomain-containing protein 4 using

molecular dynamics simulations and calculations of binding free energies. *Chemical Physics Letters, 736*(September), 136785. http://dx.doi.org/10.1016/j.cplett.2019.136785.
Wellaway, C. R., Amans, D., Bamborough, P., Barnett, H., Bit, R. A., Brown, J. A., et al. (2020). Discovery of a bromodomain and extraterminal inhibitor with a low predicted human dose through synergistic use of encoded library technology and fragment screening. *Journal of Medicinal Chemistry, 63*(2), 714−746. http://dx.doi.org/10.1021/acs.jmedchem.9b01670.
Wereszczynski, J., & McCammon, J. A. (2012). Thermodynamic properties of biomolecular recognition. *Quarterly Reviews of Biophysics, 45*(1), 1−25. http://dx.doi.org/10.1017/S0033583511000096.Statistical.
Willett, P. (2008). A history of chemoinformatics. In *Handbook of chemoinformatics* (Vol. 1, pp. 6−20). Weinheim, Germany: Wiley-VCH Verlag GmbH. http://dx.doi.org/10.1002/9783527618279.ch2.
Willett, P. (2013). Combination of similarity rankings using data fusion. *Journal of Chemical Information and Modeling, 53*(1), 1−10. http://dx.doi.org/10.1021/ci300547g. American Chemical Society.
Willett, P. (2014). The calculation of molecular structural similarity: Principles and practice. *Molecular Informatics, 33*(6−7), 403−413. http://dx.doi.org/10.1002/minf.201400024.
Wu, Q., Peng, Z., Zhang, Y., & Yang, J. (2018). COACH-D: Improved protein-ligand binding sites prediction with refined ligand-binding poses through molecular docking. *Nucleic Acids Research, 46*(W1), W438−W442. http://dx.doi.org/10.1093/nar/gky439.
Xing, J., Lu, W., Liu, R., Wang, Y., Xie, Y., Zhang, H., et al. (2017). Machine-learning-assisted approach for discovering novel inhibitors targeting bromodomain-containing protein 4. *Journal of Chemical Information and Modeling, 57*(7), 1677−1690. http://dx.doi.org/10.1021/acs.jcim.7b00098.
Xue, X., Zhang, Y., Liu, Z., Song, M., Xing, Y., Xiang, Q., et al. (2016). Discovery of benzo[cd]indol-2(1H)-ones as potent and specific BET bromodomain inhibitors: Structure-based virtual screening, optimization, and biological evaluation. *Journal of Medicinal Chemistry, 59*(4), 1565−1579. http://dx.doi.org/10.1021/acs.jmedchem.5b01511.
Yang, X. J. (2004). Lysine acetylation and the bromodomain: A new partnership for signaling. *BioEssays, 26*(10), 1076−1087. http://dx.doi.org/10.1002/bies.20104.
Yang, Y., Chen, P., Zhao, L., Zhang, F., Zhang, B., Xu, C., et al. (2019). Exploiting the 7-methylimidazo[1,5-a]pyrazin-8(7H)-one scaffold for the development of novel chemical inhibitors for Bromodomain and Extraterminal Domain (BET) family. *Bioorganic Chemistry, 90*(June), 103044. http://dx.doi.org/10.1016/j.bioorg.2019.103044.
Ye, F., Huang, J., Wang, H., Luo, C., & Zhao, K. (2019). Targeting epigenetic machinery: Emerging novel allosteric inhibitors. *Pharmacology & Therapeutics, 204*. http://dx.doi.org/10.1016/j.pharmthera.2019.107406.
Zanette, C., Bannan, C. C., Bayly, C. I., Fass, J., Gilson, M. K., Shirts, M. R., et al. (2019). Toward learned chemical perception of force field typing rules. *Journal of Chemical Theory and Computation, 15*(1), 402−423. http://dx.doi.org/10.1021/acs.jctc.8b00821.
Zaware, N., & Zhou, M. M. (2017). Chemical modulators for epigenome reader domains as emerging epigenetic therapies for cancer and inflammation. *Current Opinion in Chemical Biology, 39*, 116−125. http://dx.doi.org/10.1016/j.cbpa.2017.06.012.
Zhang, Z., Gu, L., Wang, B., Huang, W., Zhang, Y., Ma, Z., et al. (2019). Discovery of novel coumarin derivatives as potent and orally bioavailable BRD4 inhibitors based on scaffold hopping. *Journal of Enzyme Inhibition and Medicinal Chemistry, 34*(1), 808−817. http://dx.doi.org/10.1080/14756366.2019.1587417.
Zhang, G., Plotnikov, A. N., Rusinova, E., Shen, T., Morohashi, K., Joshua, J., et al. (2013). Structure-guided design of potent diazobenzene inhibitors for the BET bromodomains. *Journal of Medicinal Chemistry, 56*(22), 9251−9264. http://dx.doi.org/10.1021/jm401334s.

Zhang, G., Smith, S. G., & Zhou, M.-M. (2015). Discovery of chemical inhibitors of human bromodomains. *Chemical Reviews, 115*(21), 11625–11668. http://dx.doi.org/10.1021/acs.chemrev.5b00205.

Zhang, L., & Xu, Z. (2019). Coumarin-containing hybrids and their anticancer activities. *European Journal of Medicinal Chemistry, 181*, 111587. http://dx.doi.org/10.1016/j.ejmech.2019.111587.

Zhao, Y., Bai, L., Liu, L., McEachern, D., Stuckey, J. A., Meagher, J. L., et al. (2017). Structure-based discovery of 4-(6-Methoxy-2-methyl-4-(quinolin-4-yl)-9H-pyrimido[4,5-b]indol-7-yl)-3,5-dimethylisoxazole (CD161) as a potent and orally bioavailable BET bromodomain inhibitor. *Journal of Medicinal Chemistry, 60*(9), 3887–3901. http://dx.doi.org/10.1021/acs.jmedchem.7b00193.

Zhao, L., Cao, D., Chen, T., Wang, Y., Miao, Z., Xu, Y., et al. (2013). Fragment-based drug discovery of 2-thiazolidinones as inhibitors of the histone reader BRD4 bromodomain. *Journal of Medicinal Chemistry, 56*(10), 3833–3851. http://dx.doi.org/10.1021/jm301793a.

Zhao, H., Gartenmann, L., Dong, J., Spiliotopoulos, D., & Caflisch, A. (2014). Discovery of BRD4 bromodomain inhibitors by fragment-based high-throughput docking. *Bioorganic & Medicinal Chemistry Letters, 24*(11), 2493–2496. http://dx.doi.org/10.1016/j.bmcl.2014.04.017.

Zhi, Y., Wang, S., Huang, W., Zeng, S., Liang, M., Zhang, C., et al. (2019). Novel phenanthridin-6(5H)-one derivatives as potent and selective BET bromodomain inhibitors: Rational design, synthesis and biological evaluation. *European Journal of Medicinal Chemistry, 179*, 502–514. http://dx.doi.org/10.1016/j.ejmech.2019.06.067.

Zhou, Y., Hou, Y., Shen, J., Huang, Y., Martin, W., & Cheng, F. (2020). Network-based drug repurposing for novel coronavirus 2019-nCoV/SARS-CoV-2. *Cell Discovery, 6*(1), 1–18. http://dx.doi.org/10.1038/s41421-020-0153-3.

Zhu, S. (2019). Validation of the generalized force fields GAFF, CGenFF, OPLS-AA, and PRODRGFF by testing against experimental osmotic coefficient data for small drug-like molecules. *Journal of Chemical Information and Modeling, 59*, 4239–4247. http://dx.doi.org/10.1021/acs.jcim.9b00552. research-article.

Zwanzig, R. W. (1954). High-temperature equation of state by a perturbation method. I. Nonpolar gases. *The Journal of Chemical Physics, 22*(8), 1420–1426. http://dx.doi.org/10.1063/1.1740409.

CHAPTER SIX

# How to make an undruggable enzyme druggable: lessons from ras proteins

**Abdallah Sayyed-Ahmad[a,*] and Alemayehu A. Gorfe[b,*]**
[a]Department of Physics, Birzeit University, Birzeit, West Bank, Palestine
[b]McGovern Medical School, University of Texas Health Science Center at Houston, Department of Integrative Biology & Pharmacology, Houston, TX, United States
[*]Corresponding authors: E-mail: asayyeda@birzeit.edu; Alemayehu.G.Abebe@uth.tmc.edu

## Contents



## Abstract

Significant advances have been made toward discovering allosteric inhibitors for challenging drug targets such as the Ras family of membrane-associated signaling proteins. Malfunction of Ras proteins due to somatic mutations is associated with up to a quarter of all human cancers. Computational techniques have played critical roles in identifying and characterizing allosteric ligand-binding sites on these proteins, and to screen ligand libraries against those sites. These efforts, combined with a wide range of biophysical, structural, biochemical and cell biological experiments, are beginning to yield promising inhibitors to treat malignancies associated with mutated Ras proteins. In this chapter, we discuss some of these developments and how the lessons learned from Ras might be applied to similar other challenging drug targets.

*Advances in Protein Chemistry and Structural Biology*, Volume 122
ISSN 1876-1623
https://doi.org/10.1016/bs.apcsb.2020.05.003

## 1. Introduction

Malfunction of enzymes due to mutation, misfolding or other structural or biochemical alterations can cause various human diseases such as diabetes, neurological disorders, and cancer (Arencibia, Pastor-Flores, Bauer, Schulze, & Biondi, 2013; Bos et al., 1987; Butters, Dwek, & Platt, 2003; Cohen, 2014; Davies et al., 2002; Góth, Rass, & Páy, 2004; Hollstein, Sidransky, Vogelstein, & Harris, 1991; Karakas, Bachman, & Park, 2006; Steinberg & Kemp, 2009). A common approach to developing inhibitors of malfunctioning enzymes is to target the catalytic active site by small-molecule ligands (Das & David Kimball, 1995; Fan & Ishii, 2007; Hyde & Carmichael, 1991; Jin et al., 2018; Manetti & Botta, 2003; Wittwer et al., 2007; Wu, Nielsen, & Clausen, 2015; Zorn & Wells, 2010). However, this may not be a viable strategy for some critical drug targets. A typical example is the Ras family of guanine triphosphate (GTP)-hydrolyzing enzymes whose malfunction due to somatic mutations is associated with about 25% of all cancers. By cycling between GTP-bound active and guanine diphosphate (GDP)-bound inactive conformations, the plasma membrane-associated Ras GTPases act as molecular switches to regulate a variety of signaling pathways controlling cell growth, differentiation, and proliferation (Rajalingam, Schreck, Rapp, & Albert, 2007). Mutations in the switch regions of these proteins are linked to many cancers including pancreatic and lung cancers (Bos et al., 1987; Karachaliou et al., 2013; Mulcahy et al., 1998; Pellegata et al., 1994; Prior, Lewis, & Mattos, 2012; Riely, Marks, & Pao, 2009; Rodenhuis & Slebos, 1992; Westcott et al., 2015). Attempts in the past decades to develop drugs to treat Ras-driven cancers have failed to bear fruit in part because the Ras active site has a very high affinity (picomolar) for GTP and GDP that exist in the cell at high concentrations (~0.5 mM), making competitive inhibition essentially impractical. Furthermore, Ras proteins are ubiquitously expressed in human tissues and share structural similarities with many other small GTPases, making selective inhibition difficult.

An alternative strategy to targeting the active site would be allosteric inhibition. However, crystal structures of ligand-free Ras proteins typically depict a relatively smooth molecular surface with no obvious hydrophobic pockets suitable for structure-based drug design. This led to the pursuit of indirect Ras inhibition strategies. These included targeting downstream Ras effectors or proteins involved in its membrane targeting (Johnston,

2001; Reiss, Goldstein, Seabra, Casey, & Brown, 1990). However, since Ras proteins are major signaling hubs that integrate multiple interaction networks, many drug candidates ended up being unsuccessful often due to unintended adverse effects (Anforth, Fernandez-Peñas, & Long, 2013; Lito, Rosen, & Solit, 2013; Manousaridis, Mavridou, Goerdt, Leverkus, & Utikal, 2013; Ravoet et al., 2008). Recent studies have shown that focusing on the dynamic nature of Ras proteins, including their dynamic interactions with membranes, hold promises for discovering direct allosteric inhibitors. This chapter focuses on these developments, and is organized as follows. We first review the key studies that have established Ras as an allosteric enzyme, and then survey the computational methods that have been used to identify and characterize allosteric ligand binding sites in Ras proteins. After briefly describing some of the major discoveries regarding the allosteric effects of membrane binding and its role in the regulation of Ras functions, we conclude by a brief perspective on how lessons from Ras proteins may be applied to similar other challenging therapeutic targets.

## 2. Ras proteins as anti-cancer drug targets

Humans express three Ras isoforms: H-Ras, N-Ras and K-Ras4b (and its slice variant K-Ras4a). As illustrated in Fig. 1, the structure of Ras consists of a soluble catalytic domain (residues 1–166) (Milburn et al., 1990) and a hypervariable region (HVR, residues 167-185/6). The catalytic domain is highly conserved among the Ras isoforms and contains an effector lobe (residues 1–86) that binds to effectors and exchange factors, and an allosteric lobe (residues 87–166) that is involved in membrane interaction (Johnson et al., 2017; Abankwa, Gorfe, Inder, & Hancock, 2010) and oligomer formation (Sarkar-Banerjee et al., 2017; Prakash et al., 2017). The lipid-modified C-terminus at the HVR anchors Ras to the plasma membrane (Fig. 1), and is believed to be involved in the differential segregation of Ras isoforms into distinct membrane microdomains.

The nucleotide- and effector-interacting switch 1 (residues 25–40) and switch 2 (residues 60–75) regions undergo conformational changes upon GDP/GTP exchange and GTP hydrolysis (Milburn et al., 1990). Nucleotide exchange and GTP hydrolysis reactions are facilitated by guanine nucleotide exchange factors (GEFs) and GTPase activating proteins (GAPs) (Rajalingam et al., 2007), respectively (see Fig. 2). Somatic mutations, typically at codons 12, 13 and 61 (Prior et al., 2012), impair the GTPase activity

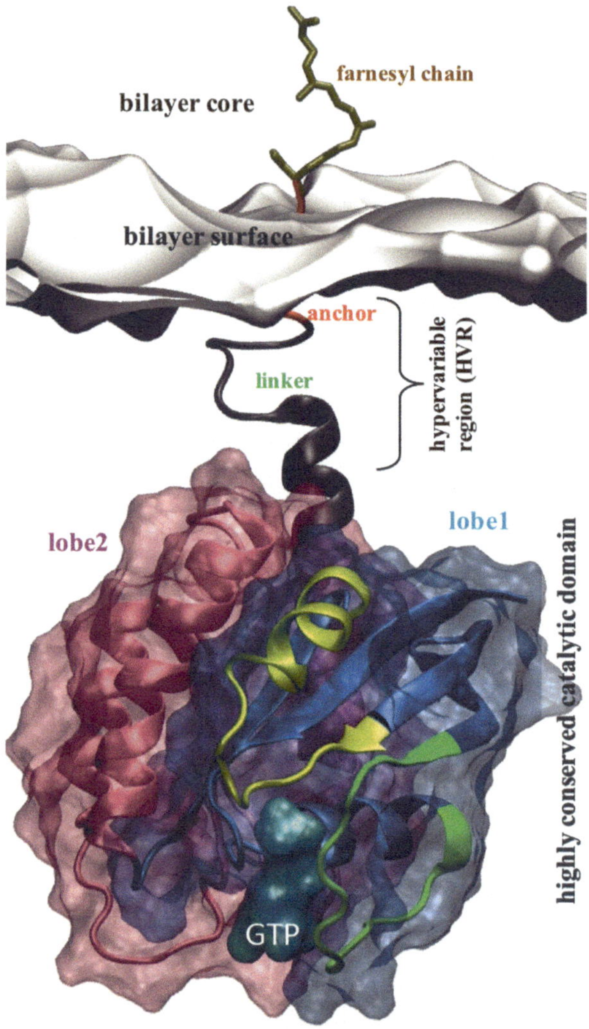


**Fig. 1** Domain architecture of K-Ras. The bi-lobed catalytic domain (residues 1–166) is highly conserved among RAS proteins. The hypervariable region (HVR; residues 167–185) harbors the membrane-interacting farnesyl (yellow) and polybasic segment (red), plus a flexible linker. Lobe 1, also called the effector lobe, (residues 1–86) is in purple while lobe 2 or the allosteric lobe (residues 87–166) is in blue. Switch 1 (residues 25–40) is shown in green and switch 2 (residues 60–75) in yellow. *This figure is adopted from Prakash, P., & Gorfe, A. A. (2013). Lessons from computer simulations of Ras proteins in solution and in membrane.* Biochim. Biophys. Acta Gen. Subj., 1830*(11), 5211–5218.*

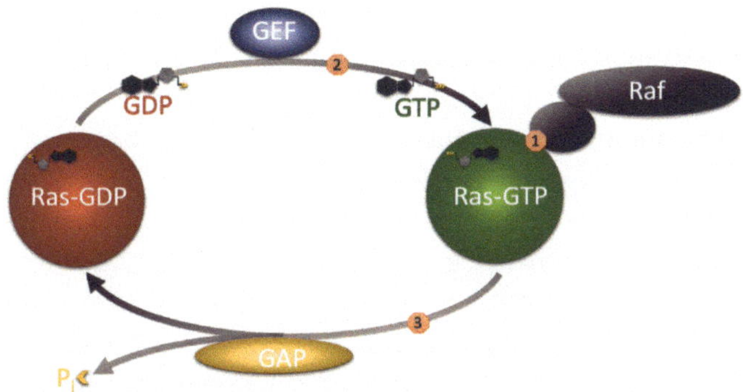


**Fig. 2** The GTPase cycle of Ras and potential mechanisms by which an allosteric inhibitor may impact its functions: (1) inhibiting effector binding; (2) preventing activation by GEFs; (3) accelerating GTP hydrolysis.

of Ras and render it constitutively active, resulting in uncontrolled cell growth/proliferation and cancer. As a result, H-Ras mutations are commonly found in thyroid and bladder cancers, K-Ras mutations in colorectal (Bos et al., 1987), lung (Karachaliou et al., 2013; Riely et al., 2009; Rodenhuis & Slebos, 1992; Westcott et al., 2015) and pancreatic (Fleming, Shen, Holloway, Davis, & Brekken, 2005; Mulcahy et al., 1998; Pellegata et al., 1994) cancers, and N-Ras mutations in myeloid leukemia (Farr, Saiki, Erlich, McCormick, & Marshall, 1988) and melanomas (Jakob et al., 2012). Therefore, finding a drug that inhibits any or all oncogenic Ras proteins would be a disruptive innovation in cancer therapy (Kloog & Cox, 2000; Spencer-Smith & O'Bryan, 2019; Wang, Fang, & Rudolph, 2012). Among the Ras isoforms, however, K-Ras4b (hereafter K-Ras) is the most important and arguably the most sought-after drug target because its mutations account for 85% of all cancer-causing Ras mutations. Therefore, although we occasionally refer to H-Ras— and N-Ras, much of the discussion in this chapter is focused on K-Ras.

## 3. Identifying and characterizing allosteric ligand binding sites on ras proteins

Efforts to indirectly inhibit aberrant Ras functions, such as by dislodging it from the plasma membrane (Johnston, 2001; Reiss et al., 1990) or inhibiting its partner proteins (Lito et al., 2013; Vigil, Cherfils, Rossman, & Der, 2010), have so far been unsuccessful due to alternative mechanisms of membrane targeting or issues of toxicity (Anforth et al., 2013; Lito et al.,

2013; Ravoet et al., 2008). Direct inhibition of mutant Ras has long been thought a long shot (Cox, Fesik, Kimmelman, Luo, & Der, 2014), because the nucleotide-binding site, which would have been the first logical target, is impractical for the reasons outlined in Introduction. Allosteric inhibitors that could overcome many of the challenges associated with targeting the active site were not considered until recently because Ras was not thought to be an allosteric enzyme. The first clue about Ras allostery emerged from a molecular dynamics (MD) simulation study of H-Ras (Gorfe, Hanzal-Bayer, Abankwa, Hancock, & McCammon, 2007) in a 1,2-dimyristoylglycero3-phosphocholine (DMPC) bilayer, which suggested that the way in which the catalytic domain interacts with membrane lipids might be modulated by the nature of the bound nucleotide or mutations. Subsequent studies provided additional evidence for the allosteric modulation of Ras dynamics and its membrane interactions (Buhrman, Kumar, Cirit, Haugh, & Mattos, 2011; Lukman, Grant, Gorfe, Grant, & McCammon, 2010; Prakash, Sayyed-Ahmad, & Gorfe, 2012a; Kearney, Johnson, Roberts, Swartz, & Mattos, 2014; Sayyed-Ahmad, Prakash, & Gorfe, 2017; Khaled, Gorfe, & Sayyed-Ahmad, 2019). This led to the idea that Ras may be inhibited by allosteric mechanisms (Gorfe, 2010). Prior to the discovery of allosteric Ras ligands (Grant et al., 2011; Sun et al., 2012; Hocker et al., 2013; Shima et al., 2013; Ostrem, Peters, Sos, Wells, & Shokat, 2013; Lim et al., 2014; McCarthy et al., 2019; Gupta et al., 2019), studies based on MD simulations and other computational techniques (Grant et al., 2011; Buhrman et al., 2010, 2011b), have led to the prediction of four allosteric ligand binding sites on Ras proteins: pockets p1, p2, p3 and p4 (see Fig. 3). Using probe-based MD (pMD) simulation, Prakash, Hancock and Gorfe (2015a) showed that the distribution of high-density probe binding regions on the surface of K-Ras agreed well with the previously characterized ligand binding pockets. Additional surface patches with high probe-binding potential were also observed in the pMD simulations, and these surfaces were proposed to be involved in membrane binding or participate in dimerization (Prakash et al., 2017; Sayyed-Ahmad, Cho, Hancock, & Gorfe, 2016; Lee, Fang, Enomoto, Seabrook, Zheng, Koide, Ikura, & Marshall., 2020) or the formation of higher order oligomers (Sarkar-Banerjee et al., 2017).

Druggability of a given pocket of an enzyme depends on its chemical and geometric features, which in turn are affected by backbone and sidechain conformational fluctuations (Sayyed-Ahmad & Gorfe, 2017). Accounting for such fluctuations increases the success rate of structure-based drug design (Durant, Leland, Henry, & Nourse, 2002; Ganesan, Coote, & Barakat,

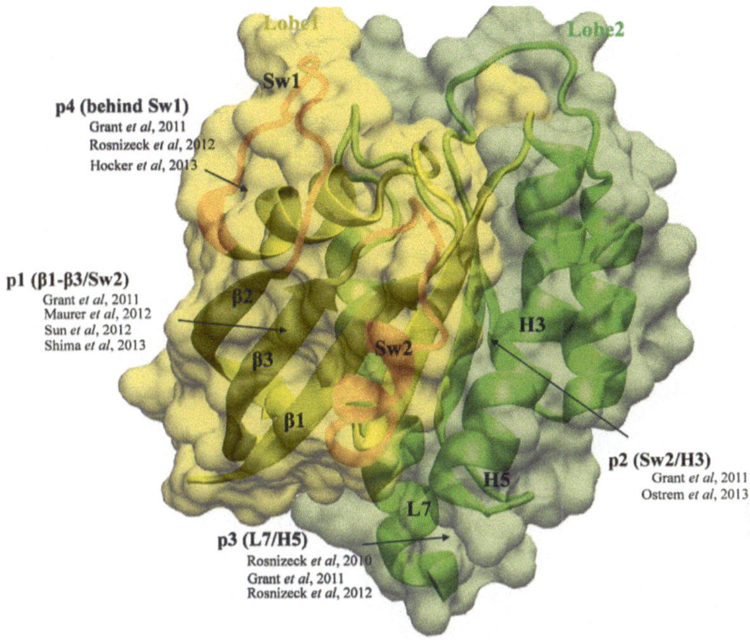


**Fig. 3** Allosteric ligand binding sites on K-Ras. The catalytic domain is shown in surface and ribbon representations with lobe 1 (residues 1–86) in yellow and lobe 2 (residues 87–167) in green. The functionally important switch regions sw1 and sw2 are highlighted. Arrows indicate the location of the four previously characterized allosteric ligand-binding sites, along with the key earlier reports that described them. *The figure is reprinted from Prakash, P., Hancock, J. F., Gorfe, A. A. (2015). Binding Hotspots on K-Ras: Consensus Ligand Binding Sites and Other Reactive Regions from Probe-Based Molecular Dynamics Analysis.* Proteins, 83*(5), 898–909.*

2017), in part by allowing access to transiently open pockets (Stank, Kokh, Fuller, & Wade, 2016). Therefore, Gupta et al. (Gupta et al., 2019) used MD simulations of wild type, G12D, G12V, G13D and Q61H K-Ras to capture conformations in which each of the four previously characterized pockets have become solvent accessible. Based on pocket volume distributions, 8–12 representative structures were selected for docking and virtual screening of ligand libraries on each protein variant and each pocket. Prior to docking, the average number of hydrogen bond acceptors/donors, the polar/non-polar surface areas, and the compactness or shape (i.e. volume-to-surface area ratio) were measured. It was found that 60%, 53%, 46% and 54% of the solvent-accessible surface area in p1, p2, p3 and p4, respectively, is non-polar. These and other observations suggested that the order of druggability of the pockets is: p1 > p2 > p4 > p3. Another remarkable observation was the significant variation in a given pocket across the K-Ras mutants. For example, the shape of p2 was found to be tubular in WT K-Ras and spherical in G12V K-Ras and Q61H K-Ras, while the non-polar surface area of p3 varies between 46% and 54% of the total solvent accessible area. Together, these observations highlight the major impact of dynamics on the geometry and chemical features, and hence druggability, of each pocket.

## 4. Allosteric inhibitors of K-Ras: light at the end of the tunnel?

The identification and detailed characterization of the four allosteric ligand binding sites described above (Buhrman et al., 2011a; Grant et al., 2011; Prakash et al., 2015a, Prakash, Sayyed-Ahmad, Gorfe, 2015b; Sayyed-Ahmad & Gorfe, 2017) and the subsequent discovery of ligands that bind to those pockets (Gupta et al., 2019; Maurer et al., 2012; McCarthy et al., 2019; Ostrem et al., 2013; Rosnizeck et al., 2010, 2012) means that development of allosteric K-Ras inhibitors is now a promising strategy to treat Ras-driven cancers. Indeed, various ligands including small-molecules, peptidomimetics (Niida et al., 2017; Sakamoto et al., 2017; Trinh, Upadhyaya, Qian, & Pei, 2016; Upadhyaya, Qian, Habir, & Pei, 2014) and monobodies (Spencer-Smith et al., 2017) have been reported that allosterically inhibit Ras interactions with its effectors or activators possibly by, for example, shifting the distribution of the conformational ensemble toward the inactive state. Currently available direct allosteric small molecule inhibitors of Ras can be divided into two groups: non-covalent and covalent inhibitors. Sulfhydryl-rich covalent inhibitors are mainly designed to target G12C K-Ras by forming a disulfide bond with the cysteine residue in the P-loop region (Fell et al., 2018; Nnadi et al., 2018; Ostrem et al., 2013; Rudolph & Stokoe, 2014; Zeng et al., 2017). Such inhibitors exhibit high potency and significantly reduce the signaling activities of G12C K-Ras. This class of compounds appear to have a better chance of successfully treating a few cancer types such as small cell lung cancer (Romero, 2020). Indeed, three such inhibitors are currently in clinical trial: AMG-510 (Amgen), MRTX849 (Mirati) and ARS-3248 (Araxes) (Nagasaka et al., 2020). However, in parallel to the development of covalent inhibitors, development of non-covalent allosteric inhibitors will continue to be needed to treat many diverse cancers caused by other critical Ras mutants such as G12D, G12V, G13D and Q61H. These four mutations together account for more than 78% of K-Rasassociated cancers (Stephen, Esposito, Bagni, & McCormick, 2014), and are common in some of the most lethal malignancies such as colon, lung, and pancreatic (Prior et al., 2012; Stephen et al., 2014) cancer.

To the best of our knowledge, the first attempt at discovering a non-covalent Ras inhibitor was reported in 2011 by Grant et al. (2011). Utilizing a multi-level computational approach combining bioinformatic analyses with advanced molecular simulations and ensemble docking, the authors identified dozens of initial hits targeting each of the four allosteric pockets shown in Fig. 3. Initial tests of the top four ligands in glioblastoma cells suggested limited inhibition of Ras signaling. In the subsequent three years, several groups reported low-affinity allosteric K-Ras inhibitors (Maurer et al., 2012; Rosnizeck et al., 2012; Shima et al., 2013; Sun et al., 2012). For example, Shima et al. (2013) exploited the pocket found in the crystal structure of M-Ras-GTP carrying a P40D mutation to develop small-molecule Ras inhibitors. They found two compounds that exhibited inhibitory activity toward the interaction of GTP-bound H-Ras with Raf1 *in vitro,* as well as inhibition of cell growth in anchorage-dependent and -independent growth conditions. These ligands also induced apoptosis of G12D H-Ras— transformed NIH 3T3 cells and down-regulated signaling via MEK/ERK, Akt and RalA pathways. Recently, McCarthy et al. (McCarthy et al., 2019) reported a pyrazolopyrimidine-based compound (2[4-(8-methyl-3,9-diphenyl-2,6,7-triazabicyclo[4.3.0]nona-2,4,7,9-tetraen-5yl)-piperazin1-yl]ethanol) that binds to GTP-K-Ras with sub-μM affinity, disrupts effector binding, and reduces the rates of intrinsic and SOS-mediated GDP/GTP exchange reactions. The ligand reduced signal transduction through K-Ras and inhibited cancer cell growth. Using a similar workflow, Gupta et al. (2019) reported a ligand that binds K-Ras with low μM affinity, although the inhibitory activities of this compound in cancer cells has not been demonstrated.

It has been shown previously that inhibiting Ras GEFs such as son-of-seven-less (SOS) is a valid approach to inhibiting oncogenic K-Ras signaling (Hocker et al., 2013). Along this line, Sun et al. (2012) used NMR fragment-based screening and structure-based design to discover ligands that bind to K-Ras and exhibit inhibitory activity in a SOS-catalyzed nucleotide exchange assay. In a subsequent study (Burns et al., 2014), the same group discovered ligands that bind to a hydrophobic pocket in SOS adjacent to the Switch II region of Ras. These ligands were shown to increase the rate of SOS-catalyzed nucleotide exchange *in vitro* and modulate MAPK and PI3K signaling in cells at low micromolar concentrations. Evelyn et al. combined ensemble-based virtual screening, high throughput experimental screening and fluorescent-based guanine nucleotide exchange assays to find two ligands that dose-dependently inhibit SOS's catalytic activity and Ras signaling (Evelyn et al., 2015). These and many other studies on SOS (e.g. (Hodges et al., 2018; Kessler et al., 2019)) laid the ground for the emergence of perhaps one of the most promising SOS-binding pan-Ras inhibitors, Boehringer's BI 1701963, which is currently in clinical trial. These and other examples (Burns et al., 2018; Casique-Aguirre et al., 2018; Evelyn et al., 2014; Jayakanthan et al., 2009; Kopra et al., 2017) highlight some of the successes in inhibiting oncogenic Ras (particularly K-Ras) signaling using non-covalent binders of Ras or its activator SOS (note: throughout this chapter, SOS refers to isoform 1 or SOS-1). The remaining challenge is overcoming the persistent problems of weak affinity and pan-Ras activity (Quevedo et al., 2018; Welsch et al., 2017). In this regard, paying special attention to subtle differences in the structure and especially in the dynamics of Ras isoforms (Johnson et al., 2017; Sayyed-Ahmad et al., 2017) may aid in the future development of isoform-selective Ras inhibitors. As discussed below, a deeper understanding of Ras-membrane interactions may also play a critical role in the development of isoform- and potentially even mutation-specific Ras inhibitors.

## 5. Ras membrane binding, lateral segregation, and reorientation

Similar to many other crucial drug targets that require membrane binding for function, Ras proteins need to be targeted to the inner surface of the plasma membrane for their biological activity; cytosolic Ras is not able to transduce extracellular signals to the nucleus. At the plasma membrane, activated (GTP-bound) Ras binds to many diverse downstream kinases to activate different signaling pathways (Laude & Prior, 2008). Ras isoforms have different preferences for effector kinases (e.g. K-Ras preferentially interacts with Raf kinases while H-Ras has more preference for PI3K (Prior & Hancock, 2001)). Although precisely how different Ras isoforms preferentially interact with different effectors is not clear, it is believed to be linked, at least in part, to their differential membrane localization/interaction with lipids. For example, activated K-Ras and H-Ras localize at liquid-disordered and liquid-ordered membrane domains, respectively, likely as a result of their difference in post-translational lipid-modification and in the sequence of the hypervariable region. This is supported by the observation that the farnesylated polybasic membrane anchor of K-Ras and its farnesylated and palmitoylated counterpart of H-Ras localize in disordered and ordered lipid phases of the plasma membrane, respectively (Prior, Muncke, Parton, & Hancock, 2003).

A key question has been whether membrane binding of Ras proteins involves the lipidated C-terminus alone or also the catalytic domain. The first evidence for the involvement of the catalytic domain in membrane binding emerged from MD simulations of G12V H-Ras in a DMPC bilayer (Gorfe et al., 2007). The results suggested that the catalytic domain of GDP-bound H-Ras is largely solvent exposed and rarely interacts with lipids. In the GTP-bound form, the catalytic domain interacts with lipids through R128 and R135 from helix α4 as well as other polar side chains from helix α5. Complemented by subsequent cell signaling and imaging studies (Abankwa et al., 2010), these observations led to the prediction that the switching function of Ras may be modulated by its orientational dynamics with respect to the membrane plane, opening up new potential avenues of developing Ras inhibitors (Gorfe, 2010). In a subsequent study, Prakash, Zhou, Liang, Hancock and Gorfe (2016) used large-scale MD simulations of GTP-bound G12D K-Ras in a 1-palmitoyl-2-oleoyl-sn-glycero-3-phosphocholine/palmitoyl-oleoyl phosphatidylserine (POPC/POPS) bilayer to demonstrate that the catalytic domain samples at least two distinct membrane orientations. In the first orientation (OS1), the effector binding region is exposed while helices α3 and α4 in the C-terminal lobe are engaged in interaction with the membrane. In the second orientation (OS2), the effector binding region is occluded by the membrane since switch 2 and strands β1 − β3 face the membrane. Most recently, it has been shown that G12V K-Ras samples three distinct orientations based on long (20μs) atomistic MD simulations and single-molecule fluorescence resonance energy transfer (smFRET) measurements in native lipid nanodiscs. Similar conclusions were derived from 20μs-long simulations of G12D and Q61H K-Ras (Prakash & Gorfe, 2019). These studies suggested that small energy barriers separate the different orientation states, and that OS1, the signaling-competent orientation state with an accessible effector loop, dominates the overall population (Prakash et al., 2019). The robustness of these observations was driven home by the work of Neale and Garcia (Neale & García, 2020) who carried out extensive (1.45 ms aggregate time) MD simulations on both GTP- and GDP-bound wild-type K-Ras in POPC and

POPC/POPS bilayers. These simulations identified both previously characterized and new orientations and confirmed the role of ionic interactions (Gregory, McLean, & Sligar, 2017; Li & Buck, 2017; McLean, Stephen, & Sligar, 2019) in modulating Ras membrane orientation. However, unlike earlier short MD simulations of G12V H-Ras, these extensive simulations of wild-type K-Ras did not indicate nucleotide-dependent orientation preferences, perhaps suggesting potential differences among Ras isoforms in terms of orientational dynamics.

## 6. Ras membrane binding and druggability

What is the significance of membrane orientation dynamics to inhibiting aberrant Ras signaling? To be effective in the cell, any potential Ras inhibitor would have to be able to bind to the membrane-bound form of the protein. Subtle structural differences between Ras in solution and membrane environments could therefore impact drug discovery efforts. Even from a purely steric consideration, it can be surmised that the effector (and exchange factor) binding region may be occluded by the membrane in some orientations, such as OS2. This may limit the effectiveness of a potential inhibitor that targets a pocket in the occluded region of Ras. Conversely, a ligand that stabilizes the occluded orientation would likely have inhibitory effects; two recent reports provided initial support for this hypothesis (Fang et al., 2018; Jansen et al., 2017).

Many common computational methods cannot be readily used to assess the druggability of membrane-bound targets in general and surface-bound ones in particular. For example, drug fragments such as those used in probe-based MD (pMD) simulations, a widely used MD-based ligand binding site identification technique that has been applied to a variety of soluble proteins, affect the interaction of the target protein with lipids and also partition into the hydrophobic core of lipid membranes (Prakash et al., 2012b, 2015b; Sayyed-Ahmad, Lichtenberger, & Gorfe, 2010). To address this issue, we introduced a technique called pMD-membrane (Prakash et al., 2015b). This technique extended the applicability of probe-based MD to membrane-bound drug targets by preventing probe molecules from partitioning into the membrane core. This was achieved by modifying selected pairwise vdW interactions between the probe and lipid molecules. Application of the technique on membrane-bound G12D K-RAS not only identified the known allosteric ligand binding sites but also suggested variations among the druggable pockets in probe saturation as a result of conformational changes caused by membrane binding. The technique was then used to compare two systems in which K-Ras adopted different orientations with respect to the bilayer plane (OS1 and OS2 orientations). The results indicated that pocket accessibility is modulated by the differential dynamics and orientation. We expanded pMD-membrane to probe mixtures featuring charged, polar, hydrophobic and other chemical functional groups representing moieties commonly found in marketed drugs (Sayyed-Ahmad & Gorfe, 2017). We also introduced a projection technique for determining probe densities on the protein surface, as well as a technique to construct surface topography maps directly from the probe-binding propensity of surface residues. This approach revealed geometric features that helped in filtering out probe-binding hotspots lacking pocket-like characteristics. It also robustly recognized the known allosteric ligand binding sites, and indicated that the accessibility of some pockets is modulated by mutation or differential dynamics, including differential membrane orientation. For example, surface hotspots are found at the flat α3α4 surface in the OS2 orientation but not in the OS1 orientation (Fig. 4). Conversely, probe binding

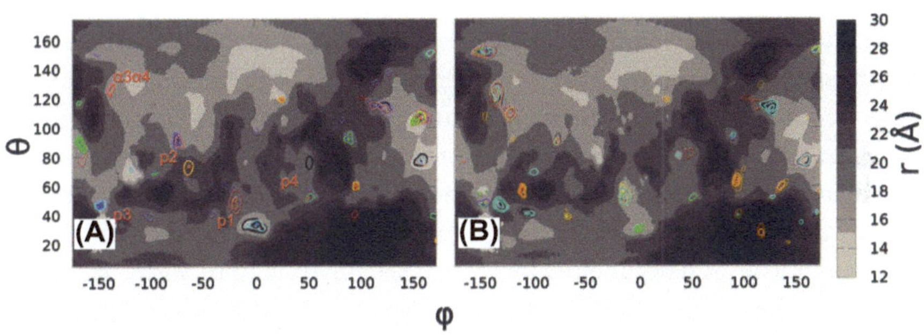


**Fig. 4** Probe densities overlaid on the topography map of G12D K-Ras in orientation state OS1 (left panel) and OS2 (right panel). High probe surface density regions are highlighted in colored contour lines (isobutane, isopropyl alcohol, acetamide, acetate, acetone, DMSO, and urea are shown in black, cyan, blue, orange, red, yellow, and green), and pockets p1-p4 are labeled. The three parameters $r$, $\theta$, $\varphi$ are the conventual spherical coordinate variables used to construct a topological map of the protein surface and characterize the corresponding probe densities. The azimuthal and polar angles $\theta$ and $\varphi$ are used to map locations on the surface of the approximately spherical protein while $r$ characterizes the average distance of a probe atom from the center of mass of the protein. In the gray shaded map, locally lower values of $r$ represent a depression in the surface that can be linked to potential binding sites through probe densities (colored contours). *This figure is reprinted from Sayyed-Ahmad, A., Gorfe, A. A. (2017). Mixed-Probe Simulation and Probe-Derived Surface Topography Map Analysis for Ligand Binding Site Identification.* J. Chem. Theory Comput., 13 *(4), 1851–1861.*

propensity at α4α5 is higher in the OS1 orientation, despite its comparable solvent accessibility in OS2. Further analysis revealed unique chemical signatures of the known allosteric ligand binding sites as determined by the probe-type interaction preferences. As shown in Fig. 4, pocket p1 tends to prefer isobutane, acetamide and acetone in the OS1 orientation, while p3 was found to mainly interact with isobutane, isopropanol and urea in the OS1 orientation.

The observations described above illustrate how membrane binding of Ras and, once on the membrane, its ability to adopt multiple orientations might modulate access (and likely also affinity) of various potential inhibitors to the different allosteric pockets. It is worth noting, however, that in addition to steric effects membrane may modulate Rasligand interactions in at least two other ways: (1) by serving as a reservoir for the ligand (so that Ras reorientation serves as the means by which to tap into the reservoir), and (2) by stabilizing ligands that might simultaneously bind to Ras and membrane.

## 7. Concluding remarks and implications for other challenging drug targets

Strategies that combine a range of computational tools with experiments such as those used to uncover allosteric Ras inhibitors could be extended to develop allosteric modulators of similar other challenging drug targets, especially membrane-bound targets. These could include non-Ras small GTPases, ATPases, some kinases and other classes of proteins whose active site is not amenable for drug design, or in cases in which allosteric inhibition is desirable for improved selectivity and to allow access to a larger chemical space. Specific examples may include Rab proteins, ADP-ribosylation factor (Seixas, Barros, Seabra, & Barral, 2013), cyclin-dependent kinase 2, Heat shock protein 90 (Neckers, 2002) and BCR-Abl tyrosine kinase (Lambert, Duhme-Klair, Morgan, & Ramjee, 2013), as well as some G protein-coupled receptors (Hauser, Attwood, Rask-Andersen, Schiöth, & Gloriam, 2017) such as metabotropic glutamate receptor 5 (Jaeschke, Wettstein, Nordquist, & Spooren, 2008) or β2 adrenergic receptor (Kolb et al., 2009).

The first step toward making an undruggable enzyme druggable through allosteric mechanisms may be to probe the surface of the protein for putative allosteric ligand binding sites. In principle, such an allosteric site could have different physicochemical properties from that of the catalytic site or another orthosteric ligand binding site. Because allosteric transitions inherently entail conformational fluctuations, it is difficult to detect allosteric sites from a static crystallographic structure alone or even from an ensemble of NMR structures. Computational approaches that allow for generating conformational ensembles can help find allosteric sites that may only be visible in transiently sampled conformations. Atomistic MD simulations can be used to map protein conformational states, and in some cases to determine the relative stability of those states and the transition mechanisms among them. MD can also be used to explore long-range allosteric communications and coupled motions that may involve networks of residue contacts linking the active site with distal allosteric sites. The predicted allosteric sites can be characterized based on shape, volume and chemical fingerprint of the pocket. This can be done using geometric methods (Liang, Woodward, & Edelsbrunner, 1998) or with the aid of probe-based MD simulations (Ghanakota & Carlson, 2016; Sayyed-Ahmad, 2018) utilizing chemically diverse probes. Once a potentially druggable allosteric site is identified, a combination of pharmacophore modeling (Sun, 2008) and molecular docking (Lyne, 2002), as well as high throughput virtual screening of chemical libraries, can be used to identify initial hits targeting that site. Typically, compounds with the highest binding scores are then subjected to experimental validation using, for instance, binding affinity measurements with a variety of biophysical techniques or appropriate functional assays. Examples of common biophysical techniques for validation may include isothermal titration calorimetry (Weber & Salemme, 2003), fluorescence resonance energy transfer (Rodems et al., 2002), and nuclear magnetic resonance (Pellecchia, Sem, & Wüthrich, 2002) spin relaxation. These and other biochemical and cell biological techniques may be used to check if a ligand binds to the intended target and has the intended allosteric mode of action.

## Acknowledgments

This work was supported in part by the National Institutes of Health grant R01GM124233 and the Cancer Prevention and Research Institute of Texas (CPRIT) grant RP190366. We thank the Texas Advanced Computing Center (TACC) for computational resources.

## References

Abankwa, D., Gorfe, A. A., Inder, K., & Hancock, J. F. (2010). Ras membrane orientation and nanodomain localization generate isoform diversity. *Proceedings of the National Academy of Sciences, 107*(3), 1130.

Anforth, R., Fernandez-Peñas, P., & Long, G. V. (2013). Cutaneous toxicities of RAF inhibitors. *The Lancet Oncology, 14*(1), e11–e18.

Arencibia, J. M., Pastor-Flores, D., Bauer, A. F., Schulze, J. O., & Biondi, R. M. (2013). AGC protein kinases: From structural mechanism of regulation to allosteric drug development for the treatment of human diseases. *Biochim. Biophys. Acta BBA - Proteins Proteomics, 1834*(7), 1302–1321.

Bos, J. L., Fearon, E. R., Hamilton, S. R., Vries, M. V., van Boom, J. H., van der Eb, A. J., et al. (1987). Prevalence of ras gene mutations in human colorectal cancers. *Nature, 327*(6120), 293–297.

Buhrman, G., Holzapfel, G., Fetics, S., & Mattos, C. (2010). Allosteric modulation of ras positions Q61 for a direct role in catalysis. *Proceedings of the National Academy of Sciences, 107*(11), 4931–4936.

Buhrman, G., Kumar, V. S. S., Cirit, M., Haugh, J. M., & Mattos, C. (2011). Allosteric modulation of ras-GTP is linked to signal transduction through RAF kinase. *Journal of Biological Chemistry, 286*(5), 3323–3331.

Buhrman, G., O'Connor, C., Zerbe, B., Kearney, B. M., Napoleon, R., Kovrigina, E. A., et al. (2011). Analysis of binding site hot spots on the surface of ras GTPase. *Journal of Molecular Biology, 413*(4), 773–789.

Burns, M. C., Howes, J. E., Sun, Q., Little, A. J., Camper, D. V., Abbott, J. R., et al. (2018). High-throughput screening identifies small molecules that bind to the RAS:SOS:RAS complex and perturb RAS signaling. *Analytical Biochemistry, 548*, 44–52.

Burns, M. C., Sun, Q., Daniels, R. N., Camper, D., Kennedy, J. P., Phan, J., et al. (2014). Approach for targeting ras with small molecules that activate SOS-mediated nucleotide exchange. *Proceedings of the National Academy of Sciences, 111*(9), 3401–3406.

Butters, T. D., Dwek, R. A., & Platt, F. M. (2003). Therapeutic applications of imino sugars in lysosomal storage disorders. *Current Topics in Medicinal Chemistry, 3*(5).

Casique-Aguirre, D., Briseño-Díaz, P., García-Gutiérrez, P., la Rosa, C. H. G., QuinteroBarceinas, R. S., Rojo-Domínguez, A., et al. (2018). KRas4B-PDE6δ complex stabilization by small molecules obtained by virtual screening affects ras signaling in pancreatic cancer. *BMC Cancer, 18*(1), 1299.

Cohen, P. (2014). Immune diseases caused by mutations in kinases and components of the ubiquitin system. *Nature Immunology, 15*(6), 521–529.

Cox, A. D., Fesik, S. W., Kimmelman, A. C., Luo, J., & Der, C. J. (2014). Drugging the undruggable RAS: Mission possible? *Nature Reviews Drug Discovery, 13*(11), 828–851.

Das, J., & David Kimball, S. (1995). Thrombin active site inhibitors. *Bioorganic & Medicinal Chemistry, 3*(8), 999–1007.

Davies, H., Bignell, G. R., Cox, C., Stephens, P., Edkins, S., Clegg, S., et al. (2002). Mutations of the BRAF gene in human cancer. *Nature, 417*(6892), 949–954.

Durant, J. L., Leland, B. A., Henry, D. R., & Nourse, J. G. (2002). Reoptimization of MDL keys for use in drug discovery. *Journal of Chemical Information and Computer Sciences, 42*(6), 1273–1280.

Evelyn, C. R., Biesiada, J., Duan, X., Tang, H., Shang, X., Papoian, R., et al. (2015). Combined rational design and a high throughput screening platform for identifying chemical inhibitors of a ras-activating enzyme. *Journal of Biological Chemistry, 290*(20), 12879–12898.

Evelyn, C. R., Duan, X., Biesiada, J., Seibel, W. L., Meller, J., & Zheng, Y. (2014). Rational design of small molecule inhibitors targeting the ras GEF, SOS1. *Chemistry & Biology, 21*(12), 1618–1628.

Fang, Z., Marshall, C. B., Nishikawa, T., Gossert, A. D., Jansen, J. M., Jahnke, W., et al. (2018). Inhibition of K-RAS4B by a unique mechanism of action: Stabilizing MembraneDependent occlusion of the effector-binding site. *Cell Chemical Biology, 25*(11), 1327–1336.e4.

Fan, J.-Q., & Ishii, S. (2007). Active-site-specific chaperone therapy for fabry disease. *FEBS Journal, 274*(19), 4962–4971.

Farr, C. J., Saiki, R. K., Erlich, H. A., McCormick, F., & Marshall, C. J. (1988). Analysis of RAS gene mutations in acute myeloid leukemia by polymerase chain reaction and oligonucleotide probes. *Proceedings of the National Academy of Sciences, 85*(5), 1629.

Fell, J. B., Fischer, J. P., Baer, B. R., Ballard, J., Blake, J. F., Bouhana, K., et al. (2018). Discovery of tetrahydropyridopyrimidines as irreversible covalent inhibitors of KRAS-G12C with in vivo activity. *ACS Medicinal Chemistry Letters, 9*(12), 1230–1234.

Fleming, J. B., Shen, G.-L., Holloway, S. E., Davis, M., & Brekken, R. A. (2005). Molecular consequences of silencing mutant K-ras in pancreatic cancer cells: Justification for KRas–directed therapy. *Molecular Cancer Research, 3*(7), 413.

Ganesan, A., Coote, M. L., & Barakat, K. (2017). Molecular dynamics-driven drug discovery: Leaping forward with confidence. *Drug Discovery Today, 22*(2), 249–269.

Ghanakota, P., & Carlson, H. A. (2016). Driving structure-based drug discovery through cosolvent molecular dynamics. *Journal of Medicinal Chemistry, 59*(23), 10383–10399.

Gorfe, A. A. (2010). Mechanisms of allostery and membrane attachment in ras GTPases: Implications for anti-cancer drug discovery. *Current Medicinal Chemistry, 17*(1), 1–9.

Gorfe, A. A., Hanzal-Bayer, M., Abankwa, D., Hancock, J. F., & McCammon, J. A. (2007). Structure and dynamics of the full-length lipid-modified H-ras protein in a 1,2-dimyristoylglycero3-phosphocholine bilayer. *Journal of Medicinal Chemistry, 50*(4), 674–684.

Góth, L., Rass, P., & Páy, A. (2004). Catalase enzyme mutations and their association with diseases. *Molecular Diagnosis, 8*(3), 141–149.

Grant, B. J., Lukman, S., Hocker, H. J., Sayyah, J., Brown, J. H., McCammon, J. A., et al. (2011). Novel allosteric sites on ras for lead generation. *PLoS One, 6*(10), e25711.

Gregory, M. C., McLean, M. A., & Sligar, S. G. (2017). Interaction of KRas4b with anionic membranes: A special role for PIP2. *Biochemical and Biophysical Research Communications, 487*(2), 351–355.

Gupta, A. K., Wang, X., Pagba, C. V., Prakash, P., Sarkar-Banerjee, S., Putkey, J., et al. (2019). Multi-target, ensemble-based virtual screening yields novel allosteric KRAS inhibitors at high success rate. *Chemical Biology & Drug Design, 94*(2), 1441–1456.

Hauser, A. S., Attwood, M. M., Rask-Andersen, M., Schiöth, H. B., & Gloriam, D. E. (2017). Trends in GPCR drug discovery: New agents, targets and indications. *Nature Reviews Drug Discovery, 16*(12), 829–842.

Hocker, H. J., Cho, K.-J., Chen, C.-Y. K., Rambahal, N., Sagineedu, S. R., Shaari, K., et al. (2013). Andrographolide derivatives inhibit guanine nucleotide exchange and abrogate oncogenic ras function. *Proceedings of the National Academy of Sciences, 110*(25), 10201–10206.

Hodges, T. R., Abbott, J. R., Little, A. J., Sarkar, D., Salovich, J. M., Howes, J. E., et al. (2018). Discovery and structure-based optimization of benzimidazole-derived activators of SOS1-mediated nucleotide exchange on RAS. *Journal of Medicinal Chemistry, 61*(19), 8875–8894.

Hollstein, M., Sidransky, D., Vogelstein, B., & Harris, C. C. (1991). P53 mutations in human cancers. *Science, 253*(5015), 49–53.

Hyde, E. G., & Carmichael, W. W. (1991). Anatoxin-A(S), a naturally occurring organophosphate, is an irreversible active site-directed inhibitor of acetylcholinesterase (EC 3.1.1.7). *Journal of Biochemical Toxicology, 6*(3), 195–201.

Jaeschke, G., Wettstein, J. G., Nordquist, R. E., & Spooren, W. (2008). MGlu5 receptor antagonists and their therapeutic potential. *Expert Opinion on Therapeutic Patents, 18*(2), 123–142.

Jakob, J. A., Bassett, R. L., Jr., Ng, C. S., Curry, J. L., Joseph, R. W., Alvarado, G. C., et al. (2012). NRAS mutation status is an independent prognostic factor in metastatic melanoma. *Cancer, 118*(16), 4014–4023.

Jansen, J. M., Wartchow, C., Jahnke, W., Fong, S., Tsang, T., Pfister, K., et al. (2017). Inhibition of prenylated KRAS in a lipid environment. *PLoS One, 12*(4), e0174706.

Jayakanthan, M., Wadhwa, G., Mohan, T. M., Arul, L., Balasubramanian, P., & Sundar, D. (2009). Computer-aided drug design for cancer-causing H-ras P21 mutant protein. *Letters in Drug Design and Discovery, 6*(1).

Jin, Y., Li, D., Lu, S., Zhao, H., Chen, Z., Hou, W., et al. (2018). Ebselen reversibly inhibits human glutamate dehydrogenase at the catalytic site. *Assay and Drug Development Technologies, 16*(2), 115–122.

Johnson, C. W., Reid, D., Parker, J. A., Salter, S., Knihtila, R., Kuzmic, P., et al. (2017). The small GTPases K-ras, N-ras, and H-ras have distinct biochemical properties determined by allosteric effects. *Journal of Biological Chemistry, 292*(31), 12981–12993.

Johnston, S. R. (2001). Farnesyl transferase inhibitors: A novel targeted therapy for cancer. *The Lancet Oncology, 2*(1), 18–26.

Karachaliou, N., Mayo, C., Costa, C., Magrí, I., Gimenez-Capitan, A., Molina-Vila, M. A., et al. (2013). KRAS mutations in lung cancer. *Clinical Lung Cancer, 14*(3), 205–214.

Karakas, B., Bachman, K. E., & Park, B. H. (2006). Mutation of the PIK3CA oncogene in human cancers. *British Journal of Cancer, 94*(4), 455–459.

Kearney, B. M., Johnson, C. W., Roberts, D. M., Swartz, P., & Mattos, C. (2014). DRoP: A water analysis program identifies ras-GTP-specific pathway of communication between membrane-interacting regions and the active site. *Journal of Molecular Biology, 426*(3), 611–629.

Kessler, D., Gmachl, M., Mantoulidis, A., Martin, L. J., Zoephel, A., Mayer, M., et al. (2019). Drugging an undruggable pocket on KRAS. *Proceedings of the National Academy of Sciences, 116*(32), 15823.

Khaled, M., Gorfe, A., & Sayyed-Ahmad, A. (2019). Conformational and dynamical effects of Tyr32 phosphorylation in K-ras: Molecular dynamics simulation and markov state models analysis. *The Journal of Physical Chemistry B, 123*(36), 7667–7675.

Kloog, Y., & Cox, A. D. (2000). RAS inhibitors: Potential for cancer therapeutics. *Molecular Medicine Today, 6*(10), 398–402.

Kolb, P., Rosenbaum, D. M., Irwin, J. J., Fung, J. J., Kobilka, B. K., & Shoichet, B. K. (2009). Structure-based discovery of β2-adrenergic receptor ligands. *Proceedings of the National Academy of Sciences, 106*(16), 6843–6848.

Kopra, K., van Adrichem, A. J., Salo-Ahen, O. M. H., Peltonen, J., Wennerberg, K., & Härmä, H. (2017). High-throughput dual screening method for ras activities and inhibitors. *Analytical Chemistry, 89*(8), 4508–4516.

Lambert, G. K., Duhme-Klair, A.-K., Morgan, T., & Ramjee, M. K. (2013). The background, discovery and clinical development of BCR-ABL inhibitors. *Drug Discovery Today, 18*(19), 992–1000.

Laude, A. J., & Prior, I. A. (2008). Palmitoylation and localisation of RAS isoforms are modulated by the hypervariable linker domain. *Journal of Cell Science, 121*(4), 421–427.

Lee, K.-Y., Fang, Z., Enomoto, M., Seabrook, G. G., Zheng, L., Koide, S., … Marshall, C. B. (2020). Two distinct structures of membrane-associated homodimers of GTP- and GDPBound KRAS4B revealed by paramagnetic relaxation enhancement. *Angewandte Chemie*. http://dx.doi.org/10.1002/ange.202001758.

Liang, J., Woodward, C., & Edelsbrunner, H. (1998). Anatomy of protein pockets and cavities: Measurement of binding site geometry and implications for ligand design. *Protein Science, 7*(9), 1884–1897.

Li, Z.-L., & Buck, M. (2017). Computational modeling reveals that signaling lipids modulate the orientation of K-Ras4A at the membrane reflecting protein topology. *Structure, 25*(4), 679–689.e2.

Lim, S. M., Westover, K. D., Ficarro, S. B., Harrison, R. A., Choi, H. G., Pacold, M. E., et al. (2014). Therapeutic targeting of oncogenic K-ras by a covalent catalytic site inhibitor. *Angewandte Chemie International Edition, 53*(1), 199–204.

Lito, P., Rosen, N., & Solit, D. B. (2013). Tumor adaptation and resistance to RAF inhibitors. *Nature Medicine, 19*(11), 1401–1409.

Lukman, S., Grant, B. J., Gorfe, A. A., Grant, G. H., & McCammon, J. A. (2010). The distinct conformational dynamics of K-ras and H-ras A59G. *PLoS Computational Biology, 6*(9), e1000922.

Lyne, P. D. (2002). Structure-based virtual screening: An overview. *Drug Discovery Today, 7*(20), 1047–1055.

Manetti, F., & Botta, M. (2003). Small-molecule inhibitors of fibroblast growth factor receptor (FGFR) tyrosine kinases (TK). *Current Pharmaceutical Design, 9*(7), 567–581.

Manousaridis, I., Mavridou, S., Goerdt, S., Leverkus, M., & Utikal, J. (2013). Cutaneous side effects of inhibitors of the RAS/RAF/MEK/ERK signalling pathway and their management. *Journal of the European Academy of Dermatology and Venereology, 27*(1), 11–18.

Maurer, T., Garrenton, L. S., Oh, A., Pitts, K., Anderson, D. J., Skelton, N. J., et al. (2012). SmallMolecule ligands bind to a distinct pocket in ras and inhibit SOS-mediated nucleotide exchange activity. *Proceedings of the National Academy of Sciences, 109*(14), 5299–5304.

McCarthy, M. J., Pagba, C. V., Prakash, P., Naji, A. K., van der Hoeven, D., Liang, H., et al. (2019). Discovery of high-affinity noncovalent allosteric KRAS inhibitors that disrupt effector binding. *ACS Omega, 4*(2), 2921–2930.

McLean, M. A., Stephen, A. G., & Sligar, S. G. (2019). PIP2 influences the conformational dynamics of membrane-bound KRAS4b. *Biochemistry, 58*(33), 3537–3545.

Milburn, M., Tong, L., deVos, A., Brunger, A., Yamaizumi, Z., Nishimura, S., et al. (1990). Molecular switch for signal transduction: Structural differences between active and inactive forms of protooncogenic ras proteins. *Science, 247*(4945), 939.

Mulcahy, H. E., Lyautey, J., Lederrey, C., qi Chen, X., Anker, P., Alstead, E. M., et al. (1998). A prospective study of K-ras mutations in the plasma of pancreatic cancer patients. *Clinical Cancer Research, 4*(2), 271.

Nagasaka, M., Li, Y., Sukari, A., Ou, S.-H. I., Al-Hallak, M. N., & Azmi, A. S. (2020). KRAS G12C game of thrones, which direct KRAS inhibitor will claim the iron throne? *Cancer Treatment Reviews, 84*, 101974.

Neale, C., & García, A. E. (2020). The plasma membrane as a competitive inhibitor and positive allosteric modulator of KRas4B signaling. *Biophysical Journal, 118*(5), 1129–1141.

Neckers, L. (2002). Hsp90 inhibitors as novel cancer chemotherapeutic agents. *Trends in Molecular Medicine, 8*(4), S55–S61.

Niida, A., Sasaki, S., Yonemori, K., Sameshima, T., Yaguchi, M., Asami, T., et al. (2017). Investigation of the structural requirements of K-Ras(G12D) selective inhibitory peptide KRpep-2d using alanine scans and cysteine bridging. *Bioorganic & Medicinal Chemistry Letters, 27*(12), 2757–2761.

Nnadi, C. I., Jenkins, M. L., Gentile, D. R., Bateman, L. A., Zaidman, D., Balius, T. E., et al. (2018). Novel K-ras G12C switch-II covalent binders destabilize ras and accelerate nucleotide exchange. *Journal of Chemical Information and Modeling, 58*(2), 464–471.

Ostrem, J. M., Peters, U., Sos, M. L., Wells, J. A., & Shokat, K. M. (2013). K-Ras(G12C) inhibitors allosterically control GTP affinity and effector interactions. *Nature, 503*(7477), 548–551.

Pellecchia, M., Sem, D. S., & Wüthrich, K. (2002). Nmr in drug discovery. *Nature Reviews Drug Discovery, 1*(3), 211–219.

Pellegata, N. S., Sessa, F., Renault, B., Bonato, M., Leone, B. E., Solcia, E., et al. (1994). K-ras and P53 gene mutations in pancreatic cancer: Ductal and nonductal tumors progress through different genetic lesions. *Cancer Research, 54*(6), 1556.

Prakash, P., & Gorfe, A. A. (2019). Probing the conformational and energy landscapes of KRAS membrane orientation. *The Journal of Physical Chemistry B, 123*(41), 8644–8652.

Prakash, P., Hancock, J. F., & Gorfe, A. A. (2015). Binding hotspots on K-ras: Consensus ligand binding sites and other reactive regions from probe-based molecular dynamics analysis. *Proteins Struct. Funct. Bioinforma., 83*(5), 898–909.

Prakash, P., Litwin, D., Liang, H., Sarkar-Banerjee, S., Dolino, D., Zhou, Y., et al. (2019). Dynamics of membrane-bound G12V-KRAS from simulations and single-molecule FRET in native nanodiscs. *Biophysical Journal, 116*(2), 179–183.

Prakash, P., Sayyed-Ahmad, A., Cho, K.-J., Dolino, D. M., Chen, W., Li, H., et al. (2017). Computational and biochemical characterization of two partially overlapping interfaces and multiple weak-affinity K-ras dimers. *Scientific Reports,* 7, 40109.

Prakash, P., Sayyed-Ahmad, A., & Gorfe, A. A. (2012). The role of conserved waters in conformational transitions of Q61H K-ras. *PLoS Computational Biology, 8*(2), e1002394.

Prakash, P., Sayyed-Ahmad, A., & Gorfe, A. A. (2015). PMD-membrane: A method for ligand binding site identification in membrane-bound proteins. *PLoS Computational Biology, 11*(10), e1004469.

Prakash, P., Sayyed-Ahmad, A., Zhou, Y., Volk, D. E., Gorenstein, D. G., Dial, E., et al. (2012). Aggregation behavior of ibuprofen, cholic acid and dodecylphosphocholine micelles. *Biochimica et Biophysica Acta - Biomembranes, 1818*(12), 3040–3047.

Prakash, P., Zhou, Y., Liang, H., Hancock, J. F., & Gorfe, A. A. (2016). Oncogenic K-ras binds to an anionic membrane in two distinct orientations: A molecular dynamics analysis. *Biophysical Journal, 110*(5), 1125–1138.

Prior, I. A., & Hancock, J. F. (2001). Compartmentalization of ras proteins. *Journal of Cell Science, 114*(9), 1603–1608.

Prior, I. A., Lewis, P. D., & Mattos, C. (2012). A comprehensive survey of ras mutations in cancer. *Cancer Research, 72*(10), 2457.

Prior, I. A., Muncke, C., Parton, R. G., & Hancock, J. F. (2003). Direct visualization of ras proteins in spatially distinct cell surface microdomains. *The Journal of Cell Biology, 160*(2), 165–170.

Quevedo, C. E., Cruz-Migoni, A., Bery, N., Miller, A., Tanaka, T., Petch, D., et al. (2018). Small molecule inhibitors of RAS-effector protein interactions derived using an intracellular antibody fragment. *Nature Communications, 9*(1), 1–12.

Rajalingam, K., Schreck, R., Rapp, U. R., & Albert, Š. (2007). Ras oncogenes and their downstream targets. *Biochimica et Biophysica Acta - Molecular Cell Research, 1773*(8), 1177–1195.

Ravoet, C., Mineur, P., Robin, V., Debusscher, L., Bosly, A., André, M., et al. (2008). Farnesyl transferase inhibitor (lonafarnib) in patients with myelodysplastic syndrome or secondary acute myeloid leukaemia: A phase II study. *Annals of Hematology, 87*(11), 881–885.

Reiss, Y., Goldstein, J. L., Seabra, M. C., Casey, P. J., & Brown, M. S. (1990). Inhibition of purified P21ras farnesyl:protein transferase by cys-AAX tetrapeptides. *Cell, 62*(1), 81–88.

Riely, G. J., Marks, J., & Pao, W. (2009). KRAS mutations in non–small cell lung cancer. *Proceedings of the American Thoracic Society, 6*(2), 201–205.

Rodems, S. M., Hamman, B. D., Lin, C., Zhao, J., Shah, S., Heidary, D., et al. (2002). A FRET-based assay platform for ultra-high density drug screening of protein kinases and phosphatases. *Assay and Drug Development Technologies, 1*(1), 9–19.

Rodenhuis, S., & Slebos, R. J. C. (1992). Clinical significance of ras oncogene activation in human lung cancer. *Cancer Research, 52*(9 Suppl. ment), 2665s.

Romero, D. (2020). Two new agents target KRAS G12C. *Nature Reviews Clinical Oncology, 17*(1), 6–6.

Rosnizeck, I. C., Graf, T., Spoerner, M., Tränkle, J., Filchtinski, D., Herrmann, C., et al. (2010). Stabilizing a weak binding state for effectors in the human ras protein by cyclen complexes. *Angewandte Chemie International Edition, 49*(22), 3830–3833.

Rosnizeck, I. C., Spoerner, M., Harsch, T., Kreitner, S., Filchtinski, D., Herrmann, C., et al. (2012). Metal—Bis(2-Picolyl)Amine complexes as state 1(T) inhibitors of activated ras protein. *Angewandte Chemie International Edition, 51*(42), 10647—10651.

Rudolph, J., & Stokoe, D. (2014). Selective inhibition of mutant ras protein through covalent binding. *Angewandte Chemie International Edition, 53*(15), 3777—3779.

Sakamoto, K., Kamada, Y., Sameshima, T., Yaguchi, M., Niida, A., Sasaki, S., et al. (2017). K-Ras(G12D)-Selective inhibitory peptides generated by random peptide T7 phage display technology. *Biochemical and Biophysical Research Communications, 484*(3), 605—611.

Sarkar-Banerjee, S., Sayyed-Ahmad, A., Prakash, P., Cho, K.-J., Waxham, M. N., Hancock, J. F., et al. (2017). Spatiotemporal analysis of K-ras plasma membrane interactions reveals multiple high order homo-oligomeric complexes. *Journal of the American Chemical Society, 139*(38), 13466—13475.

Sayyed-Ahmad, A. (2018). Hotspot identification on protein surfaces using probe-based MD simulations: Successes and challenges. *Current Topics in Medicinal Chemistry, 18*(27), 2278—2283.

Sayyed-Ahmad, A., Cho, K.-J., Hancock, J. F., & Gorfe, A. A. (2016). Computational equilibrium thermodynamic and kinetic analysis of K-ras dimerization through an effector binding surface suggests limited functional role. *The Journal of Physical Chemistry B, 120*(33), 8547—8556.

Sayyed-Ahmad, A., & Gorfe, A. A. (2017). Mixed-probe simulation and probe-derived surface topography map analysis for ligand binding site identification. *Journal of Chemical Theory and Computation, 13*(4), 1851—1861.

Sayyed-Ahmad, A., Lichtenberger, L. M., & Gorfe, A. A. (2010). Structure and dynamics of cholic acid and Dodecylphosphocholine—Cholic acid aggregates. *Langmuir, 26*(16), 13407—13414.

Sayyed-Ahmad, A., Prakash, P., & Gorfe, A. A. (2017). Distinct dynamics and interaction patterns in H- and K-ras oncogenic P-loop mutants. *Proteins Struct. Funct. Bioinforma., 85*(9), 1618—1632.

Seixas, E., Barros, M., Seabra, M. C., & Barral, D. C. (2013). Rab and arf proteins in genetic diseases. *Traffic, 14*(8), 871—885.

Shima, F., Yoshikawa, Y., Ye, M., Araki, M., Matsumoto, S., Liao, J., et al. (2013). In silico discovery of small-molecule ras inhibitors that display antitumor activity by blocking the ras—effector interaction. *Proceedings of the National Academy of Sciences, 110*(20), 8182—8187.

Spencer-Smith, R., Koide, A., Zhou, Y., Eguchi, R. R., Sha, F., Gajwani, P., et al. (2017). Inhibition of RAS function through targeting an allosteric regulatory site. *Nature Chemical Biology, 13*(1), 62—68.

Spencer-Smith, R., & O'Bryan, J. P. (2019). Direct inhibition of RAS: Quest for the holy grail? *Some chinks ras armor nov. Cancer Therapy Horizons, 54*, 138—148.

Stank, A., Kokh, D. B., Fuller, J. C., & Wade, R. C. (2016). Protein binding pocket dynamics. *Accounts of Chemical Research, 49*(5), 809—815.

Steinberg, G. R., & Kemp, B. E. (2009). AMPK in Health and disease. *Physiological Reviews, 89*(3), 1025—1078.

Stephen, A. G., Esposito, D., Bagni, R. K., & McCormick, F. (2014). Dragging ras back in the ring. *Cancer Cell, 25*(3), 272—281.

Sun, H. (2008). Pharmacophore-based virtual screening. *Current Medicinal Chemistry, 15*(10), 1018—1024.

Sun, Q., Burke, J. P., Phan, J., Burns, M. C., Olejniczak, E. T., Waterson, A. G., et al. (2012). Discovery of small molecules that bind to K-ras and inhibit sos-mediated activation. *Angewandte Chemie International Edition, 51*(25), 6140—6143.

Trinh, T. B., Upadhyaya, P., Qian, Z., & Pei, D. (2016). Discovery of a direct ras inhibitor by screening a combinatorial library of cell-permeable bicyclic peptides. *ACS Combinatorial Science, 18*(1), 75–85.

Upadhyaya, P., Qian, Z., Habir, N. A. A., & Pei, D. (2014). Direct ras inhibitors identified from a structurally rigidified bicyclic peptide library. *Tetrahedron, 70*(42), 7714–7720.

Vigil, D., Cherfils, J., Rossman, K. L., & Der, C. J. (2010). Ras superfamily GEFs and GAPs: Validated and tractable targets for cancer therapy? *Nature Reviews Cancer, 10*(12), 842–857.

Wang, W., Fang, G., & Rudolph, J. (2012). Ras inhibition via direct ras binding—is there a path forward? *Bioorganic & Medicinal Chemistry Letters, 22*(18), 5766–5776.

Weber, P. C., & Salemme, F. R. (2003). Applications of calorimetric methods to drug discovery and the study of protein interactions. *Current Opinion in Structural Biology, 13*(1), 115–121.

Welsch, M. E., Kaplan, A., Chambers, J. M., Stokes, M. E., Bos, P. H., Zask, A., et al. (2017). Multivalent small-molecule pan-RAS inhibitors. *Cell, 168*(5), 878–889.e29.

Westcott, P. M. K., Halliwill, K. D., To, M. D., Rashid, M., Rust, A. G., Keane, T. M., et al. (2015). The mutational landscapes of genetic and chemical models of kras-driven lung cancer. *Nature, 517*(7535), 489–492.

Wittwer, A. J., Hills, R. L., Keith, R. H., Munie, G. E., Arner, E. C., Anglin, C. P., et al. (2007). Substrate-dependent inhibition kinetics of an active site-directed inhibitor of ADAMTS-4 (aggrecanase 1). *Biochemistry, 46*(21), 6393–6401.

Wu, P., Nielsen, T. E., & Clausen, M. H. (2015). FDA-approved small-molecule kinase inhibitors. *Trends in Pharmacological Sciences, 36*(7), 422–439.

Zeng, M., Lu, J., Li, L., Feru, F., Quan, C., Gero, T. W., et al. (2017). Potent and selective covalent quinazoline inhibitors of KRAS G12C. *Cell Chem. Biol., 24*(8), 1005–1016.e3.

Zorn, J. A., & Wells, J. A. (2010). Turning enzymes ON with small molecules. *Nature Chemical Biology, 6*(3), 179–188.

CHAPTER SEVEN

# *In silico* tools to study molecular targets of neglected diseases: inhibition of TcSir2rp3, an epigenetic enzyme of *Trypanosoma cruzi*

**Edgar López-López[a,b], Carolina Barrientos-Salcedo[c], Fernando D. Prieto-Martínez[a] and José L. Medina-Franco[a,*]**

[a]Department of Pharmacy, School of Chemistry, National Autonomous University of Mexico, Mexico City, Mexico

[b]Department of Pharmacology, Center of Research and Advanced Studies of the National Polytechnic Institute (CINVESTAV), Mexico City, Mexico

[c]Medicinal Chemistry Laboratory, School of Bioanalysis, University of Veracruz, Veracruz, Mexico

[*]Corresponding author: E-mails: medinajl@unam.mx; jose.medina.franco@gmail.com

## Contents



*Advances in Protein Chemistry and Structural Biology*, Volume 122
ISSN 1876-1623
https://doi.org/10.1016/bs.apcsb.2020.04.001

## Abstract

There is a growing interest to study and address neglected tropical diseases (NTD). To this end, *in silico* methods can serve as the bridge that connects academy and industry, encouraging the development of future treatments against these diseases. This chapter discusses current challenges in the development of new therapies, available computational methods and successful cases in computer-aided design with particular focus on human trypanosomiasis. Novel targets are also discussed. As a case study, we identify amentoflavone as a potential inhibitor of TcSir2rp3 (sirtuine) from *Trypanosoma cruzi* (20.03 μM) with a workflow that integrates chemoinformatic approaches, molecular modeling, and theoretical affinity calculations, as well as *in vitro* assays.

## 1. Introduction

The term "neglected tropical disease" (NTD) refers to tropical diseases caused by bacteria, protozoa, helminths, and viruses, which normally affect vulnerable populations of low socioeconomic status, without access to effective treatments. For this reason, NTDs are considered important factors in the stability of public health and in the world economy, affecting 2 billion people in 149 countries, generating 150,000 deaths per year, and more than 25 billion dollars in losses due to Disability-Adjusted Life Year (DALY) (WHO, 2020).

For centuries NTD has been considered one of the most important causes of death and disability in developing countries. Moreover, globalization and the poor efforts of the pharmaceutical industry to address these types of diseases have aggravated their globally (Nii-Trebi, 2017). As a reference, recent approvals and drug development, investment around NTD accounts for only 1% of total investment (compared to other types of conditions) (Aerts, Sunyoto, Tediosi, & Sicuri, 2017). Approximately 90% of the investment in research and development (in NTD) comes from public and non-profit organizations (Moran et al., 2009). In this scenario, the rational design using computational resources (*in silico* methods) could allow efficient use of the available economic resources. In addition, this can bring down development times significantly and provide valuable insights into the molecular mechanisms involved.

Currently, there are large multidisciplinary efforts to propose novel strategies for the treatment of NTD. However, other biological aspects that include incubation, transmission, and development cycles of this type of disease make it difficult for the development of efficient treatments. Nowadays, the discovery of alternative therapeutic targets, high-performance assays, and

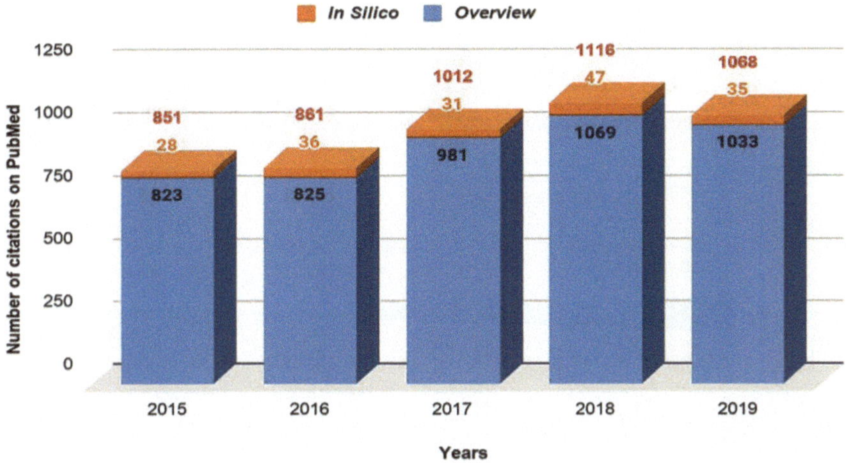


**Fig. 1** Frequency of the keywords "neglected diseases" and "*in silico*" in PubMed. The total results linked to the keyword "neglected disease" are shown in blue. The results associated with the search "neglected disease + *in silico*" per year are shown in orange.

molecular understanding of these diseases opens up new areas of opportunity. Fig. 1 shows the increasing number of citations linked to the keywords "neglected disease" and "*in silico*" on the PubMed platform. It is noted that only 3%—4% of citations per year are linked to computational approaches (mainly in bioinformatics and chemoinformatics areas).

After the implementation of "high-performance techniques" for the synthesis (combinatorial chemistry) and biological evaluation (high-throughput screening) of molecules, the need to screen a large number of candidates arose. Similarly, in the academic setting is often needed to prioritize compounds with the most promising bioactive potential. Computer-aided drug design (CADD) is a discipline that brings together chemical-molecular and quantum approaches that allow the discovery, design, and development of novel chemical entities and the repositioning of approved drugs for a new therapeutic use. CADD has revolutionized the conventional process of drug discovery and development (Fig. 2), significantly shortening the time and costs associated with the development of therapies (Martin, Hutchens, Hawkins, & Radnov, 2017).

In general, the drug discovery process has three main stages (Fig. 2): (1) the discovery phase, in which the possible targets of interest for the subsequent identification of active molecules (*hits*) are identified; (2) the development phase, where hits molecules are evaluated and optimized (*leads*) using *in vivo* and *in vitro* models (this phase includes the preclinical, clinical

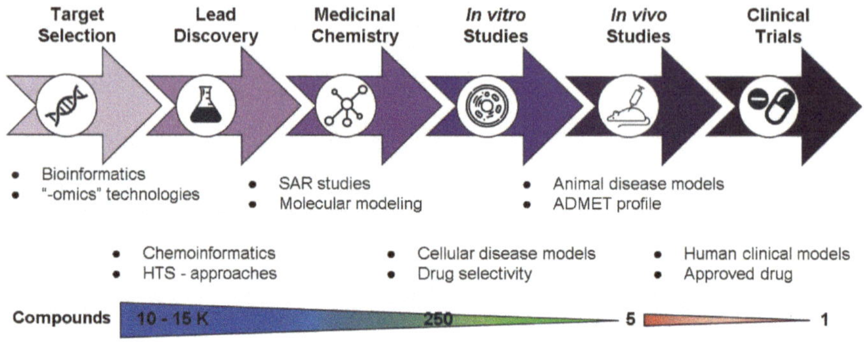


**Fig. 2** Overview of a classic workflow of drug discovery and drug development.

I, II, and III subphases); and (3) the phase that allows the commercialization and study of drugs in a controlled and systematic way in different human populations. It has been estimated that the average preclinical cost in this process is 3.4 million dollars, which increases to 8.6 and 21.4 million dollars, respectively, in clinical phases II and III (Martin et al., 2017).

There are few drugs approved as the first line of treatment for NTD (Fig. 3). However, most of them have several problems such as resistance, low efficacy and selectivity, and multiple side effects due to long dosage periods.

It is noteworthy that most of these approved drugs (Fig. 3) were discovered several decades ago when the molecular and genomic understanding of NTDs was limited. These drugs have a broad range of mechanisms of action (now known) among which are: tubulin inhibition (Albendazole and Triclabendazole), membrane disruption (Praziquantel, Amphotericin B, Miltefosine), free radical toxicity (Benznidazole), DNA chelator (Clofazimine), and others (Swinney & Pollastri, 2019).

Efforts to discover and develop new agents with potential activity against NTD have increased significantly. An attractive strategy is "drug repurposing", which consists of "recycling" or redirecting the initial clinical indication and have completed the drug development process (Fig. 2). These compounds represent a promising alternative to treat NTD (Ferreira & Andricopulo, 2016). Representative successful cases of drug repurposing are outlined in Fig. 3: Moxidectin, approved for the treatment of onchocerciasis (2018) and Fexinidazole, approved for the treatment of human African trypanosomiasis (HAT) (in mid-2019).

In addition to synthetic compounds, natural products are a fundamental part of the development and innovation of new therapeutic agents. Natural

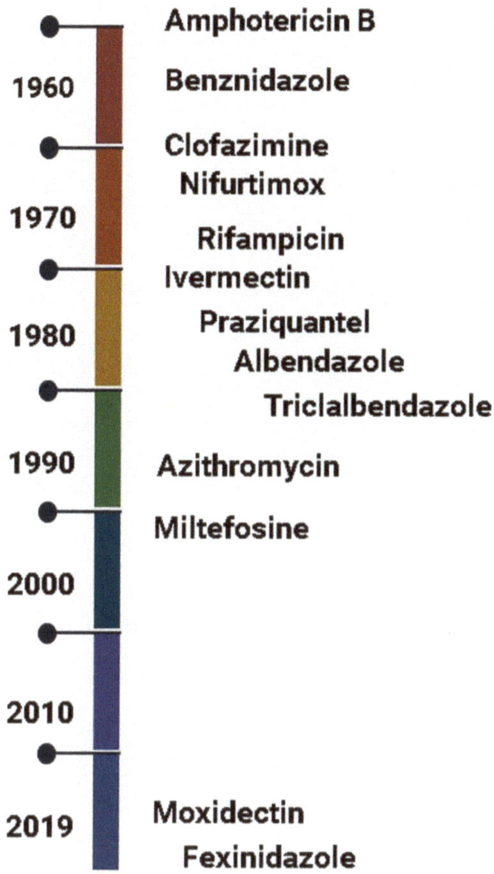


**Fig. 3** Representative drugs used in neglected tropical diseases.

products have had a positive impact in the search of active molecules to treat diseases such as HAT (or sleeping sickness), leishmaniasis, schistosomiasis, and lymphatic filariasis (Cheuka, Mayoka, Mutai, & Chibale, 2016). This shows that natural compounds continue to be a direct or indirect source of new compounds (Cragg & Newman, 2013).

## 2. Overview of current *in silico* approaches used for the study of neglected diseases

There is a broad range of computational tools that allow the systematic, and exhaustive study of the large diversity of information and scientific content available in different databases (related to "big data"), transforming

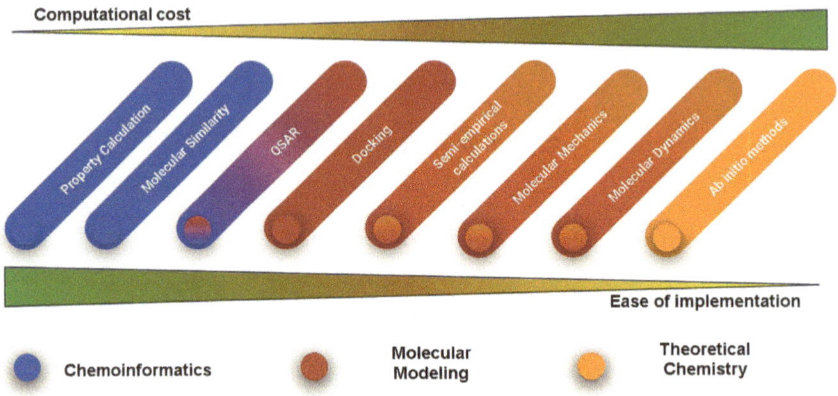


**Fig. 4** Relationship computational cost-ease of implementation of different computational approaches commonly used in drug discovery.

data into knowledge. Computational tools have revolutionized the drug development process *per se* (Fig. 2), and have facilitated the study of complex biological systems, from a molecular perspective to an atomistic one. Fig. 4 schematically summarizes the common computational approaches in CADD.

*In silico* approaches can be classified into three main areas of study (Fig. 4): (1) chemoinformatics, which is a mix of information resources used to transform (chemical) data into information and information into knowledge. These resources are used to make better and faster decisions in lead identification and optimization (Naveja, Oviedo-Osornio, Trujillo-Minero, & Medina-Franco, 2018). The most common chemoinformatic tools are those that allow the calculation of chemical and structural descriptors (analysis of scaffolds and their substituents in chemical compounds), as well as predictions of structure-activity relationships (SAR) (Naveja, Norinder, López-López, & Medina-Franco, 2018). The main advantage of this approach is: efficient handling of large amounts of chemical information and its straightforward implementation for particular case studies. Note, however, that chemoinformatic approaches heavily rely on chemical information availability; (2) Molecular modeling, based on several computational and mathematical techniques for two and three-dimensional descriptions of molecules (Prieto-Martínez, Arciniega, & Medina-Franco, 2018). The ultimate goal of these approaches is to approximate interactions between proteins and ligands in a rigid (molecular docking) or dynamic (molecular dynamics) environment. Among the advantages of molecular modeling approaches is the computational cost-implementation relationship. Usually, such studies are enriched when there is crystallographic information about the targets. If there is not a reference structure, it is possible to build a "chimeric protein" (homology modeling).

One of the disadvantages of methods such as docking and dynamics is their applicability to systems like metalloproteins, protein-protein complexes, and supramolecular structures. This is mainly due to technical limitations that have not been solved so far (Lane, Shukla, Beauchamp, & Pande, 2013; Onawole, Sulaiman, Adegoke, & Kolapo, 2017). The third area of study (Fig. 4) is (3) Theoretical chemistry that uses physical laws to answer chemical questions about the reactivity of chemical compounds, their binding mechanism, molecular structure and spectrometry (Tomasi, 1988). Advantages of theoretical chemistry are the high structural resolution of the generated models, and the atomistic explanation of the phenomena studied. The largest disadvantage is the overall high computational cost and, therefore, the limited number of chemical compounds and biological systems that can be analyzed in a reasonable time.

Hence, in CADD the availability of chemical information is perhaps the fundamental aspect when deciding which tool should be used. In NTD, the available information is in particular limited, since there are very few reported crystallographic structures of the etiological agents of this type of diseases (which limits the approaches based on the structure of the targets). Another limitation often found is the imbalance of information in biological reports; only "active" compounds are usually reported, and not the inactive molecules (limiting ligand-based approaches). This "slant" of information technically limits the study of this type of disease, where each laboratory or pharmaceutical company must generate its data, which generates a "vicious cycle" of low scientific production around these problems. Despite these limitations, there are successful cases where computational methods have contributed significantly to the development of new therapeutic agents (Table 1) (Belllera, Sbaraglini, Alberca, Alice, & Talevi, 2019).

Given the complexity of the systems to be studied in NTDs, an integral study of the molecular phenomena involved in the therapeutic effect generated by each molecule results in both difficult and expensive. Therefore, structure-phenotype relationship (SPhR) studies are increasingly pursued (Swinney & Pollastri, 2019). Phenotypic assays refer to the analysis of the morphology of a biological system (cell culture or animal model) from a reductionist perspective, with the idea of identifying physiological and morphological patterns that predict a possible therapeutic effect of molecule studied, for example, a decrease in proliferation or cell death (Aulner, Danckaert, Ihm, Shum, & Shorte, 2019).

**Table 1** Representative examples of drug repositioning for NTD by *in silico* methods.

| Computational method | NTD studied | Approach | Reference |
|---|---|---|---|
| Comparative homology modeling | Filarial nematodes | Molecular modeling | Galvin et al. (2014) |
| SAR | Trypanosomiasis | Chemoinformatics | Blaazer et al. (2018) |
| Homology modeling | Schistosomes | Molecular modeling | Weeks et al. (2018) |
| Virtual Screening | Leishmaniasis | Molecular modeling and chemoinformatics | Alberca et al. (2016) |
| Molecular docking and Dynamics | Dengue virus | Molecular modeling | Botta, Rivara, Zuliani and Radi (2018) |
| Target-based chemogenic screening | Schistosomes | Phenotypic studies and chemoinformatics | Neves et al. (2016) |
| Network analysis | Trypanosomiasis | – | Berenstein, Magariños, Chernomoretz and Agüero (2016) |

Adapted of Bellera L.

## 3. Technological trends applied to CADD

The drug discovery landscape is constantly changing, with new trends and developments being introduced each year. There are little explored technologies for NTD that have proven widely useful in CADD. Table 2 summarizes emerging technologies that the authors believe should be incorporated into future studies (Prieto-Martínez, López-López, Eurídice Juárez-Mercado, & Medina-Franco, 2019).

### 3.1 Network analysis

Network analysis is one of the most promising tools in CADD. It allows the study of large amounts of data from different perspectives (available for free in public databases), and these approaches are currently optimized through the implementation of learning algorithms (machine learning, deep learning,

**Table 2** Selected trending technologies in Computer-Aided Drug Design.

| Technology | Description | Reference |
|---|---|---|
| Machine learning | Algorithms and models that may be self fed, toward optimization and improvement of data. | Khamis, Gomaa and Ahmed (2015) |
| Deep learning | Implementation of machine learning based on artificial neural networks with nonlinear processing units. | Zhang, Tan, Han and Zhu (2017) |
| Artificial intelligence | Refers to an algorithm or machine, capable of mimicking cognitive functions without supervision or user input. | Duch, Swaminathan and Meller (2007) |

and artificial intelligence) (Norinder, Naveja, López-López, Mucs, & Medina-Franco, 2019). A "network" is a graph, which represents the relationship between its nodes. Each node could represent various biological and chemical parameters of the compounds evaluated, the targets studied, or the post-treatment morphological effects (Vidal, Cusick, & Barabási, 2011). However, the nature and quality of the data with which the models are trained on impacts significantly on their predictions.

Fig. 5 shows an example of exploratory analysis by networks of active compounds against the target cruzipain present in the parasite trypanosoma (*Trypanosoma cruzi* and *Trypanosoma brucei*) etiologic agent of NTD (Chagas disease and HAT, respectively). 82 representative compounds with reported

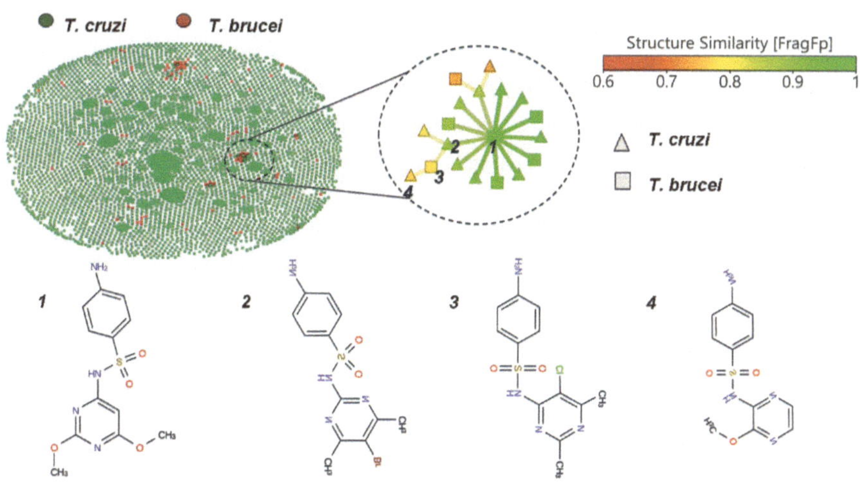


**Fig. 5** Analysis of molecular similarity by networks of representative bioactive molecules against *Trypanosoma cruzi* and *Trypanosoma brucei*. Compounds: 1 (CHEMBL2107253); 2 (CHEMBL2111078); 3 (CHEMBL2107588); 4 (CHEMBL1525826).

activity for *T. brucei*, and 3966 bioactive compounds for *T. cruzi*, obtained from ChEMBL database (https://www.ebi.ac.uk/chembl/) were analyzed. A similarity network based on the fragmentation algorithm (FragP) implemented in the free DataWarrior software was generated (López-López, Naveja, & Medina-Franco, 2019). Of the entire network, a single cluster of compounds tested with both *T. cruzi* and *T. brucei* was analyzed.

One of the first uses of the networks is the inference of the possible target: the compounds were evaluated in different tests, for example, compounds **1**, **2**, and **4** (Fig. 5) were evaluated directly in the target cruzipain. In contrast, compound **3** was evaluated in a phenotypic assay (cell culture). The network in Fig. 5 leads to the hypothesis that compound **3** is active against cruzipain. A second example of the use of the network is the prediction of power (bioactivity): in the database, the bioactivity values reported in different units ($IC_{50}$, $K_i$, and potency). Since the compounds come from different tests, the extrapolation of results is not correct, because each test is performed under different biological conditions. However, the network allows to "approximate" SAR: for instance, compound **1** has a reported potency of 6.3 μM. Then, following the similarity principle, the other compounds are expected to have a similar potency (assuming there are no activity cliffs) (Medina-Franco, 2013). A third use of the networks is the identification of possible repositioning, for example, compounds **3** and **4** (Fig. 5), were also tested in *Mycobacterium tuberculosis* (phenotypic assay) having important inhibitory results.

## 3.2 Polypharmacology and NTD

A classical paradigm in drug discovery is optimization, iterative SAR cycles which result in highly selective compounds. Thus, the main goal is single target recognition (one drug, one target, one disease approach) to reduce toxicity or off-target effects. However, combined treatments have recently been proposed, where more than one molecule is administered to inhibit more than one target. In this sense, why not look for a molecule with the potential to inhibit two or more targets? This train of thought gives rise to polypharmacology (Chaudhari, Tan, & Zhang, 2017).

For example, citing a case of NTD, malaria is treated with a combination of amodiaquine (HMNT inhibitor) and artesunate (EXP1 inhibitor) (Mutabingwa et al., 2005). The objective would be identifying if the different molecular targets have common binding cavities (models of comparative homology or comparative crystallographic structures) or if the drugs share chemical characteristics (pharmacophore models or structural similarity analysis).

In the near future, such polypharmacological approach will have greater importance. However, there are still significant challenges, such as the optimization for dual activity in a set of molecules (López-López, Prieto-Martínez, & Medina-Franco, 2018). As well as the analysis of structure-multiple activity relationships (SmART), to preserve the selectivity of compounds by their targets without generating side effects (Saldívar-González, Naveja, Palomino-Hernández, & Medina-Franco, 2017) (Medina-Franco, Naveja, & López-López, 2020).

In the case of NTDs, polypharmacological approaches are particularly interesting for several reasons: (1) The development of inhibitors of different targets for the same etiologic agent would tentatively decrease the cases of resistance generation; (2) It would be possible to inhibit targets that are expressed in different stages of the etiologic agent during the disease (depending on the life cycle of each agent, these may pass through extra- and intra-cellular stages, which limits the classical approach to drug development); (3) It would enhance drug repurposing, directly collaborating in the identification of new therapeutic targets and possible treatments for existing ones (Chaudhari, Tan, Huang et al., 2017).

## 4. Novel targets and treatments for NTD

As discussed thus far, computational methods not only facilitate the optimization of resources and time. CADD also offers the identification and prediction of potential new targets. The basic science around the understanding of the molecular aspects linked to infection, life cycle, and survival of the etiological agents of NTDs, has allowed a gradual advance in the identification of new targets, which is essential to reconsider the pharmacological paradigms. That is, to propose new and better possible solutions to old problems. In this section, examples of novel targets identified for NTD care are discussed.

### 4.1 Leishmaniasis

A broad range of compounds with therapeutic potential have has recently studied. Some examples are nitroaromatic compounds and oxaborols. At the same time, joint efforts between academia and industry have made large progress in repositioning with high-performance techniques of aminopyrazoles and pyrazolopyrimidines (Ortalli et al., 2019; Reguera et al., 2019).

From the phenotypic perspective, quinolines have been studied extensively around the inhibition of multiple biological systems, however, within current approaches, designs oriented to the generation of mitochondrial alteration and bioenergetic collapse in parasites stand out. This class of compounds has proved interesting since they are compounds with high selectivity on parasitic targets, low toxicity of uninfected cells and micromolar activity in different parasitic stages (Staderini et al., 2019).

Novel studies for identifying new therapeutic alternatives in protozoan parasites (such as *Leishmania spp* and *Trypanosoma spp*) are exploring the potential of molecules that generate alterations in proteasomes (Winzeler & Ottilie, 2019).

In addition, the identification of a novel target involved in the transport of the heme group (essential for the survival of *L. spp*) has recently been identified (LmFLVCRb), opening the possibility of a new approach in target-oriented design (Cabello-Donayre et al., 2020).

## 4.2 Trypanosomiasis

There is a large list of inhibitors reported against trypanosomes (*T. cruzi* and *T brucei*). The most studied have effects on mitochondrial, glycosomal and oxidative metabolism. For example, tripanotation reductase inhibitors and inducers of reactive oxygen species (ROS). One of the most interesting targets explored by multidisciplinary groups in academia and industry is C14$\alpha$-sterol demethylase (CYP51), an approach that has led to studies in the clinical phase. Other types of inhibitors seek to attack the integrity of the parasitic membrane, by inhibiting pathways involved in the synthesis of sterols or squalene (cholesterol precursor) and protein traffic. Within the large collection of molecular targets, one of the most studied is cruzain (a cysteine protease), involved in parasitic infection and therefore in the cycle of natural infection (André Vannier-Santos et al., 2019; Singh et al., 2016).

Less studied targets but with great potential have been recently described. Examples are targets present in the nucleus and kinetoplast of the parasite, also known as chromatin remodelers or epigenetic targets (*vide infra*) (André Vannier-Santos et al., 2019).

## 4.3 Malaria

Phenotypic assays have demonstrated the activity of quinine on the *Plasmodium falciparum* parasite. However, until recently the possible molecular mechanism linked in its action was unknown. *In silico* studies allowed linking the biological effects with the inhibition of the falcipain-II target, which opens up novel possibilities for repositioning, design, and optimization of known inhibitors. The same case occurs with the *P. falciparum* orthologue calcium ATPase (PfATP6), as well as with other targets linked to parasitic membrane homeostasis such as PfNCR1, with alterations of the CTP biosynthetic pathway (studied in apicomplexan and kinetoplastids parasites) and with bioenergetic abnormalities in the parasite (Bhatnagar, Nicklas, Morrisey, Goldberg, & Vaidya, 2019; Kaiser, 2019; Nunes et al., 2019; Wang et al., 2019).

## 4.4 Tuberculosis

Investigation in tuberculosis has led identify a large diversity of promising molecular targets. Among the recently discovered targets are transcriptional regulatory types such as EthR and the enzymes Guab2 (an inosine monophosphate dehydrogenase) and PyrG (CTP-Synthase) (Yamurai Bishi, Sundeep Chaitanya Vedithi, Blundell, & Mugumbate, 2019).

In addition, computational approaches have identified potential new inhibitors against proteasome, topoisomerase I and signaling pathways generated by serine-threonine kinases (pknB) of *M. tuberculosis*. (Brindha et al., 2017; Ekins et al., 2017; Lone, Kumar, Athar, & Jha, 2018).

## 4.5 Filariasis

Recently, a series of analogs derived from tylosin scaffold, a repositioned drug for veterinary use, currently in clinical phase I. These compounds have shown an improved activity and pharmacokinetic profile with respect to the inhibitors reported to date (von Geldern et al., 2019).

Inhibitors of FtSZ (a protein involved in binary fission reproduction of *Wolbachia spp*) have recently been identified such inhibitors are metabolites of albenzadole (Serbus, Landman, Lindley, White, & Sullivan, 2018).

## 4.6 Epigenetic targets for NTD

Over the past few years, the trends to identify new "druggable" targets have evolved. With no hesitation, the new "-omic" technologies have changed the way in which new paths for the rational design of drugs are identified. A clear example is the epigenetic targets (López-López, Rabal, Oyarzabal, & Medina-Franco, 2020). An epigenetic target can be defined as the one that modifies the chromatin compaction without altering the sequence of the genetic material (DNA). Among the best-known targets are DNA methyltransferases (DNMTs), histone deacetyl/acetyl transferases (HDACs/HATs), chromatin remodeling factors (e.g. bromodomains), and non-coding RNAs (Dueñas-González, Naveja, & Medina-Franco, 2016).

**Table 3** Representative druggable epigenetic targets of NTD.

| NTD and etiological agent | Epigenetic target class | Targets | Reference |
|---|---|---|---|
| *T. brucei* (Tripanosomiasis Humana Africana) | HDAC | DAC1, DAC3 | Ingram and Horn (2002) |
| | HAT | HAT1, HAT2 | Kawahara et al. (2008) |
| | HMT | DOT1B | Duraisingh and Horn (2016) |
| | Bromodomain | TbBDF2, TbBDF3 | Schulz et al. (2015) |
| *T. cruzi* (Tripanosomiasis Americana/Chagas disease) | Sirtuin | TcSIR2RP1, TcSIR2RP3 | Moretti et al. (2015) |
| *L. infantum* (Leishmaniasis) | Sirtuin | LiSIR2RP1 | Tavares et al. (2008) |
| *P. falciparum* (Malaria) | HDAC | PfHDAC1, PfHDAC2, PFHDAC3 | Coleman et al. (2014) |
| *S. mansoni* (Schitossomiasis) | HDAC | SmHDACC8 | Marek et al. (2013) |
| | HAT | SmGCN5, SmCBP | Carneiro et al. (2014) |
| *M. tuberculosis* (Tuberculosis) | HMT | Rv1988 | Khosla, Sharma and Yaseen (2016) |

Adapted to Pierce and Khalife (2019).

The complex cycles of infection, survival, and virulence of the etiological agents of NTDs are the product of changes in genetic expression, and therefore in the compaction and decompaction of their chromatin. In this sense, it is reasonable to think that the inhibition of the enzymes involved in these functions causes the death of infectious agents, opening the way to highly selective therapeutic approaches (Pierce & Khalife, 2019) (Table 3).

## 5. Case study: *In silico* study of epigenetic target on *Trypanosoma cruzi*

In this section, we discuss a workflow (Fig. 6) to study a novel target against the NTD, Chagas. As illustrated in the workflow, the integration of different computational tools is recommended to generate more efficient "screening" with pharmacological (Lipinski rules), biological, and chemical criteria that is expected to select the best candidates for biological evaluation.

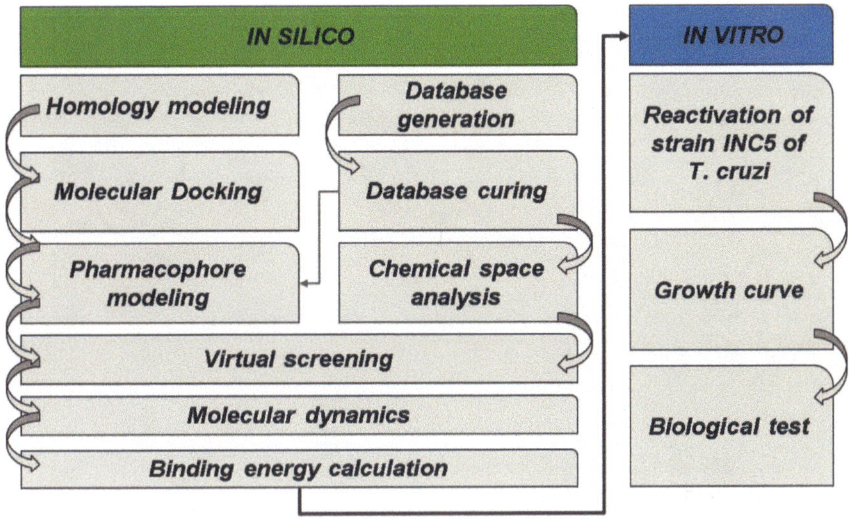


**Fig. 6** Proposed workflow for the study of the TcSir2rp3 of *Trypanosoma cruzi*.

The model of the sirtuin TcSir2rp3 (UniProtKB/Swiss-Prot: Q4DP02) was generated using YASARA software v.18.8.9 and as principal templates in NAD-binding state (reported in Protein Data Bank - PDB) (PDB IDs: 3ZGV, 4I5I). (Krieger & Vriend, 2014) The missing loops were completed by *ab initio* calculations and subsequently optimized (Canutescu, Shelenkov, & Dunbrack, 2003; Michalsky, Goede, & Preissner, 2003). The models generated were qualified and validated with PROCHECK based on the parameters: Root-mean-square deviation (RMSD), percentage of identity with the template, alignment coverage, and normality of the dihedral angles. For all tools, the default parameters were used (Laskowski, MacArthur, Moss, & Thornton, 1993).

Molecular docking was performed using the homology model. Gasteiger charges were assigned to the protein and ligands, the grid was centered in the NAD-binding site with a size of 15 $Å^3$. The compounds were subjected (*vide infra*) to 50 search steps with RMSD clusters of 5 Å. Additional parameters were set by default. The best binding poses of the different clusters were identified (Trott & Olson, 2010). Docked ligands (previously minimized with MMFF94x force field) were NAD (natural co-substrate), nicotinamide (NAD-dependent inhibitor), and SRT1720 (NAD - not dependent inhibitor) (Nguyen, Schaefer, Gertz, Weyand, & Steegborn, 2013; Schuetz et al., 2007) (Cho, Guallar, Berne, & Friesner, 2005). Fig. 7 shows for each ligand the key interactions with the amino acids in the binding site of TcSir2rp3.

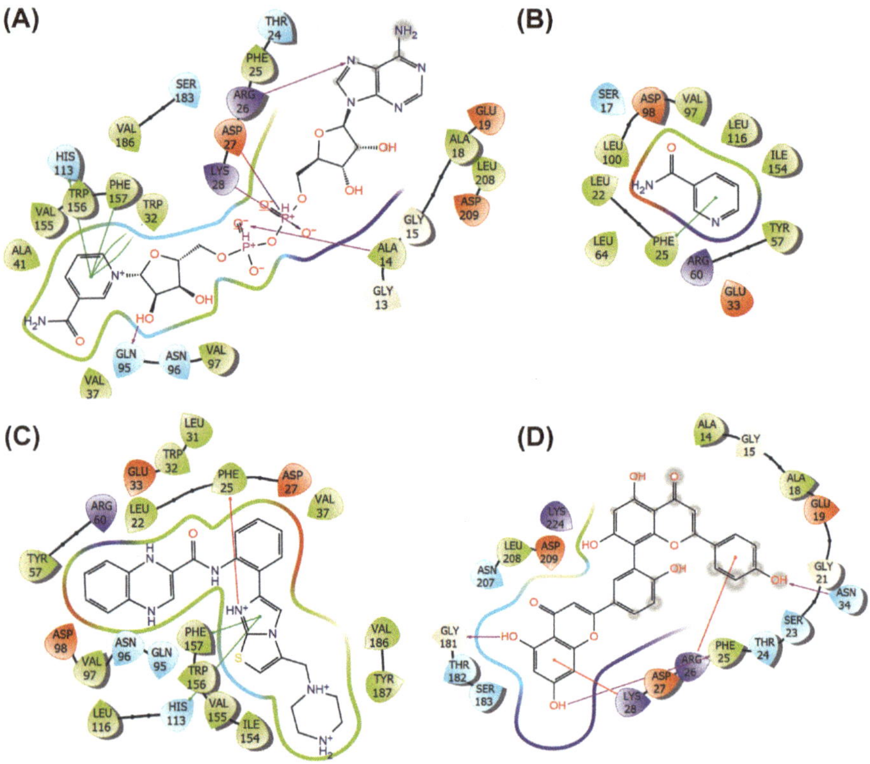


**Fig. 7** The binding pose of inhibitors reported and amentoflavone against TcSir2rp3. (A) NAD, (B) Nicotinamide, (C) SRT1720, and (D) Amentoflavone.

In parallel, using the Swiss Similarity platform (www.swisssimilaruty.ch/), and using the endogenous agonist of the sirtuin TcSir2rp3 (NAD) and the ligands reported with activity against this target as a structural reference (Fig. 7), 10,432 compounds were identified with high structural similarity (>80%). For the similarity studies, MACCS keys fingerprint (166-bits) and the Tanimoto coefficient were used as a metric. This number of compounds were screened with the generated pharmacophore model (with the molecular docking poses shown in Fig. 7) and with a molecular similarity analysis (chemical space) based on the network algorithm (Frag-P), identifying five candidate molecules with potential of enzymatic inhibition of TcSir2rp3. The five molecules selected where further evaluated using molecular docking: DB03168 (+25.0), 94186 - Fluka (-4.44), 66042 - Fluka (-4.09), DB03613 (+2.56), and amentoflavone (-8.11) (López-López et al., 2019). The best candidate (amentoflavone) was selected for additional testing (Fig. 8A).

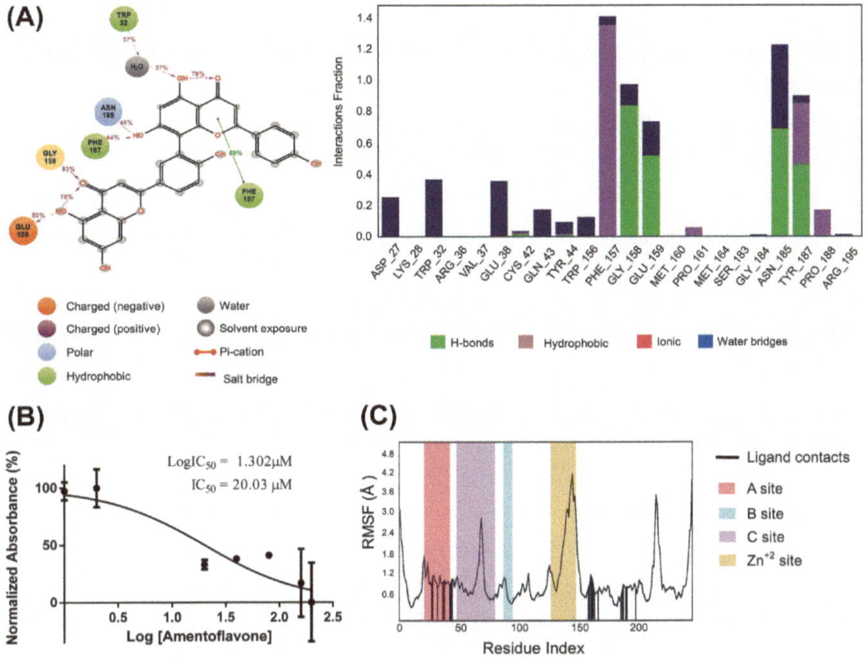


**Fig. 8** Overview of TcSir2rp3 analysis. (A) TcSir2rp3—Amentoflavone contact analysis during the 30 ns simulations. The diagrams show the fraction and type of interactions with representative ligands during the simulation; (B) Amentoflavone activity in INC5 strain of *Trypanosoma cruzi*; (C) Sites of catalytic and structural importance in sirtuin TcSir2rp3. Results of molecular dynamics of sirtuine against amentoflavone (ligand) corresponding to 30 ns.

The most representative binding mode on the molecular docking for amentoflavone was used as a starting point for molecular dynamics (MD) simulations. MD studies were performed with Desmond software (Schrödinger) with the OPLS 2005 force field (Bowers et al., 2006). The complexes were prepared with the System Builder Utility in an orthomobic box (10 × 10 × 10 Å; buffering the protein), using the transferable intermolecular potential with a 3-point model for water (TIP3P). The complexes were neutralized and NaCl was added in a 0.15 M concentration.

Complexes were minimized using the steep-descent (SD) algorithm followed by the low memory Broyden-Fletcher-Goldfarb-Shanno (L-BFGS) optimization in three stages. In the first stage water-heavy atoms were restrained, with a force constant of 1000 kcal $mol^{-1}$ $Å^{-2}$ for 5000 steps (1000 SD, 4000 L-BFGS) with a convergence criterion of 50 kcal $mol^{-1}$ $Å^{-2}$; Following a second stage, the protein backbone was constrained with a

10 kcal $mol^{-1}$ Å$^{-2}$ force constant, using a convergence criterion of 10 kcal $mol^{-1}$ Å$^{-2}$ for 2000 steps (1000 SD, 1000 L-BFGS); Finally, for the third stage, the systems were minimized with no restraints for 1000 steps (750 SD, 250 L-BFGS) with a convergence criterion of 1 kcal $mol^{-1}$ Å$^{-2}$.

Equilibration was carried out in several steps, using Desmond's default protocol, reported elsewhere (Ozen, Sherman, & Schiffer, 2013). Subsequently, the system was submitted to 30 ns of production runs, under NPT ensemble at 1 bar using the Martyna-Tuckerman-Klein (MTK) barostat and 310 K using the Nose-Hoover thermostat. Electrostatic forces were calculated with the smooth PME method using a 9 Å cut-off, while constraints were enforced with the M-SHAKE algorithm (Kräutler, van Gunsteren, & Hünenberger, 2001). Integration was done every 1.2 fs, with a recording interval of 50 ps. The quality of the simulation and trajectory analyses were carried out with the tools implemented in the Maestro-GUI tool (Schrödinger).

Fig. 8A describes the interactions established between amentoflavone and TcSir2rp3, highlighting the well-conserved interactions with PHE 157, GLY 158, GLU 159, ASN 185 and TYR 187. The estimation of binding energy (MM/PBSA method) was -52.35 ($\pm$4.2) kcal/mol. Similar interactions have been described for bichalcones and human sirtuins, which showed that the main component of binding is hydrophobic (Karaman et al., 2018). This may explain the micromolar potency found for polyphenolic compounds and sirtuins (Rahnasto-Rilla et al., 2018).

*In vitro* evaluation was carried out according to the protocol described Sülsen et al., 2007, with minor modifications. The concentration of parasites was quantified by Neubauer method for 20 consecutive days, identifying the number of parasites per milliliter of medium (per triplicate), allowing the generation of a standard growth curve (Sülsen et al., 2007).

In a sterile 96-well plate, 10,000 parasites (per well) was deposited adjusted to a volume of 100 μL, maintaining negative control (Phosphate Buffered Saline or PBS), positive controls (nifurtimox and benznidazole, $IC_{50}$ 0.31 mM and 0.69 mM, respectively), technician control (parasites without treatment) and amentoflavone (with different concentrations: 2, 20, 40, 80, 160 and 200 μM) (Díaz-Chiguer et al., 2012). Final concentration of dimethylsulfoxide (DMSO) in the medium was set below 1%. Following the standard protocol of the MTT test (Thermo Fisher Scientific), it was incubated for 4 h with the MTT reagent, isopropanol was used as a detergent solution and the absorbance was read at 570 nm. Which allowed identifying the parasitic viability, according to the production of formazan crystals (Díaz-Chiguer et al., 2012). Trypanocidal effect was determined by comparing the remaining trypomastigotes in each concentration with the negative control group. Each assay was performed in triplicate. The $IC_{50}$ (Fig. 8B) of amentoflavone was determined (20.03 μM).

The TcSir2 family of sirtuins is a set of emerging epigenetic targets in the discovery of anti-*T. cruzi* drugs (Religa & Waters, 2012). Particularly, the inhibition of TcSir2rp3 is associated with death due to mitochondrial alterations in the parasite. Its homologous protein in humans differs widely in the catalytic site, therefore, unlike other targets, it offers a novel venue for the development of selective ligands (Soares et al., 2012). However, due to the limited structural information, as well as the limited list of selective compounds against this target, a comprehensive study is not possible yet.

Biflavonoids have been broadly described as bioactive agents, with several reports of antiprotozoal activity (Tasdemir et al., 2006). In particular, amentoflavone has been evaluated against epigenetically relevant targets, showing positive results, which postulates it as a candidate for optimization processes (Tasdemir et al., 2006). Additionally, amentoflavone shows other advantages when compared to other antiprotozoal compounds. For example, its ability to act in privileged sites, e.g., it's capacity to cross the blood-brain membrane, which would prove useful during the chronic phase of Chagas disease (Gutmann et al., 2002).

From the morphological context, some reports link certain post-treatment morphology characteristics with the type of cell death (*T. cruzi)* (Jiménez-Ruiz et al., 2010; Mosquillo et al., 2018). Biflavonoids (e.g. amentoflavone) have shown an affinity for the SIRT2 family of proteins, coupled with the fact that post-treatment morphology (INC5 strain) suggests mitochondrial abnormalities, we suggest an interaction with the TcSir2rp3 (a mitochondrial sirtuin of the SIRT2 family) of *T. cruzi*.

MD studies of different sirtuins have allowed describing their three-dimensional structures (Sakkiah, Arooj, Cao, & Lee, 2013). In general, they have three sites of conformational importance: A, B, and C (binding to NAD adenine, NAD ribose, and acetyl-lysine, respectively). The reported inhibitors are capable of allosterically modulating the different sites of conformational importance in sirtuins (Sanders, Jackson, & Marmorstein, 2010). As illustrated in Fig. 8C, the herein MD with amentoflavone and TcSir2rp3 suggested conformational variations at the C site, in addition to the binding site to their cofactor $Zn^{+2}$.

## 6. Challenges and emerging problems

It is well-known the limited interest of private initiatives and large pharmaceutical companies to investigate neglected diseases (Nwaka & Ridley, 2003). In this scenario, the role of the academic community is to contribute to providing shorter and more efficient ways that can be optimized by pharmaceutical companies. Research in countries with endemic NTD plays a key role in reducing time in the generation of basic research and in the promotion of therapeutic alternatives, as well as in the establishment of policies that allow the direct treatment of these conditions. One of the most important challenges but at the same time opportunities is increasing the multidisciplinary efforts among groups that communicate to the academy, industry, and government allowing the necessary studies and financing to combat NTD. Otherwise, and as until now, the "globalization" factor will contribute to the dissemination of this type of disease, increasing the rates of infection, death, resistance to treatments, and disability associated with these diseases.

From a chemical perspective, there are repeated structural scaffolds in molecules with bioactivity against NTD (e.g., privileged structures) (Leite, Espíndola, de Oliveira Cardoso, & de Oliveira Filho, 2019). These structural and molecular patterns have been little explored, although we now have the technology to generate more efficient structure-activity-phenotype approaches, as well as the prediction and proposal of shared targets among these diseases. These privileged structures could be an excellent starting point for designing new therapeutic agents, where synthetic feasibility could be evaluated prior to biological analysis.

As discussed throughout this chapter, drug repositioning is one of the most promising alternatives in NTD, being economic and more versatile approaches (Farha & Brown, 2019). However, obtaining false positives is very common, as evidently complex biological systems are very difficult to model.

Another current challenge is the generation of models to predict ADMETox (Absorption, Distribution, Metabolism, Excretion, and Toxicology) properties.

We also consider that reporting "negative data" (e.g., inactive compounds) will have a positive impact on the training and generation of predictive models.

## 7. Conclusions

This chapter discusses the different *in silico* tools available to study NTD, successful cases, problems, emerging challenges, as well as the possible new targets to explore. However, there will be no significant progress in the development of more effective, powerful and safe therapeutic alternatives to the point where "multidisciplinary" approaches are generated aimed exclusively at combating these problems, where the academic and private sectors manage to merge properly.

As discussed throughout this chapter, phenotypic approaches have proven very useful in detecting new possible treatments for NTD. These approaches to being "reductionists" do not allow to demonstrate the effect of drugs on a particular target, at the molecular level which limits the information available to computational chemists to develop predictive models. From now on, it will be considered necessary to complement phenotypic tests with more directed inhibition tests that allow to better hypothesize possible targets.

The key factor to adequately treat NTDs is the generation of more efficient administrative and research paths supported by the academic, industrial and government sectors. To this end, a joint development of efficient public health policies to combat NTD is a major first step.

## Acknowledgements

F.D.P.-M thanks the PhD scholarship grant from CONACyT No. 660465/576637. We also thank the Programa de Nuevas Alternativas de Tratamiento para Enfermedades Infecciosas (NUATEI-IIB-UNAM).

### Formatting of funding sources

This work was funded by the School of Chemistry of the *Universidad Nacional Autónoma de México* (UNAM), the Programa de Apoyo a Proyectos de Investigación e Innovación Tecnológica (PAPIIT) grant number IA203718, UNAM and the Consejo Nacional de Ciencia y Tecnología (CONACyT) grant number 282785.

## References

Aerts, C., Sunyoto, T., Tediosi, F., & Sicuri, E. (2017). Are public-private partnerships the solution to tackle neglected tropical diseases? A systematic review of the literature. *Health Policy, 121*(7), 745–754. http://dx.doi.org/10.1016/j.healthpol.2017.05.005.

Alberca, L. N., Sbaraglini, M. L., Balcazar, D., Fraccaroli, L., Carrillo, C., Medeiros, A., et al. (2016). Discovery of novel polyamine analogs with anti-protozoal activity by computer guided drug repositioning. *Journal of Computer-Aided Molecular Design, 30*(4), 305–321. http://dx.doi.org/10.1007/s10822-016-9903-6.

André Vannier-Santos, M., Brunoro, G. V., M de Nazaré, C. S., DeCastro, S. L., & Menna-Barreto, R. F. S. (2019). Parasite, compartments, and molecules: Trick versus treatment on Chagas disease. In W. De Souza (Ed.), *Biology of Trypanosoma cruzi*. IntechOpen. http://dx.doi.org/10.5772/intechopen.84472.

Aulner, N., Danckaert, A., Ihm, J., Shum, D., & Shorte, S. L. (2019). Next-generation phenotypic screening in early drug discovery for infectious diseases. *Trends in Parasitology, 35*(7), 559–570. http://dx.doi.org/10.1016/j.pt.2019.05.004.

Belllera, C. L., Sbaraglini, M. L., Alberca, L. N., Alice, J. I., & Talevi, A. (2019). In silico modeling of FDA-approved drugs for discovery of therapies against neglected diseases: A drug repurposing approach. In *In silico drug design* (pp. 625–648). Elsevier. http://dx.doi.org/10.1016/B978-0-12-816125-8.00021-3.

Berenstein, A. J., Magariños, M. P., Chernomoretz, A., & Agüero, F. (2016). A multilayer network approach for guiding drug repositioning in neglected diseases. *PLoS Neglected Tropical Diseases, 10*(1), e0004300. http://dx.doi.org/10.1371/journal.pntd.0004300.

Bhatnagar, S., Nicklas, S., Morrisey, J. M., Goldberg, D. E., & Vaidya, A. B. (2019). Diverse chemical compounds target plasmodium falciparum plasma membrane lipid homeostasis. *ACS Infectious Diseases, 5*(4), 550–558. http://dx.doi.org/10.1021/acsinfecdis.8b00277.

Blaazer, A. R., Singh, A. K., de Heuvel, E., Edink, E., Orrling, K. M., Veerman, J. J. N., et al. (2018). Targeting a subpocket in trypanosoma brucei phosphodiesterase B1 (TbrPDEB1) enables the structure-based discovery of selective inhibitors with trypanocidal activity. *Journal of Medicinal Chemistry, 61*(9), 3870–3888. http://dx.doi.org/10.1021/acs.jmedchem.7b01670.

Botta, L., Rivara, M., Zuliani, V., & Radi, M. (2018). Drug repurposing approaches to fight Dengue virus infection and related diseases. *Frontiers in Bioscience, 23*, 997–1019. http://dx.doi.org/10.2741/4630.

Bowers, K. J., Chow, D. E., Xu, H., Dror, R. O., Eastwood, M. P., Gregersen, B. A., et al. (November 17, 2006). Scalable Algorithms for molecular dynamics Simulations on commodity clusters. In *Proceedings of the 2006 ACM/IEEE conference on supercomputing*.

Brindha, S., Vincent, S., Velmurugan, D., Ananthakrishnan, D., Sundaramurthi, J. C., & Gnanadoss, J. J. (2017). Bioinformatics approach to prioritize known drugs towards repurposing for tuberculosis. *Medical Hypotheses, 103*, 39–45. http://dx.doi.org/10.1016/j.mehy.2017.04.005.

Cabello-Donayre, M., Orrego, L. M., Herráez, E., Vargas, P., Martínez-García, M., Campos-Salinas, J., et al. (2020). Leishmania heme uptake involves LmFLVCRb, a novel porphyrin transporter essential for the parasite. *Cellular and Molecular Life Sciences, 77*, 1827–1845. http://dx.doi.org/10.1007/s00018-019-03258-3.

Canutescu, A. A., Shelenkov, A. A., & Dunbrack, R. L. (2003). A graph-theory algorithm for rapid protein side-chain prediction. *Protein Science, 12*(9), 2001–2014. http://dx.doi.org/10.1110/ps.03154503.

Carneiro, V. C., de Abreu da Silva, I. C., Torres, E. J. L., Caby, S., Lancelot, J., Vanderstraete, M., et al. (2014). Epigenetic changes modulate schistosome egg formation and are a novel target for reducing transmission of schistosomiasis. *PLoS Pathogens, 10*(5), e1004116. http://dx.doi.org/10.1371/journal.ppat.1004116.

Chaudhari, R., Tan, Z., Huang, B., & Zhang, S. (2017). Computational polypharmacology: A new paradigm for drug discovery. *Expert Opinion on Drug Discovery, 12*(3), 279–291. http://dx.doi.org/10.1080/17460441.2017.1280024.

Chaudhari, R., Tan, Z., & Zhang, S. (2017). Overview of drug polypharmacology and multitargeted molecular design. In *Comprehensive medicinal chemistry* (Vol. III, pp. 259–275). Elsevier. http://dx.doi.org/10.1016/B978-0-12-409547-2.12323-6.

Cheuka, P. M., Mayoka, G., Mutai, P., & Chibale, K. (2016). The role of natural products in drug discovery and development against neglected tropical diseases. *Molecules, 22*(1). http://dx.doi.org/10.3390/molecules22010058.

Cho, A. E., Guallar, V., Berne, B. J., & Friesner, R. (2005). Importance of accurate charges in molecular docking: Quantum mechanical/molecular mechanical (QM/MM) approach. *Journal of Computational Chemistry, 26*(9), 915–931. http://dx.doi.org/10.1002/jcc.20222.

Coleman, B. I., Skillman, K. M., Jiang, R. H. Y., Childs, L. M., Altenhofen, L. M., Ganter, M., et al. (2014). A Plasmodium falciparum histone deacetylase regulates antigenic variation and gametocyte conversion. *Cell Host & Microbe, 16*(2), 177–186. http://dx.doi.org/10.1016/j.chom.2014.06.014.

Cragg, G. M., & Newman, D. J. (2013). Natural products: A continuing source of novel drug leads. *Biochimica et Biophysica Acta, 1830*(6), 3670–3695. http://dx.doi.org/10.1016/j.bbagen.2013.02.008.

Díaz-Chiguer, D. L., Márquez-Navarro, A., Nogueda-Torres, B., de la Luz León-Ávila, G., Pérez-Villanueva, J., Hernández-Campos, A., et al. (2012). In vitro and in vivo trypanocidal activity of some benzimidazole derivatives against two strains of Trypanosoma cruzi. *Acta Tropica, 122*(1), 108–112. http://dx.doi.org/10.1016/j.actatropica.2011.12.009.

Duch, W., Swaminathan, K., & Meller, J. (2007). Artificial intelligence approaches for rational drug design and discovery. *Current Pharmaceutical Design, 13*(14), 1497–1508. http://dx.doi.org/10.2174/138161207780765954.

Dueñas-González, A., Naveja, J. J., & Medina-Franco, J. L. (2016). Introduction of epigenetic targets in drug discovery and current status of epi-drugs and epi-probes. In *Epi-informatics* (pp. 1–20). Elsevier. http://dx.doi.org/10.1016/B978-0-12-802808-7.00001-0.

Duraisingh, M. T., & Horn, D. (2016). Epigenetic regulation of virulence gene expression in parasitic protozoa. *Cell Host & Microbe, 19*(5), 629–640. http://dx.doi.org/10.1016/j.chom.2016.04.020.

Ekins, S., Godbole, A. A., Kéri, G., Orfi, L., Pato, J., Bhat, R. S., et al. (2017). Machine learning and docking models for *Mycobacterium tuberculosis* topoisomerase I. *Tuberculosis, 103*, 52–60. http://dx.doi.org/10.1016/j.tube.2017.01.005.

Farha, M. A., & Brown, E. D. (2019). Drug repurposing for antimicrobial discovery. *Nature Microbiology, 4*(4), 565–577. http://dx.doi.org/10.1038/s41564-019-0357-1.

Ferreira, L. G., & Andricopulo, A. D. (2016). Drug repositioning approaches to parasitic diseases: A medicinal chemistry perspective. *Drug Discovery Today, 21*(10), 1699–1710. http://dx.doi.org/10.1016/j.drudis.2016.06.021.

Galvin, B. D., Li, Z., Villemaine, E., Poole, C. B., Chapman, M. S., Pollastri, M. P., et al. (2014). A target repurposing approach identifies N-myristoyltransferase as a new candidate drug target in filarial nematodes. *PLoS Neglected Tropical Diseases, 8*(9), e3145. http://dx.doi.org/10.1371/journal.pntd.0003145.

von Geldern, T. W., Morton, H. E., Clark, R. F., Brown, B. S., Johnston, K. L., Ford, L., et al. (2019). Discovery of ABBV-4083, a novel analog of Tylosin A that has potent anti-Wolbachia and anti-filarial activity. *PLoS Neglected Tropical Diseases* (Vol. 13,(2), e0007159. http://dx.doi.org/10.1371/journal.pntd.0007159.

Gutmann, H., Bruggisser, R., Schaffner, W., Bogman, K., Botomino, A., & Drewe, J. (2002). Transport of amentoflavone across the blood-brain barrier in vitro. *Planta Medica, 68*(9), 804–807. http://dx.doi.org/10.1055/s-2002-34401.

Ingram, A. K., & Horn, D. (2002). Histone deacetylases in trypanosoma brucei: Two are essential and another is required for normal cell cycle progression. *Molecular Microbiology, 45*(1), 89–97. http://dx.doi.org/10.1046/j.1365-2958.2002.03018.x.

Jiménez-Ruiz, A., Alzate, J. F., Macleod, E. T., Lüder, C. G. K., Fasel, N., & Hurd, H. (2010). Apoptotic markers in protozoan parasites. *Parasites & Vectors, 3*, 104. http://dx.doi.org/10.1186/1756-3305-3-104.

Kaiser, A. (2019). Druggable targets in cyclic nucleotide signaling pathways in apicomplexan parasites and kinetoplastids against disabling Protozoan diseases in humans. *International Journal of Molecular Sciences, 20*(1). http://dx.doi.org/10.3390/ijms20010138.

Karaman, B., Alhalabi, Z., Swyter, S., Mihigo, S. O., Andrae-Marobela, K., Jung, M., et al. (2018). Identification of bichalcones as sirtuin inhibitors by virtual screening and in vitro testing. *Molecules, 23*(2). http://dx.doi.org/10.3390/molecules23020416.

Kawahara, T., Siegel, T. N., Ingram, A. K., Alsford, S., Cross, G. A. M., & Horn, D. (2008). Two essential MYST-family proteins display distinct roles in histone H4K10 acetylation and telomeric silencing in trypanosomes. *Molecular Microbiology, 69*(4), 1054–1068. http://dx.doi.org/10.1111/j.1365-2958.2008.06346.x.

Khamis, M. A., Gomaa, W., & Ahmed, W. F. (2015). Machine learning in computational docking. *Artificial Intelligence in Medicine, 63*(3), 135–152. http://dx.doi.org/10.1016/j.artmed.2015.02.002.

Khosla, S., Sharma, G., & Yaseen, I. (2016). Learning epigenetic regulation from mycobacteria. *Microbial Cell, 3*(2), 92–94. http://dx.doi.org/10.15698/mic2016.02.480.

Kräutler, V., van Gunsteren, W. F., & Hünenberger, P. H. (2001). A fast SHAKE algorithm to solve distance constraint equations for small molecules in molecular dynamics simulations. *Journal of Computational Chemistry, 22*(5), 501–508. http://dx.doi.org/10.1002/1096-987X(20010415)22:5<501::AID-JCC1021>3.0.CO;2-V.

Krieger, E., & Vriend, G. (2014). YASARA View - molecular graphics for all devices - from smartphones to workstations. *Bioinformatics, 30*(20), 2981–2982. http://dx.doi.org/10.1093/bioinformatics/btu426.

Lane, T. J., Shukla, D., Beauchamp, K. A., & Pande, V. S. (2013). To milliseconds and beyond: Challenges in the simulation of protein folding. *Current Opinion in Structural Biology, 23*(1), 58–65. http://dx.doi.org/10.1016/j.sbi.2012.11.002.

Laskowski, R. A., MacArthur, M. W., Moss, D. S., & Thornton, J. M. (1993). Procheck: A program to check the stereochemical quality of protein structures. *Journal of Applied Crystallography, 26*(2), 283–291. http://dx.doi.org/10.1107/S0021889892009944.

Leite, A. C. L., Espíndola, J. W. P., de Oliveira Cardoso, M. V., & de Oliveira Filho, G. B. (2019). Privileged structures in the design of potential drug candidates for neglected diseases. *Current Medicinal Chemistry, 26*(23), 4323–4354. http://dx.doi.org/10.2174/0929867324666171023163752.

Lone, M. Y., Kumar, S. P., Athar, M., & Jha, P. C. (2018). Exploration of *Mycobacterium tuberculosis* structural proteome: An in-silico approach. *Journal of Theoretical Biology, 439*, 14–23. http://dx.doi.org/10.1016/j.jtbi.2017.11.021.

López-López, E., Naveja, J. J., & Medina-Franco, J. L. (2019). DataWarrior: An evaluation of the open-source drug discovery tool. *Expert Opinion on Drug Discovery, 14*(4), 335–341. http://dx.doi.org/10.1080/17460441.2019.1581170.

López-López, E., Prieto-Martínez, F. D., & Medina-Franco, J. L. (2018). Activity landscape and molecular modeling to explore the SAR of dual epigenetic inhibitors: A focus on G9a and DNMT1. *Molecules, 23*(12), 3282. http://dx.doi.org/10.3390/molecules23123282.

López-López, E., Rabal, O., Oyarzabal, J., & Medina-Franco, J. L. (2020). Towards the understanding of the activity of G9a inhibitors: an activity landscape and molecular modeling approach. *Journal of Computer-Aided Molecular Design, 34*, 659–669. http://dx.doi.org/10.1007/s10822-020-00298-x.

Marek, M., Kannan, S., Hauser, A.-T., Moraes Mourão, M., Caby, S., Cura, V., et al. (2013). Structural basis for the inhibition of histone deacetylase 8 (HDAC8), a key epigenetic player in the blood fluke Schistosoma mansoni. *PLoS Pathogens, 9*(9), e1003645. http://dx.doi.org/10.1371/journal.ppat.1003645.

Martin, L., Hutchens, M., Hawkins, C., & Radnov, A. (2017). How much do clinical trials cost? *Nature Reviews Drug Discovery, 16*(6), 381–382. http://dx.doi.org/10.1038/nrd.2017.70.

Medina-Franco, J. L. (2013). Activity cliffs: Facts or artifacts? *Chemical Biology & Drug Design, 81*(5), 553–556. http://dx.doi.org/10.1111/cbdd.12115.

Medina-Franco, J. L., Naveja, J. J., & López-López, E. (2020). Reaching for the bright StARs in chemical space. *Drug Discovery Today, 24*(11), 2162–2169. http://dx.doi.org/10.1016/j.drudis.2019.09.013.

Michalsky, E., Goede, A., & Preissner, R. (2003). Loops in Proteins (LIP)–a comprehensive loop database for homology modelling. *Protein Engineering, 16*(12), 979–985. http://dx.doi.org/10.1093/protein/gzg119.

Moran, M., Guzman, J., Ropars, A.-L., McDonald, A., Jameson, N., Omune, B., et al. (2009). Neglected disease research and development: How much are we really spending? *PLoS Medicine, 6*(2), e30. http://dx.doi.org/10.1371/journal.pmed.1000030.

Moretti, N. S., da Silva Augusto, L., Clemente, T. M., Antunes, R. P. P., Yoshida, N., Torrecilhas, A. C., et al. (2015). Characterization of trypanosoma cruzi sirtuins as possible drug targets for Chagas disease. *Antimicrobial Agents and Chemotherapy, 59*(8), 4669–4679. http://dx.doi.org/10.1128/AAC.04694-14.

Mosquillo, M. F., Bilbao, L., Hernández, F., Tissot, F., Gambino, D., Garat, B., et al. (2018). Trypanosoma cruzi biochemical changes and cell death induced by an organometallic platinum-based compound. *Chemical Biology & Drug Design, 92*(3), 1657–1669. http://dx.doi.org/10.1111/cbdd.13332.

Mutabingwa, T. K., Anthony, D., Heller, A., Hallett, R., Ahmed, J., Drakeley, C., et al. (2005). Amodiaquine alone, amodiaquine+sulfadoxine-pyrimethamine, amodiaquine+artesunate, and artemether-lumefantrine for outpatient treatment of malaria in Tanzanian children: A four-arm randomised effectiveness trial. *The Lancet, 365*(9469), 1474–1480. http://dx.doi.org/10.1016/S0140-6736(05)66417-3.

Naveja, J. J., Norinder, U., López-López, E., & Medina-Franco, J. L. (2018). Chemical space, diversity and activity landscape analysis of estrogen receptor binders. *RSC Advances, 8*, 38229–38237. http://dx.doi.org/10.1039/C8RA07604A.

Naveja, J. J., Oviedo-Osornio, C. I., Trujillo-Minero, N. N., & Medina-Franco, J. L. (2018). Chemoinformatics: A perspective from an academic setting in Latin America. *Molecular Diversity, 22*(1), 247–258. http://dx.doi.org/10.1007/s11030-017-9802-3.

Neves, B. J., Dantas, R. F., Senger, M. R., Melo-Filho, C. C., Valente, W. C. G., de Almeida, A. C. M., et al. (2016). Discovery of new anti-schistosomal hits by integration of QSAR-based virtual screening and high content screening. *Journal of Medicinal Chemistry, 59*(15), 7075–7088. http://dx.doi.org/10.1021/acs.jmedchem.5b02038.

Nguyen, G. T. T., Schaefer, S., Gertz, M., Weyand, M., & Steegborn, C. (2013). Structures of human sirtuin 3 complexes with ADP-ribose and with carba-NAD+ and SRT1720: Binding details and inhibition mechanism. *Acta Crystallographica Section D Biological Crystallography, 69*(Pt 8), 1423–1432. http://dx.doi.org/10.1107/S0907444913015448.

Nii-Trebi, N. I. (2017). Emerging and neglected infectious diseases: Insights, advances, and challenges. *BioMed Research International, 2017*, 5245021. http://dx.doi.org/10.1155/2017/5245021.

Norinder, U., Naveja, J. J., López-López, E., Mucs, D., & Medina-Franco, J. L. (2019). Conformal prediction of HDAC inhibitors. *SAR and QSAR in Environmental Research, 30*(4), 265–277. http://dx.doi.org/10.1080/1062936X.2019.1591503.

Nunes, R. R., Fonseca, A. L. da, Pinto, A. C. de S., Maia, E. H. B., Silva, A. M. da, Varotti, F. de P., et al. (2019). Brazilian malaria molecular targets (BraMMT): Selected receptors for virtual high-throughput screening experiments. *Memorias Do Instituto Oswaldo Cruz, 114*, e180465. http://dx.doi.org/10.1590/0074-02760180465.

Nwaka, S., & Ridley, R. G. (2003). Virtual drug discovery and development for neglected diseases through public-private partnerships. *Nature Reviews Drug Discovery, 2*(11), 919–928. http://dx.doi.org/10.1038/nrd1230.

Onawole, A. T., Sulaiman, K. O., Adegoke, R. O., & Kolapo, T. U. (2017). Identification of potential inhibitors against the Zika virus using consensus scoring. *Journal of Molecular Graphics and Modelling, 73*, 54–61. http://dx.doi.org/10.1016/j.jmgm.2017.01.018.

Ortalli, M., Varani, S., Rosso, C., Quintavalla, A., Lombardo, M., & Trombini, C. (2019). Evaluation of synthetic substituted 1,2-dioxanes as novel agents against human leishmaniasis. *European Journal of Medicinal Chemistry, 170*, 126–140. http://dx.doi.org/10.1016/j.ejmech.2019.02.070.

Ozen, A., Sherman, W., & Schiffer, C. A. (2013). Improving the resistance profile of hepatitis C NS3/4A inhibitors: Dynamic substrate envelope guided design. *Journal of Chemical Theory and Computation, 9*(12), 5693–5705. http://dx.doi.org/10.1021/ct400603p.

Pierce, R. J., & Khalife, J. (2019). Parasite epigenetic targets. In W. Sippl, & M. Jung (Eds.), *Epigenetic drug discovery* (pp. 457–475). Wiley. http://dx.doi.org/10.1002/9783527809257.ch16.

Prieto-Martínez, F. D., Arciniega, M., & Medina-Franco, J. L. (2018). Acoplamiento molecular: Avances recientes y retos. *Tip Revista Especializada en Ciencias Químico-Biológicas, 21*. http://dx.doi.org/10.22201/fesz.23958723e.2018.0.143.

Prieto-Martínez, F. D., López-López, E., Eurídice Juárez-Mercado, K., & Medina-Franco, J. L. (2019). Computational drug design methods—current and future perspectives. In *In silico drug design* (pp. 19–44). Elsevier. http://dx.doi.org/10.1016/B978-0-12-816125-8.00002-X.

Rahnasto-Rilla, M., Tyni, J., Huovinen, M., Jarho, E., Kulikowicz, T., Ravichandran, S., et al. (2018). Natural polyphenols as sirtuin 6 modulators. *Scientific Reports, 8*(1), 4163. http://dx.doi.org/10.1038/s41598-018-22388-5.

Reguera, R. M., Pérez-Pertejo, Y., Gutiérrez-Corbo, C., Domínguez-Asenjo, B., Ordóñez, C., García-Estrada, C., et al. (2019). Current and promising novel drug candidates against visceral leishmaniasis. *Pure and Applied Chemistry, 91*(8), 1385–1404. http://dx.doi.org/10.1515/pac-2018-1102.

Religa, A. A., & Waters, A. P. (2012). Sirtuins of parasitic protozoa: In search of function(s). *Molecular and Biochemical Parasitology, 185*(2), 71–88. http://dx.doi.org/10.1016/j.molbiopara.2012.08.003.

Sakkiah, S., Arooj, M., Cao, G. P., & Lee, K. W. (2013). Insight the C-site pocket conformational changes responsible for sirtuin 2 activity using molecular dynamics simulations. *PloS One, 8*(3), e59278. http://dx.doi.org/10.1371/journal.pone.0059278.

Saldívar-González, F. I., Naveja, J. J., Palomino-Hernández, O., & Medina-Franco, J. L. (2017). Getting SMARt in drug discovery: Chemoinformatics approaches for mining structure–multiple activity relationships. *RSC Advances, 7*(2), 632–641. http://dx.doi.org/10.1039/C6RA26230A.

Sanders, B. D., Jackson, B., & Marmorstein, R. (2010). Structural basis for sirtuin function: What we know and what we don't. *Biochimica et Biophysica Acta, 1804*(8), 1604–1616. http://dx.doi.org/10.1016/j.bbapap.2009.09.009.

Schuetz, A., Min, J., Antoshenko, T., Wang, C.-L., Allali-Hassani, A., Dong, A., et al. (2007). Structural basis of inhibition of the human NAD+-dependent deacetylase SIRT5 by suramin. *Structure, 15*(3), 377–389. http://dx.doi.org/10.1016/j.str.2007.02.002.

Schulz, D., Mugnier, M. R., Paulsen, E.-M., Kim, H.-S., Chung, C. W., Tough, D. F., et al. (2015). Bromodomain proteins contribute to maintenance of bloodstream form stage identity in the african trypanosome. *PLoS Biology, 13*(12), e1002316. http://dx.doi.org/10.1371/journal.pbio.1002316.

Serbus, L., Landman, F., Lindley, C., White, P., & Sullivan, W. (2018). *Compositions and methods employing Wolbachia FtsZ as a target for Albendazole sulfone* (Patent No. US 10 , 220 , 022 B2).

Singh, N., Shah, P., Dwivedi, H., Mishra, S., Tripathi, R., Sahasrabuddhe, A. A., et al. (2016). Integrated machine learning, molecular docking and 3D-QSAR based approach for identification of potential inhibitors of trypanosomal N-myristoyltransferase. *Molecular BioSystems, 12*(12), 3711–3723. http://dx.doi.org/10.1039/c6mb00574h.

Soares, M. B. P., Silva, C. V., Bastos, T. M., Guimarães, E. T., Figueira, C. P., Smirlis, D., et al. (2012). Anti-Trypanosoma cruzi activity of nicotinamide. *Acta Tropica, 122*(2), 224–229. http://dx.doi.org/10.1016/j.actatropica.2012.01.001.

Staderini, M., Piquero, M., Abengózar, M.Á., Nachér-Vázquez, M., Romanelli, G., López-Alvarado, P., et al. (2019). Structure-activity relationships and mechanistic studies of novel mitochondria-targeted, leishmanicidal derivatives of the 4-aminostyrylquinoline scaffold. *European Journal of Medicinal Chemistry, 171*, 38–53. http://dx.doi.org/10.1016/j.ejmech.2019.03.007.

Sülsen, V. P., Cazorla, S. I., Frank, F. M., Redko, F. C., Anesini, C. A., Coussio, J. D., et al. (2007). Trypanocidal and leishmanicidal activities of flavonoids from Argentine medicinal plants. *The American Journal of Tropical Medicine and Hygiene,* 77(4), 654–659. http://dx.doi.org/10.4269/ajtmh.2007.77.654.

Swinney, D. C., & Pollastri, M. P. (2019). Drug discovery strategies for neglected tropical diseases: Repurposing knowledge, mechanisms and therapeutics to increase discovery efficiency. In D. Swinney, & M. Pollastri (Eds.), *Neglected tropical diseases: Drug discovery and development* (pp. 1–13). Wiley. http://dx.doi.org/10.1002/9783527808656.ch1.

Tasdemir, D., Kaiser, M., Brun, R., Yardley, V., Schmidt, T. J., Tosun, F., et al. (2006). Antitrypanosomal and antileishmanial activities of flavonoids and their analogues: In vitro, in vivo, structure-activity relationship, and quantitative structure-activity relationship studies. *Antimicrobial Agents and Chemotherapy, 50*(4), 1352–1364. http://dx.doi.org/10.1128/AAC.50.4.1352-1364.2006.

Tavares, J., Ouaissi, A., Santarém, N., Sereno, D., Vergnes, B., Sampaio, P., et al. (2008). The Leishmania infantum cytosolic SIR2-related protein 1 (LiSIR2RP1) is an NAD+ -dependent deacetylase and ADP-ribosyltransferase. *Biochemical Journal, 415*(3), 377–386. http://dx.doi.org/10.1042/BJ20080666.

Tomasi, J. (1988). Models and modeling in theoretical chemistry. *Journal of Molecular Structure: THEOCHEM, 179*(1), 273–292. http://dx.doi.org/10.1016/0166-1280(88)80128-3.

Trott, O., & Olson, A. J. (2010). AutoDock vina: Improving the speed and accuracy of docking with a new scoring function, efficient optimization, and multithreading. *Journal of Computational Chemistry, 31*(2), 455–461. http://dx.doi.org/10.1002/jcc.21334.

Vidal, M., Cusick, M. E., & Barabási, A.-L. (2011). Interactome networks and human disease. *Cell, 144*(6), 986–998. http://dx.doi.org/10.1016/j.cell.2011.02.016.

Wang, X., Miyazaki, Y., Inaoka, D. K., Hartuti, E. D., Watanabe, Y.-I., Shiba, T., et al. (2019). Identification of plasmodium falciparum mitochondrial malate: Quinone oxidoreductase inhibitors from the pathogen box. *Genes, 10*(6). http://dx.doi.org/10.3390/genes10060471.

Weeks, J. C., Roberts, W. M., Leasure, C., Suzuki, B. M., Robinson, K. J., Currey, H., et al. (2018). Sertraline, paroxetine, and chlorpromazine are rapidly acting anthelmintic drugs capable of clinical repurposing. *Scientific Reports, 8*(1), 975. http://dx.doi.org/10.1038/s41598-017-18457-w.

WHO. (2020). *Neglected tropical diseases*. https://www.who.int/neglected_diseases/diseases/.

Winzeler, E. A., & Ottilie, S. (2019). The proteasome as a target: How not tidying up can have toxic consequences for parasitic protozoa. *Proceedings of the National Academy of Sciences of the United States of America, 116*(21), 10198–10200. http://dx.doi.org/10.1073/pnas.1904694116.

Yamurai Bishi, L., Sundeep Chaitanya Vedithi, D., Blundell, T. L., & Mugumbate, G. (2019). Computational deorphaning of *Mycobacterium tuberculosis* targets. In *Tuberculosis - beyond the biomedical.* [Working Title]. IntechOpen. http://dx.doi.org/10.5772/intechopen.82374.

Zhang, L., Tan, J., Han, D., & Zhu, H. (2017). From machine learning to deep learning: Progress in machine intelligence for rational drug discovery. *Drug Discovery Today, 22*(11), 1680–1685. http://dx.doi.org/10.1016/j.drudis.2017.08.010.

CHAPTER EIGHT

# Controlling Ser/Thr protein phosphatase PP1 activity and function through interaction with regulatory subunits

**Antonio Casamayor and Joaquín Ariño***
Institut de Biotecnologia i Biomedicina & Departament de Bioquímica i Biologia Molecular, Universitat Autònoma de Barcelona, Cerdanyola, del Vallès, Spain
*Corresponding author: E-mail: Joaquin.Arino@uab.cat

## Contents



*Advances in Protein Chemistry and Structural Biology*, Volume 122
ISSN 1876-1623
https://doi.org/10.1016/bs.apcsb.2020.06.004

## Abstract

Protein phosphatase 1 is a major Ser/Thr protein phosphatase activity in eukaryotic cells. It is composed of a catalytic polypeptide (PP1C), with little substrate specificity, that interacts with a large variety of proteins of diverse structure (regulatory subunits). The diversity of holoenzymes that can be formed explain the multiplicity of cellular functions under the control of this phosphatase. In quite a few cases, regulatory subunits have an inhibitory role, downregulating the activity of the phosphatase. In this chapter we shall introduce PP1C and review the most relevant families of PP1C regulatory subunits, with particular emphasis in describing the structural basis for their interaction.

## 1. Introduction

Reversible covalent modification of proteins is a major mechanism for the modulation of these key biomolecules and, in consequence, of a myriad of cellular functions. In eukaryotes, proteins are phosphorylated mostly at Ser and Thr residues and, to a lesser extent, in Tyr. The phosphorylation status of a give protein is the results of the balancing activities of two kind of enzymes: protein kinases and protein phosphatases. In eukaryotes, as a rule, the number of polypeptides with catalytic protein kinase activity exceeds by a factor of 2–3 that of protein phosphatases. If we consider the enzymes involved in the, by far, most frequent Ser/Thr phosphorylation events, possibly accounting for 99% of cases (Sharma et al., 2014), it seems clear that Ser/Thr kinases are commonly single polypeptides, whereas in many cases Ser/Thr phosphatases activities are composed of a catalytic polypeptide accompanied by one or more non-catalytic components or subunits. Such components, which can be very numerous and varied, can specifically modulate the activity of the catalytic polypeptide, confer substrate specificity, and/or determine its subcellular localization. Therefore, the actual function of many Ser/Thr phosphatases are defined for the specific conjunction or assembly of the catalytic polypeptide and one or more of these so-called "regulatory subunits". This scenario, which is particularly akin to the subject of this chapter, Ser/Thr protein phosphatase 1, gave rise recently to the concept of "split protein phosphatase system" (Bertolotti, 2018).

The first (classical) classification of the Ser/Thr protein phosphatase (PP) activity was performed using enzymatic criteria, differentiating between PP type 1 (PP1) and 2 (PP2) depending on the ability of the enzymes to dephosphorylate the β subunit of the phosphorylase kinase. Further subclassification of type 2 was based in the requirements for divalent cation: none for PP2A, $Ca^{2+}$ (PP2B) and $Mg^{2+}$ (Ingebritsen & Cohen, 1983). It was subsequently found that PP2C catalytic subunits are structurally distant from the rest of catalytic polypeptides. Thus, the catalytic subunits of PP1, PP2A and PP2B were included in the phosphoprotein phosphatase (PPP) family, whereas PP2C comprise the protein phosphatase $Mg^{2+}/Mn^{2+}$-dependent (PPM) family.

The existence of regulatory proteins able to build up holoenzymes based on a diversity of hetero-oligomeric complexes is particularly relevant in the canonical PP1 proteins, since their catalytic subunits are relatively short polypeptides (~300 residues) mostly consisting in the catalytic moiety that spans about 90% of the entire protein, thus leaving little room for extra functions (Fig. 1).

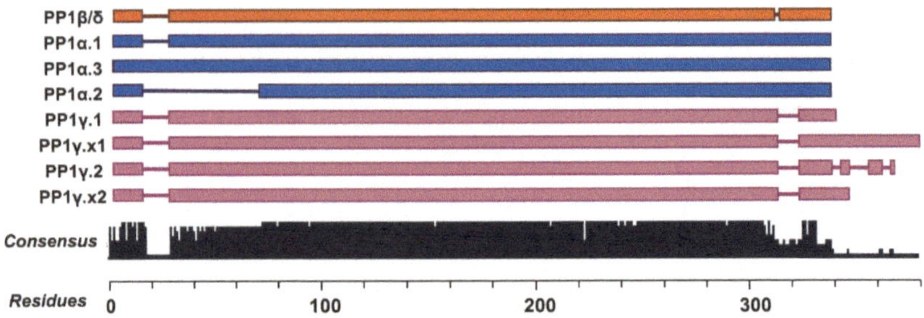


**Fig. 1** Schematic alignment of the different mammalian PP1C. Sequences were obtained from NCBI (https://www.ncbi.nlm.nih.gov/) and aligned with Clustal Omega. The alignment was loaded into JalView for calculation of the consensus sequence at each position.

## 2. The catalytic subunit of PP1 (PP1C)

PP1 is a major serine/threonine phosphatase activity that regulates a variety of cellular functions, such as glycogen metabolism, muscle contraction or cell division (Plattner & Bibb, 2012). PP1 consists of an evolutionary conserved catalytic subunit (PP1C), with little substrate specificity, and a large number of regulatory subunits (PP1R) that regulate its function by determining the subcellular localization of PP1C and substrate specificity. Three different genes encoding for four distinct PP1C are present in mammalian genomes: PP1$\alpha$, PP1$\beta/\delta$, and the two splicing forms of the PP1C$\gamma$ gene, PP1$\gamma_1$ and PP1$\gamma_2$. The genome of the budding yeast cells possesses one only gene coding for the PP1C (*GLC7*), whereas multiple variants can be found in plants (Moorhead, De Wever, Templeton, & Kerk, 2009; Smith & Walker, 1993). As shown in Fig. 1 in the case of mammalian PP1C, the short N– and C-terminal regions flanking the catalytic domain of the different PP1C isoforms present a much lower level of sequence conservation than the catalytic region. As will be described below, both tails contain sequence motifs that provide specificity to several regulatory subunits.

PP1c exhibits a compact $\alpha/\beta$ fold, whit a $\beta$ sandwich packed between two $\alpha$-helical domains. Its catalytic site is located at the center of three shallow furrows, forming a Y-shaped surface. This site contains two metal ions ($Mn^{2+}$ and $Fe2^{+}$), that are coordinated by three histidines, two aspartic acids, and one asparagine, highly conserved not only in PP1C, but also in all members of the PPP family. These metal ions are key in the catalytic process, as they are thought to bind and activate a water molecule, which initiates a nucleophilic attack on the phosphorous atom (Yigong Shi, 2009). Tumor-promoting toxins that strongly inhibit PP1C (i.e. microcystin and okadaic acid) strongly associate with the catalytic site by means of interactions with residues in its surface.

It is generally accepted that PP1C rarely exists as an individual protein in the cell and, for instance, overexpression of yeast PP1C (Glc7) has proven to be deleterious (Liu, Krizek, & Bretscher, 1992; Zhang, Guha, & Volkert, 1995), suggesting that tight control of PP1C activity is needed. In this chapter we will describe and discuss a number of regulatory subunits and the mechanism by which these subunits modulate (and often inhibit) PP1C catalytic activity.

## 3. The PP1 regulatory subunits

Multiple PP1R subunits or PP1C-binding proteins have been identified, ranging from no less of a few dozens in the yeast to about 200 in human cells. Remarkable, only a few of them are evolutionary conserved in eukaryotes, and it has been suggested that these may have a role in the biogenesis and turnover of PP1 (see below). Biochemical and structural studies have determined that many PPP1R, as other PP2A- and PP2B-binding proteins, contain intrinsically disordered regions (IDR). A common characteristic of the IDR is the presence of one or more *S*hort *Li*near *M*otifs (SLiM) by which different unstructured PPP1R interact to the PP1C to become ordered upon the binding.

### 3.1 Examples of SLiMs identified in the PPP1R proteins

**The RVXF motif:** First identified in two parallel and independent studies. Analyzing the crystal structure of PP1C in complex with a peptide from $G_M$ (PPP1R3A), the RVXF motif, where X is any residue, was identified as a responsible of the binding and identified in other PP1C-binding proteins (Egloff et al., 1997). The consensus sequence "(R/H/Q/K)−V-X-(F/W)" was also identified as the results of a screening of a random peptide display library searching form peptides that bind to PP1C "(R/H/Q/K)−V-X-(F/W)" (Zhao & Lee, 1997). Using both alignments of established RVXF sequences and site-directed mutagenesis, the consensus sequence for this site was later redefined as [RK]-X(0,1)-[VI]-{P}-[FW] where {P} is any residue but P (Wakula, Beullens, Ceulemans, Stalmans, & Bollen, 2003). This study also revealed that the binding of the RVXF motif promotes the binding of secondary interaction sites to PP1C that could affect the substrate specificity. Posterior systematic analysis of each position revealed which residues are important in each position of the motif, and the proposed new consensus motif was defined as: [HKR]-[ACHKMNQRSTV]-[V]-[CHKNQRST]-[FW], where the first position represent the key anchor point of the motif by forming a network of salt bridges hydrogen bonds (Meiselbach, Sticht, & Enz, 2006). A more recent *in silico* screening in search of novel PP1C interactors, combining the definitions described by Wakula and by Meiselbach, identified 115 new human candidates, validating most of them (78), being 63% (49) inhibitors of the PP1C activity, Nevertheless, these experiments were performed using glycogen phosphorylase *a* as a substrate and the real effects of the PP1C interactors may dependent on the specific substrate of the holoenzyme (Hendrickx et al., 2009). Analysis of this five-residue "RVXF" consensus motif in all the 143 known human PP1C interactors determined the followed percentages of occurrence by residue: $[K_{55}R_{34}][K_{28}R_{26}]$ $[V_{94}I_{6}]$\{FIMYDP\}$[F_{83}W_{17}]$, where braced residues denotes exclusion (Hendrickx et al., 2009). It should be noted the possible relevance of Ser and Thr residues in position 4 (21% and 18% respectively), whose phosphorylation could contribute to inhibit the binding of the RV[**S**/**T**]F-containing regulatory subunits to PP1C (Dent, Campbell, Caudwell, & Cohen, 1990; Liu et al., 2010).

**The SILK motif**. Conformed by the sequence [G/S]IL[RK], it was first identified as essential for binding of PP1C to Inhibitor-2. Posterior analyses demonstrated that additional regulatory subunits also contained the SILK motif (Hendrickx et al., 2009; Hurley et al., 2007).

**The MyPhoNE (myosin phosphatase N-terminal element) motif.** This motif is characterized by the RXXQ[VIL] [KR]X[YW] consensus sequence, commonly located N-terminus to the RVXF motif (Hendrickx et al., 2009). It was identified in several myosin phosphatase targeting proteins (Bollen, Peti, Ragusa, & Beullens, 2010; Hendrickx et al., 2009; Terrak, Kerff, Langseimo, Tao, & Dominguez, 2004).

**The ΦΦ motif.** This PP1C anchor domain is composed by two sequential hydrophobic residues present in the regulatory subunits that interacts with the PP1C **ΦΦ** grove, near to the RVXF-binding pocket (Choy et al., 2014; O'Connell et al., 2012). It was first observed in the spinophilin:PP1C complex, but also detected in multiple PP1C-binding proteins (O'Connell et al., 2012; Ragusa et al., 2010).

**The FXXR/KXR/K motif**: It was found in Bcl-$x_L$ and Bcl-w as required for the binding to PP1α (Ayllón, Cayla, García, Fleischer, &

Rebollo, 2002). This consensus motif was also identified, together with an RVXF motif [R/K-X(0,1)−V-X-F], in the BH domain of Bcl-2-related proteins. The simultaneous occurrence of both motifs has been proposed and validated as a signature to identify PP1C interacting proteins. Furthermore, the successive appearance of a RVXF and a FXXR/KXR/K motif in the form of a R/K-X(0,1)-V/I-X-F-X-X-R/K-X-R/K sequence has been identified as an apoptotic signature (Garcia et al., 2004; Godet, Guergnon, Maire, Croset, & Garcia, 2010).

**The KiR-Small linear motive**. Named by the Ki-67-RepoMan SLIM, IT corresponds to a stretch of residues exclusively located C-terminal to the canonical RVXF of the nuclear subunits Ki-67 and RepoMan. The consensus sequence of the KiR-SLIM determined by Kumar and collaborators is FDXXLP[P/A]N[T/S]PL[R/K] [R/K]GX[T/S]P (Kumar et al., 2016).

## 4. PP1C interacting sites for regulatory proteins

Comprehensive studies about protein phosphatases interactors and structures have been published in the last few years (Esteves, Domingues, Da Cruz E Silva, Fardilha, & Da Cruz E Silva, 2012; Sacco, Perfetto, Castagnoli, & Cesareni, 2012; Shi, 2009; Yadav et al., 2017). Here we will briefly review the regions of PP1C relevant for interaction with its multiple regulatory proteins. It is important to remember that in many cases different interacting proteins share one or more of PP1C target regions, thus generating a model of combinatorial control of the activity of the phosphatase (Bollen, 2001). Fig. 2 presents the disposition of the catalytic site together with the most relevant docking regions for specific SLIMs.

The crystal structure of different PP1C isoforms has been resolved revealing a very similar structure consisting in two tightly linked N− and C-terminal domains with three different and connected clefts (Goldberg et al., 1995; Kita et al., 2002; Terrak et al., 2004; Yu, Deng, & Xiang, 2018). Since the hydrophobic groove of PP1C were the β-strand containing the RVXF motif binds to is distant from the PP1C catalytic site, binding of the RVXF motif does not influence by itself the PP1C activity (Peti, Nairn, & Page, 2013; Terrak et al., 2004).

The C-terminal region of PPC1β/δ (301−309) possesses a binding site for the ankyrin repeats present in several regulatory subunits (Terrak et al.,

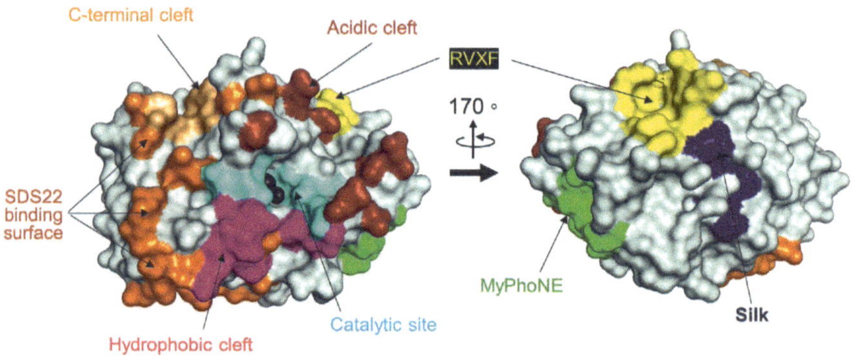


**Fig. 2** The PP1α model based in the 4MOV entry at the Protein Data Bank. Annotation was done using Pymol and data for interaction sites gathered from diverse publications, mainly (Chen et al., 2016; Choy et al., 2014, 2019). Note that diverse residues required for SDS22 binding overlap with residues from the C-terminal, the acidic, and the hydrophobic clefts and, therefore, are not represented as SDS22 binding elements.

2004). The MyPhone-binding region of this PPC1β/δ isoform also could participate providing additional specificity (Scotto-Lavino, Garcia-Diaz, Du, & Frohman, 2010).

Type 2 SH3-binding motif (PXXPXR) is present in the C-terminus of all human, but not in yeast or fly, PP1C isoforms, as well as a set of one to four basic residues on the last five N-terminal residues (conserved in the yeast Glc7 but not in other PP1C-like proteins). The basic residues following the PXXPXR motif and the length of the tail after the PXXPXR motif is what might determine the binding specificity between the PP1α isoform and the SH3 domain of the ASPP2 protein (Bertran et al., 2019).

## 5. PP1C interacting proteins and regulators

Here we will describe the nature of PP1-interacting proteins, from the point of view of multicellular, organisms (mostly mammals), with occasional incursions into unicellular organisms such as yeasts. Because the number of PP1C-binding proteins is very high, it would be out of the scope of this chapter to attempt a comprehensive review of all of them. Therefore, we will focus in a selected number as example of mechanisms for enzyme regulation (Table 1). It must be stressed that only few of these proteins are conserved among most eukaryotes, ranging from human to yeast. This has led to the proposal (Verbinnen, Ferreira, & Bollen, 2017) that at least three

**Table 1** Major regulatory subunits of the PP1C phosphatase.

| Family | Gene | Protein | Inhibitor | PP1C-binding described motifs | Other motifs |
|---|---|---|---|---|---|
| Small inhibitors | PPP1R1A | Inhibitor-1 | Yes | RVXF | IPP-1/DARPP-32 |
| | PPP1R1B | DARPP-32 | Yes | RVXF | PP-1/DARPP-32 |
| | PPP1R1C | IPP5 | Yes | RVXF | PP-1/DARPP-32 |
| | PPP1R2 | Inhibitor-2 | Yes | SILK, RVXF | IPP-2 |
| Glycogen-targeting | PPP1R3A | PP1G | No | RVXF | CBM53, TM, C—C |
| | PPP1R3B | $G_L$ | No | RVXF | CBM53 |
| | PPP1R3C | PTG | No | RVXF | CBM53 |
| | PPP1R3D | R6 | No | RVXF | CBM53 |
| | PPP1R3E | R3E | No | RVXF | CBM53 |
| | PPP1R3F | R3F | No | RVXF | CBM53, TM |
| | PPP1R3G | R3G | No | RVXF | CBM53 |
| Actin-target. | PPP1R9A | Neurabin 1 | No | RVXF, **ΦΦ** | PDZ, TelA, Tankyrase_bdg_C, SAM |
| | PPP1R9B | Spinophilin | No | RVXF, **ΦΦ** | PDZ, TBPIP, PCRF, C—C |
| Nuclear | PPP1R7 | SDS22 | Yes[a] | LRR | — |
| | PPP1R11 | Inhibitor-3 | Yes[a] | RVXF, EKPRAFGESSTES | NLS, NoLS |
| | PPP1R8 | NIPP-1 | Yes | RVXF, **ΦΦ**, NIPP-1 helix | FHA, WAHD |
| | PPP1R10 | PNUTS | Yes | RVXF, **ΦΦ** | TFS2N, 4MP0, ZnF |
| Myosin targeting | PPP1R12A | MYPT1 | No | MyPhoNE, RVXF | ANK, C—C, PRKG1_interact. |
| | PPP1R12B | MYPT2 | No | MyPhoNE, RVXF | ANK, C—C, PRKG1_interact |
| | PPP1R12C | MBS85 | No | RVXF, MyPhoNE | ANK, PRKG1_interact. |

*(Continued)*

**Table 1** Major regulatory subunits of the PP1C phosphatase.—cont'd

| Family | Gene | Protein | Inhibitor | PP1C-binding described motifs | Other motifs |
|---|---|---|---|---|---|
| | PPP1R16A | MYPT3 | Yes | RVXF, MyPhoNE | ANK, CAAX |
| | PPP1R16B | TIMAP | No | RVXF, MyPhoNE | ANK, CAAX |
| ASPP | PPP1R13A | ASPP2 | No | RVXF, SH3 | RA, C—C, 4IRV, ANK |
| | PPP1R13B | ASPP1 | No | RVXF, SH3 | RA, C—C, 4IRV, ANK |
| | PPP1R13L | iASPP | Yes | RVXF, SH3 | C—C, ANK |
| PPP1R14 family | PPP1R14A | CPI-17 | Yes | | PP1_inhibitor |
| | PPP1R14B | PHI | Yes | RVXF | PP1_inhibitor |
| | PPP1R14C | KEPI | Yes | RVXF | PP1_inhibitor |
| | PPP1R14D | GBPI-1 | Yes | RVXF | PP1_inhibitor |
| eIF2α-PPases | PPP1R15A | PGADD 34 | No | PP1c_bdg, RVXF, **ΦΦ** | |
| | PPP1R15B | CReP | No | RVXF, **ΦΦ** | CReP_N |
| Mitotic exit | PPP1R81 | RepoMan | No | RVXF, **ΦΦ**, KiR-SLIM | |
| | PPP1R105 | Ki-67 | No | RVXF, **ΦΦ**, KiR-SLIM | FHA, K167R |
| Others | PPP1R43 | AKAP149 | No | RVXF, **ΦΦ** | TM, KH, TUDOR |
| | PPP1R55 | KNL1 | No | RVXF, **ΦΦ**, SILK | 4A1G, C—C, 4NF9 |

**Other motifs:** 4A1G: PDB 4a1g; 4IRV: PDB 4irv; 4MP0: PDB 4mp0; 4NF9: PDB 4nf9; ANK: SM00248; C2: SM00239; C—C: Coiled-coil region; CAAX: Prenylation site; CBM53: PF16760; Cons_C: PF15281; CReP_N: PF10472; FHA: CD00060; FIVE_2: PF02318; IPP-1/DARPP-32: PF05395; IPP-2: PF04979; K167R: SM001295; KH: SM000322; NLS: Nuclear localization signal; NoLS: Nucleolar localization signal; NYAP_N: PF15439; PCRF: CL23943; PF15898; PDZ: SM00228; PP1_inhibitor: PF05361; PP1c_bdg: PF10488; RA: SM000314; SAM: SM00454; SH2: SM00252; SH3: SM000326; TelA: PF05816; Tankyrase_bdg_C: PF15327; TBPIP: PF07106; TFS2N: SM00509; TM: Transmembrane region; TUDOR: SM00333; WAHD: PF11945; WBP11: PF09429; ZnF_C3H1: SM000356.
[a]Proposed that *in vivo* might act also as activator in yeast.

of them (SDS22, Inhibitor-2 and Inhibitor-3) could be crucial for the biogenesis of PP1, including metal ions load and unload, subunit interchange, and turnover of the complexes, whereas other subunits would be involved in substrate selection or activity regulation.

## 5.1 Small and thermostable inhibitors

### *5.1.1 PPP1R1A (I-1)*

Human inhibitor-1, or I-1, is a small and thermostable protein of 171 amino acids that preferentially binds, and inhibits PP1C when it is phosphorylated by PKA at Thr-35 (Endo, Zhou, Connor, Wang, & Shenolikar, 1996). Alternative splicing of the I-1 rat gene results in diverse variants: the "complete" I-1 form (contains all six exons), I-1$\alpha$ (lacks exon 5), and I-1$\beta$ (lacks exons 4 and 5) mRNAs (Weiser, Sikes, Li, & Shenolikar, 2004). I-1 possesses a phosphorylated Thr-35 and a RVXF motif sequence ($^{9}$KIQF$^{12}$) that mediates the binding to the hydrophobic groove in the $\beta$12-$\beta$13 loop of PP1C (Connor et al., 1998; Connor, Kleeman, Barik, Honkanen, & Shenolikar, 1999; Egloff et al., 1997).

The interaction of I-1 with PP1C defines diverse aspects of the function of the phosphatase. For example, the C-terminal sequence of I-1 binds to the Growth Arrest and DNA Damage-Inducible Protein GADD34, a scaffolding protein that also bind PP1C. This complex interacts with and dephosphorylates the translation initiation factor eIF2$\alpha$ and, in this way, protein synthesis can be regulated (Connor, Weiser, Li, Hallenbeck, & Shenolikar, 2001; Weiser et al., 2004).

Inhibitor-1 plays an important role in mitotic regulation since it is phosphorylated by PKA in its Ser-35. This results in increased inhibitory activity, leading to a reduction in PP1C activity during mitosis. At the mitotic exit PP1C becomes active again (by a Cdc2-dependent process) and dephosphorylates I-1, which dissociates from PP1C, leading to the full stimulation of the phosphatase activity (Wu et al., 2009).

Several studies have determined that inhibition of PP1 by I-1 has an important role in muscle cells. In cardiac muscle, I-1 regulates cardiac contractility [extensively reviewed in (Nicolaou, Hajjar, & Kranias, 2009)], and it has been described as beneficial in alleviating the detrimental effects of heart failure. Inhibition of PP1C by I-1 contributes to keep phospholamban (PLN) phosphorylated at Ser16 and Thr17. PLN is a regulator of the $Ca^{2+}$ pump in cardiac muscle cells and its phosphorylation leads to the activation of the cardiac muscle sarcoplasmic reticulum $Ca^{2+}$-ATPase SERC, the transporter of $Ca^{2+}$ into the sarcoplasmic reticulum (Nicolaou,

Hajjar, et al., 2009; Nicolaou, Rodriguez, et al., 2009; Pathak et al., 2005; Pritchard et al., 2013). Accordingly, expression of the constitutively active version of I-1 (T35D) or a C-terminal truncated version (1.65) (Nicolaou, Rodriguez, et al., 2009; Pathak et al., 2005) enhanced cardiac function. On the contrary, the naturally occurring G109E mutation in I-1 increases PP1 activity, impairing sarcoplasmic reticulum $Ca^{2+}$ cycling and arrhytmogenesis (Haghighi et al., 2015). I-1 induces a cell contractile phenotype in the vascular smooth muscle, where it is highly expressed, regulating the phospholamban phosphorylation. It has been proposed that a constitutively active version of I-1 could be a promising therapeutic strategy in the prevention of vascular proliferative disorders (Lipskaia et al., 2014).

Other cellular functions such as glycogen metabolism and cell proliferation are also regulated by I-1 (Nicolaou, Hajjar, et al., 2009). It is worth noting that chemical inhibition of the PP1 activity was described long ago as inducer of tumor growth (Fujiki & Suganuma, 2009). Accordingly, expression of the I-1 gene has been recently found to be upregulated in Ewing sarcoma, regulating, together with PKA, the tumorigenic and metastatic processes (Luo et al., 2018).

### 5.1.2 PPP1R1B (DARPP-32)

DARPP-32 stands for Dopamine- and cAMP-regulated neuronal phosphoprotein-32, and it is absent in plants and fungi. It is structurally related to I-1 and, as happens with I-1, is also up-regulated by the PKA in the brain dopamine-responsive cells by phosphorylation of the key Thr-34 residue (equivalent to Thr-35 in I-1), thus promoting the binding of DARPP-32 to PP1C and subsequent PP1C inhibition. This residue, located in a $^{31}$RRP**T**PA$^{36}$ sequence that is conserved in other PPP1R1, is dephosphorylated by phosphatase PP2B (calcineurin). Phosphorylation of other residues modulate the phosphorylation status of Thr-34. Phosphorylation of Thr-75 by CDK5 prevents PKA-mediated Thr-34 phosphorylation and its positive effect on DARPP-32 as a PP1C inhibitor, whereas phosphorylation of Ser-102 enhances phosphorylation of Thr-34 by the PKA. In addition, phosphorylation of Ser-137 attenuates the dephosphorylation of Thr-34 by calcineurin (see (Avanes, Lenz, & Momand, 2019; Reis et al., 2007; Walaas, Hemmings, Greengard, & Nairn, 2011) for reviews). As inhibitor of PP1C, DARPP-32, plays a role in tumor cell proliferation and drug resistance, by activating the PI3K-Akt/PKB pathway, and is frequently found overexpressed in several cancers, (Avanes et al., 2019).

t-DARPP is a N-terminal truncated version of DARPP-32 originated by alternative splicing of the PPP1R1B gene. In consequence, it lacks the RVXF domain responsible for binding and modulating PP1C, as well as Thr-34 (El-Rifai et al., 2002). t-DARPP is also overexpressed in multiple tumoral cells, and this overexpression causes resistance to antitumoral drugs (Avanes et al., 2019).

### *5.1.3 PPP1R1C (IPP5)*

Human PPP1R1C is a small protein (116 amino acids the longest variant) found in vertebrates, first identified in 2009 by Zeng and collaborators (Zeng et al., 2009). It is highly expressed in heart and testis.

As other PPP1R1, it has in its N-terminal region the conserved RVXF motif and the phosphorylation site. Active human IPP5 has been recognized as a potential apoptosis-promoting factor since an overexpression in HeLa cells of an active truncated version (residues 8—60) with the phosphomimetic T41D point mutation in the isoform 1 was able to arrest cell cycle in G2/M and promote apoptosis, enhancing the effects of Paclitaxel (Zeng et al., 2012, 2014).

### *5.1.4 PPP1R2 (Inhibitor-2, I-2)*

In spite of their names, the aminoacidic sequences of inhibitor 1 and 2 are not related except in the N-terminal region, where they share the RVXF motif and a few blocks of residues. Inhibitor-2 was the first PP1C regulatory subunit to be identified. Rabbit I-2 was purified in 1976 and the human cDNA cloned twenty years later (Helps, Street, Elledge, & Cohen, 1994; Huang & Glinsmann, 1976a, 1976b). I-2 forms a stable complex with PP1C and can be found in plants and in some fungi as I-2-like proteins. Like PPP1R1 inhibitors, I-2 is a small and thermostable protein that exist as four different isoforms produced by alternative splicing forms from the human gene.

I-2 is a disordered protein with three binding sites: the sequence $^{12}$KGILKN$^{17}$ that binds to a surface created by amino acids 50—59 in PP1C$\gamma$; from residues 44 to 56 interacts along the RVXF binding groove through the sequence, $^{43}$KSQKW$^{47}$; and residues 130—169, forms $\alpha$-helical regions that lie across the substrate binding cleft, thus blocking the access of the substrate (Hurley et al., 2007) (Fig. 3). Phosphorylation by CK2 in three serine residues does not alter I-2 activity but facilitates its phosphorylation on Thr-72 by GSK-3, that triggers conformational changes that activate the PP1C:I-2 complex (Lin et al., 2003; Oliver & Shenolikar, 1998). I-2 acts

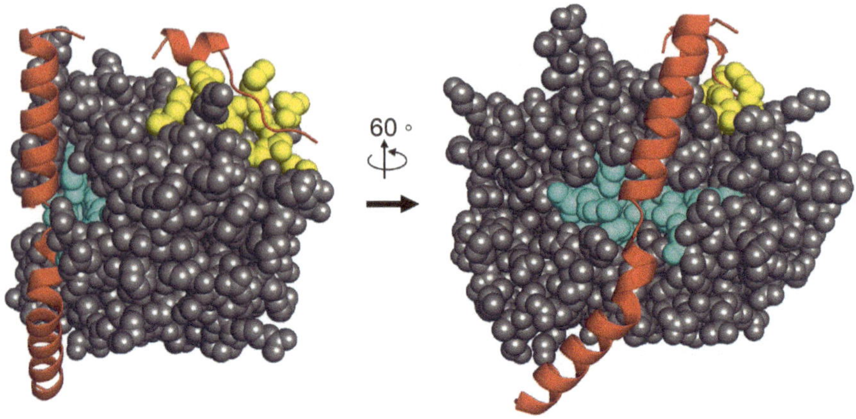


**Fig. 3** The interaction between PP1Cγ and inhibitor 2. The crystal structure of rat PP1Cγ complexed with mouse inhibitor-2 correspond to PDB 2O8A (Hurley et al., 2007) and has been prepared with Pymol. Note how the N-terminus of I-2 (red) interacts with the RVXF-binding pocket in PP1C (yellow) so the alpha-helical I-2 regions lie across the PP1C substrate-binding cleft (cyan), thus blocking substrate entry.

as a suppressor of memory formation by negatively regulating CREB-mediated gene transcription (Yang et al., 2015) and it has been described that when phosphorylated at Ser-44 by ATM in response to DNA damage (among other situations), it dissociates from the complex thus increasing PP1 activity (Tang et al., 2008).

## 5.2 Glycogen—targeting subunits

PP1C is a key activity involved in the hormonal regulation of glycogen metabolism, activating glycogen synthase and inactivating the enzymes implicated in glycogen degradation (glycogen phosphorylase and phosphorylase kinase). Subunits $G_M$ and $G_L$, the first characterized members of the PPP1R3 family (Doherty, Moorhead, Morrice, Cohen, & Cohen, 1995; Tang, Bondor, Swiderek, & DePaoli-Roach, 1991), and other PPP1R3 members, are required for the association of PP1C to the glycogen particles and proper regulation of glycogen metabolism in liver and muscle (reviewed by (Ceulemans & Bollen, 2004)). The RVXF consensus sequence has been found in all glycogenic subunits, which also share a conserved VNXXFEKXV sequence essential for the binding to the glycogen particle (Wu et al., 1998). Two independent studies have reported sequences for the binding of PP1C substrates to glycogenic subunits. Fong and collaborators postulated to occurs through a stretch of amino acids predicted to protrude from the surface (WDNNE/D) that define the conserved WDNNXGXNY consensus motif (Fong et al., 2000). Liu and collaborators, however, determined the somewhat similar WXNXGNYX(L/I) motif (Liu & Brautigan, 2000).

### 5.2.1 PPP1R3A (PP1G, $G_M$)

PP1G is a 1122 amino acids muscle glycogen-bound protein that allows regulation by PP1C of the enzymes involved in the glycogen metabolism. PP1G is likely restricted to animals. Proteins with low level of similarity exist in fungi and plants, but this similarity is limited to a carbohydrate-binding motif (Rodríguez-Sanoja, Oviedo, & Sánchez, 2005). In mammals, adrenaline-induced phosphorylation of PP1G at Ser-65 causes its dissociation from PP1C, leading to the release of PP1C from glycogen particles and the sarcoplasmic reticulum, inhibiting thus glycogen synthesis and activating glycogen phosphorylase. It is worth noting that Ser-65 in located within the RVXF PP1C binding motif ($^{63}$RV**S**F$^{66}$). Adrenaline also induces the phosphorylation of Ser-46 by PKA, which triggers glycogen synthesis by increasing PP1C-PP1G complex activity (Chen et al., 1994; Walker, Watt, & Cohen, 2000). PP1G bind to glycogen by the $^{148}$VLNVSFEKLV$^{157}$ sequence (Wu et al., 1998). Furthermore, two regions implicated in binding to glycogen synthase have been found, one between residues 77 and 117 and a second one (219–240) that contains the WXNXGXNYX(I/L) core sequence (Liu & Brautigan, 2000). An hydrophobic region near the COOH terminus has been suggested as a sarcoplasmic reticulum-binding domain (Tang et al., 1991). An isoform caused by alternative splicing lacks the N-terminal region where the RVXF and the glycogen-binding region are located. Mutations and polymorphisms in the PP1R3A gene have been respectively associated to tumorigenesis (Hayashida et al., 2005; Huang et al., 2014; Li et al., 2017; Somaiah et al., 2018) and to susceptibility to insulin resistance (Hansen et al., 1995; Sánchez-Pozos et al., 2018).

### 5.2.2 PPP1R3B (PPP1R4, $G_L$)

PPP1R3B is a 285-residues protein, closely related to PPP1R3C, that corresponds to the N-terminal region of PPP1R3A ($G_M$), comprising the carbohydrate binding domain. PPP1R3B is highly expressed in human liver, where it was first characterized as regulator of the glycogen metabolism. It was found that the amount of hepatic glycogen directly correlated with the expression level of PPP1R3B (Mehta et al., 2017; Stender et al., 2018). Consequently, genetic variants of PPP1R3B have impact on liver glycogen accumulation (Stender et al., 2018), but also in predisposition to type 2 diabetes (Dunn et al., 2006; Niazi et al., 2019) and plasmatic lipids and C-reactive protein levels (Zhang et al., 2013).

### 5.2.3 PPP1R3C (PPP1R5, PTG)

Human PPP1R3C is a 317-residues protein, containing a carbohydrate binding domain. The PTG protein interacts with the laforin-malin complex, implicated in Lafora disease (which can be considered a glycogen metabolism disorder in the brain). Laforin is a dual-specificity protein phosphatase that is usually mutated in the Lafora disease. Expression of laforin or malin induces proteasome-dependent degradation of PTG, decreasing PTG-stimulated glycogenic activity (Worby, Gentry, & Dixon, 2008). In fact, depletion of PTG in mouse or mutation of Asn-249 in the carbohydrate binding domain of the human protein, reduces glycogen synthesis and rescues or ameliorates the phenotype caused by the absence of laforin (Guerrero et al., 2011; Turnbull et al., 2014).

Hypermethylation of the PPP1R3C promoter, causing lower expression levels, were found in colorectal cancer patients (Lee et al., 2015; Takane et al., 2014). In fact, the methylation status of the promoter seems to be sensitive to glucose levels, since high glucose levels reduces methylation and this fact has been postulated as a mechanism required for cell proliferation under high glucose levels (Lee et al., 2015). It has recently been documented that suppression of hepatic gluconeogenesis by metformin-activated AMPK in mice is caused by a decrease in hepatic PPP1R3C expression (Ji et al., 2019). On the contrary, in rats with non-alcoholic fatty liver disease, expression of PPP1R3C was found elevated.

### 5.2.4 PPP1R3D (R6, PPP1R6)

Human R6 encodes a 299-aminoacids protein, first identified by Armstrong and col. (Armstrong, Browne, Cohen, & Cohen, 1997), with two putative RVXF sequences: $^{102}$RVRF$^{105}$, corresponding to the motif found in other members of the family, and the R6-specific $^{252}$RVHF$^{255}$. R6, as any member of the PPP1R3 family, possesses a carbohydrate-binding module (CBM21) in its C-terminal region, in which it has been identified a number of residues, also conserved in other PPP1R3 members, essential for the glycogenic activity. Specifically, the sequence WDNNE/D of PPP1R3D ($^{267}$WDNNDHRDY$^{275}$) is required for the binding of glycogenic targeting subunits to glycogen-related protein substrates (Rubio-Villena, Sanz, & Garcia-Gimeno, 2015). In addition, a binding site to 14-3-3 proteins has been identified involving Ser-74 located in the $^{71}$RAR**S**LP$^{76}$ sequence (Rubio-Villena et al., 2015). This phosphorylatable Ser-74 is not conserved in PPP1R3B and PPP1R3C, the closer members of the PPP1R3 family, where a proline residue (in position 57 and 33, respectively) is found, avoiding thus the binding to the 14-3-3 proteins. Phosphorylation in two other residues of serine in the same region (Ser-77 and Ser-78) that are not completely conserved in other PPP1R3 members, might play a role in the stabilization of the interaction with PP1C (Yadav et al., 2017).

As the PPP1R3C protein, R6 also interacts with the glucan phosphatase laforin allowing the ubiquitination of R6 by the E3-ubiquitin ligase malin, and targeting R6 for degradation (Rubio-Villena, Garcia-Gimeno, & Sanz, 2013). This mechanism reduces de glycogenic activity in neuron cells, where R6 is expressed, avoiding the non-desirable neuronal glycogen accumulation.

R6 also binds to the AMP-activated protein kinase AMPKβ, mediating the glucose-induced dephosphorylation of Thr-148 of AMPKβ, and thus promoting the inactivation of the kinase. This R6:AMPKβ interaction is dynamically controlled by glycogen content, being enhanced under glycogen depletion conditions (Oligschlaeger et al., 2016).

### 5.2.5 PPP1R3E (R3E)

Human R3E is a 279-residues protein involved in glycogen metabolism, structurally similar to other PP1R3 proteins, with a predicted CBM21 carbohydrate-binding domain near its C-terminal end (181–259). This protein is highly expressed in human skeletal muscle and heart but is almost absent in liver. In contrast, R3E is highly expressed in rodent liver and skeletal muscle (Munro, Ceulemans, Bollen, Diplexcito, & Cohen, 2005). Since expression levels are controlled by insulin, genetic variants of this gene are associated with diabetes (Sánchez-Pozos et al., 2018).

### 5.2.6 PPP1R3F (R3F)

R3F is present in animals. In humans, it is found as six different isoforms (1, 2, X1, X2, X3 and X4), originated by alternative splicing, ranging from 799 to 433 residues (isoform 1 and X4, respectively). It was identified in a cDNA brain library by Munro and co-workers, as a functional glycogen targeting subunit that also possesses an unusual membrane binding domain and a substrate interaction domain. It is predominantly expressed in brain and regulates glycogen synthase in response to glucose, noradrenaline and adenosine (Kelsall, Voss, Munro, Cuthbertson, & Cohen, 2011; Munro et al., 2005). R3F was found to be important for neuronal activities, and mutations of this protein have been identified in individuals with autism spectrum disorders (ASDs) (Piton et al., 2011). Subsequently, R3F has been identified as a master regulator of a regulon (177 target genes) that are associated with ASDs (Doostparast Torshizi, Duan, & Wang, 2018).

### 5.2.7 PPP1R3G (R3G)

R3G is a vertebrate regulatory subunit of PP1 with 358 amino acids in humans, where no splicing forms have been detected. This human gene was identified by genomic sequencing, and analysis of the encoded protein reveals PP1C and glycogen binding domains (Ceulemans, Stalmans, & Bollen, 2002). It is actively involved in the control of blood glucose homeostasis by regulating hepatic glycogenesis in a manner closely coordinated with the fasting—feeding cycle. The expression pattern of the R3G gene was found different from other glycogen-targeting subunits: it was found highly upregulated in the liver of fasting mice, and its mRNA level were quickly reduced after feeding (Luo et al., 2011). Stimulation of glycogenesis by R3G in hepatocytes in the absence of glucose was found dependent of its glycogen-binding motif.

## 5.3 The actin binding proteins neurabins (PPP1R9)

The neuronal actin-binding animal proteins Neurabin 1 and 2 interact to I-2-bound PP1C to form a complex involved in the regulation of the actin cytoskeleton (Terry-Lorenzo et al., 2002). The flexibility of the PP1C binding domain of neurabin-2 (also called spinophilin) allows its interaction with PP1C over an extensive surface, forming many unexpected interactions (Ragusa et al., 2010). Binding of spinophilin blocks one of the three potential PP1C substrate binding grooves without altering the PP1C active site.

### 5.3.1 PPP1R9A (Neurabin-1)

Several isoforms, originated by alternative splicing, have been identified in humans. They share both C— and N-terminus and rang from 1374 to 1090 aminoacids (isoforms 1 and 5, respectively). One PDZ in the middle of the protein (512—592) and one Sterile Alpha Motif (SAM) (1261—1327) domains are predicted, as well as two coiled-coil regions (598—627 and 697—839). Similar architecture can be found in the other 4 isoforms except for isoform 3, that lacks the C-terminal SAM domain, but an additional coiled-coil is predicted near its C-terminus. Neurabin-1 controls cytoskeleton organization and has been involved in neuronal function, being important for neuron projection development, postsynaptic actin skeleton organization, neurotransmitter secretion, and dendritic spine development, among other functions. Proper doses of neurabin-1 are important for regulation of spine structure and function as deduced from the fact that expression of PPP1R9A mRNA has been found increased in bipolar disorder subjects (Konopaske, Subburaju, Coyle, & Benes, 2015).

In the nematode *C. elegans* NeurABin-1 is a shorter protein of 527 amino acids with a predicted PDZ (located near the N-terminus), SAM (near the C-terminus) and coiled-coil domains (in the middle) that acts as an actin-binding protein involved in the synaptogenic signaling (Chia, Patel, & Shen, 2012). Its N-terminal region presents similarity with a region of human neurabins where the RXVF motif is included ($^{455}$RKIK$^{458}$ in neurabin-1, $^{447}$RKIH$^{450}$ in neurabin-2 and $^{19}$RKVS$^{22}$ in NeuaABin). Its N-terminal region ($^{9}$RKVS$^{22}$) presents similarity with a region of human neurabins where the RXVF motif is included ($^{455}$RKIK$^{458}$ in neurabin-1, $^{447}$RKIH$^{450}$ in neurabin-2) (Hsieh-Wilson, Allen, Watanabe, Nairn, & Greengard, 1999). Association of neurabin-1 with PP1C is attenuated by phosphorylation by PKA of Ser-461, adjacent to the PP1C-binding motif (Gao, Hu, Yang, & Xia, 2018).

### 5.3.2 PPP1R9B (Spinophilin, Neurabin-2)

This protein is enriched in human dendritic spines, being important for regulation of synaptic physiology by directing PP1C to substrates such as glutamate receptors. Several factors and conditions regulate expression of Spinophilin and its interaction with PP1C. Spinophilin associates with PP1Cγ1 and PP1Cβ1(Carmody, Baucum, Bass, & Colbran, 2008; MacMillan et al., 1999). It is also an unstructured protein that has sequence similarity to neurabin-1, although shorter (817 aminoacids), since it lacks the C-terminal end of the neurabin-1 isoform. Therefore, only a PDZ and a C-terminal coiled-coil domain are predicted. Binding of PP1C to Spinophilin depends on its $^{447}$RKIH$^{450}$ motif (Hsieh-Wilson et al., 1999).

Spinophilin targets PP1C to neuronal synapses. CDK5 phosphorylates Spinophilin at Ser-17 within its N-terminal actin-binding domain, enhancing its association with PP1C (Edler et al., 2018). Phosphorylation of Spinophilin at Ser-15 and Ser-205 by ERK2 modulates the spine morphogenesis by reducing the ability of Spinophilin to bind bundle actin filaments (Futter et al., 2005). Spinophilin is also phosphorylated by $Ca^{2+}$/calmodulin-dependent protein kinase II (CaMKII) in Ser-100 (an acidic residue in neurabin-1) and Ser-116 (conserved in neurabin-1), also located in the actin-binding domain (Grossman et al., 2004). All these phosphorylation events of Spinophilin may target this protein to specific locations within dendritic spines.

## 5.4 Nuclear-related subunits

### *5.4.1 PPP1R7 (SDS22)*

This is a conserved and broadly distributed regulatory subunit, present in animals, plants and fungi that is important for completion of the mitotic cycle and for targeting PP1C to mitotic kinetochores. First identified in *Schizosaccharomyces. pombe* as *sds22*$^+$, a high-dosage suppressor of defects caused by a cold sensitive mutation in PP1C (*dis2-11*) (Ohkura & Yanagida, 1991), *sds22*$^+$ was found to be an essential gene, required for completion of mitosis. Lethality of the null mutant was rescued by high dosage of the PP1C gene. Sds22 was also soon identified in *Saccharomyces cerevisiae* as a PP1C interactor that, when overexpressed, was also able to rescue mitotic defects caused by a temperature sensitive allele of the PP1C (*glc7-12*$^{ts}$) (MacKelvie, Andrews, & Stark, 1995). Characterization of the first human cDNA revealed that it encodes a protein of 360 residues, 46% identical to *S. pombe* Sds22 (Renouf et al., 1995). Sds22 proteins lack the RVXF motif but, instead, interacts to PP1C through a domain containing 11 imperfect leucine-rich repeats (LRRs) of 22 residues (Fig. 4), each of which has a hairpin-like conformation that, together, assembles into a curved superhelix important for protein-protein interaction (Ceulemans, Vulsteke, et al., 2002; Kobe & Kajava, 2001). Neither the RVXF nor the acidic groove of the yeast PP1C are essential for the binding to Sds22 (Figs. 2 and 4). Instead, Sds22 binds to the R4/R5/R6 triangle of Glc7, close to the catalytic site (Ceulemans, Vulsteke, et al., 2002). Evidences indicate that yeast Sds22 inhibits PP1 activity in different *in vivo* functions (Pedelini et al., 2007; Yao et al., 2016) and that LLR repeats 5 and 2 of the human Sds22 are enough to inhibit PP1C activity (Lesage et al., 2007).

Multiple splicing forms have been identified in the human gene giving rise to, at least, seven isoforms of PPP1R7. A central core of six exons (out of the ten existing coding exons) are shared by all seven isoforms. The two first (coding) exons in the N-terminus are variable among the different isoforms, as occurs with the last two exons of the C-terminal end, which, according to SMART analysis (Letunic I & Bork P, 2018), is predicted to encode for two LRR repeats plus the LRR cap domain, proposed to shield the hydrophobic core of the LRRs from solvent (Ceulemans, De Maeyer, Stalmans, & Bollen, 1999). The DNA sequences

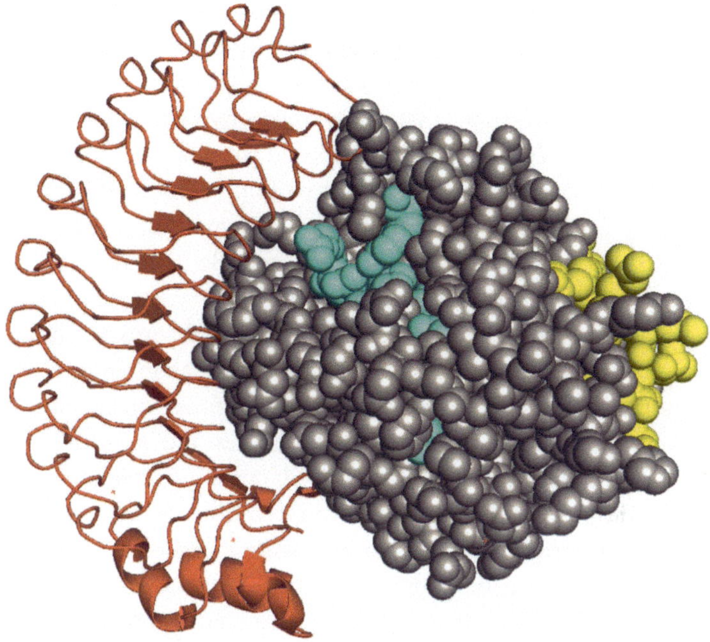

**Fig. 4** Interaction of SDS22 with human PP1Cα (PDB 6obp). Note the vicinity of the catalytic site (blue) and the position, almost opposite, of the PP1C region that binds RVXF-containing regulatory subunits (yellow).

coding for the predicted LRR are not coincident with the exons (Ceulemans, Van Eynde et al., 1999). Consequently, isoforms range from the isoform 1, containing 360 residues and 10 predicted LRR domains in addition to an LRR cap domain, to the shortest isoform 7, containing 237 residues with 8 LRR domains predicted. However, a model recently generated after crystallographic studies has determined the presence of 12 LRR domains, being the last LRR, a few residues before the LRR cap domain, an incomplete repeat (Ceulemans, De Maeyer et al., 1999; Heroes et al., 2019). The expected number of LRR domains for the 338-residues budding yeast Sds22 protein, the only PPP1R7 present in this organism, is twelve, although an LRR cap is not predicted. In the fission yeast *S. pombe* Sds22$^{+}$ is shorter (280 residues), and 11 LRR are predicted, the last of them overlapping with a possible LRR cap domain.

As mentioned above the PP1C catalytic site has two metal ions bound to it. It has been recently shown that SSD22 selectively binds a unique conformation of PP1 that contains a single metal (at position M2) at its active site and that SDS22 dissociation is accompanied by a second metal (M1) being loaded into PP1. It was observed, using the budding yeast homologs, that M1-deficient mutants cannot dissociate from SDS22. Therefore, SDS22 would acts as a "PP1C trap" that binds to inactive single metal-containing PP1C (Choy et al., 2019). It was concluded that SDS22 functions to bind and maintain PP1 in an inactive state, providing a pool of PP1 that is poised and ready for holoenzyme assembly. It was previously hypothesized that Sds22 could act as a chaperon to keep soluble the recently translated PP1C (Verbinnen, Ferreira, & Bollen, 2017).

SDS22 contributes to regulate G2/M cell cycle progression, defining PP1C localization and functions. Thus, in mammalian cells, for example, SDS22 contributes to the regulation of the mitotic kinase Aurora B and to the proper interaction between microtubules and kinetochores (Posch et al., 2010). As regulator in G2/M cell cycle progression, SDS22 has been involved in tumor suppression through induction of apoptosis by promoting the dephosphorylation (and inactivation) of PKB/Akt in Thr-308 and Ser-473 (Paul, Bargale, Rapole, Shetty, & Santra, 2019). It also participates in establishing *Drosophila* epithelial cell shape and polarity (Grusche et al., 2009). Sds22 is also implicated in the inhibition of the PP1C autodephosphorylation (Heroes et al., 2019) and it has been suggested that, in yeast cells, Sds22 functions positively with PP1C (Glc7) to promote dephosphorylation of nuclear substrates required for proper chromosomes segregation during mitosis (Peggie et al., 2002). Very recent crystallographic studies performed by Heroes and collaborators have hypothesized that Sds22 functions as an "ancillary" subunit of PP1 holoenzymes required for the binding to a "real" regulatory subunit containing IDRs (Intrinsically disordered regions). This hypothesis has been demonstrated for the case of the Bcl2-associated factor 1, which form a trimeric complex with Sds22 and PP1C (Heroes et al., 2019). SDS22 has been found to form a ternary complex with PP1C and other regulatory subunit, Inhibitor-3 (Cheng & Chen, 2015; Pedelini et al., 2007) and that this occurs with newly synthesized PP1C. The formed inactive complex needs to be disassembled by the p97 AAA-ATPase to promote exchange to other subunits that would serve, for instance, as substrate specifiers (Weith et al., 2018). Sds22, together with I-2 and I-3, participate as a regulator of PP1$\gamma$2, a PP1C isoform specifically expressed in the mammals testis and sperm that is essential for the normal function of mammalian sperm, as recently review in (Dey, Brothag, & Vijayaraghavan, 2019).

### *5.4.2 PPP1R11 (I-3, Ypi1)*

Present in eukaryotic organisms, and first identified in humans as Inhibitor-3, it is a small (126 amino acids), hydrophilic, heat-stable PP1C-specific inhibitor (Zhang, Qi, Gao, & Lee, 2008). Human I-3 bind PP1C by its $^{39}$KKVEW$^{43}$ sequence, similar to other RVXF motives. A second region of human I-3 ($^{65}$EKPRAFGESSTES$^{77}$) is required for interaction with and inhibition of PP1C activity (Zhang et al., 2008). Those motives are not well conserved in proteins from other organisms, such as in the putative yeast homolog Ypi1 (Yeast Phosphatase Inhibitor 1). A basic nuclear localization signal has been identified near the C-terminus of human I-3 ($^{94}$HRKGRRR$^{100}$) (Huang, Pozarowski, Gao, Darzynkiewicz, & Lee, 2005), which is conserved in the yeast protein ($^{120}$ERRHRK$^{125}$). In contrast, neither the N-terminal nuclear targeting signal ($^{33}$RKRK$^{36}$) nor a putative caspase-3 cleavage site detected in human I-3 ($^{46}$DTVD$^{49}$) are found in Ypi1 (Huang et al., 2005).

Multiple nuclear functions for inhibitor-3 have been identified. The nuclear-localized ternary complex of PP1C with Sds22 and Inhibitor-3 could be required for PP1C to antagonize the action of Aurora B kinase in the kinetochores for faithful chromosome segregation (Eiteneuer et al., 2014). I-3 associates with PP1C$\gamma$ in the nucleoli and with PP1C$\alpha$ in the centrosome, but it does not immunoprecipitate with the P1C$\beta$/$\delta$ isoform (Huang et al., 2005). I-3 is degraded during apoptosis in a caspase-3-depenent manner, and this degradation is important for the apoptotic process (Huang & Lee, 2008). Recent evidences point to a role of inhibitor-3 in regulation of the immune response. Inhibitor-3 has been found phosphorylated upon T cell activation, having an effect on shaping T cell signaling probably by affecting the phosphoinositide-dependent pathways, suggesting that inhibitor-3 could be a negative regulator of T cell activation-induced cytokine expression (Joshi et al., 2017, 2019).

In yeast cells, the I-3 ortholog Ypi1 is a 155-aminoacid essential protein that was identified as interactor of both, Glc7, the canonical PP1C in yeast cells, and the PP1-ralated phosphatase Ppz1. Yeast Ypi1 forms stable ternary complexes with Glc7 and Sds22 and the inhibitory effect of Ypi1 on PP1C is potentiated, *in vitro*, by the addition of Sds22, displaying additive Glc7 inhibitory capacity (García-Gimeno, Muñoz, Ariño, & Sanz, 2003; Pedelini et al., 2007). Ypi1 is important for cation homeostasis (Marquina, González, et al., 2012) and is required to maintain the nuclear localization of Glc7 (Pedelini et al., 2007). Consequently, different nuclear functions such as the G2/M cell cycle transition and the morphogenesis and spindle checkpoints are dependent on Ypi1 (Bharucha, Larson, Gao, Daves, & Tatchell, 2008; Marquina, Queralt, Casamayor, & Ariño, 2012). Loss of Ypi1 causes cell growth arrest, with aberrant mitotic spindle, similar phenotypes to those described for the loss of Sds22 (Marquina, Queralt, et al., 2012; Pedelini et al., 2007). However, other observations have been taken as evidence that Ypi1 and Sds22 are, *in vivo*, positive regulators of Glc7 (Bharucha et al., 2008).

### 5.4.3 PPP1R8 (NIPP1)

First isolated from bovine thymus as an small peptide with specific and potent capacity to inhibit PP1 activity (Beullens, Van Eynde, Stalmans, & Bollen, 1992), cloning of its cDNA revealed that its C-terminal region corresponded to the Ard1 protein, involved in RNA processing (Van Eynde et al., 1995; Wang & Cohen, 1994). It has been identified in plants and animals, and there are three different isoforms in human, consequence of alternative splicing. Isoform α, the largest, has 350 amino acids, and comprises the sequences of the other two isoforms. Isoforms β/δ and γ contain, respectively, the last 209 and 127 residues of isoform α. A forkhead associated (FHA) domain (Hofmann & Bucher, 1995), predicted in the N-terminal region of isoform α (48–101), would be absent in the other isoforms. The central domain of NIPP1, an intrinsically disordered protein, anchors PP1C in a RVXF ($^{200}$RVTF$^{203}$) motif-dependent manner (O'Connell et al., 2012). In addition to the RVXF motif, an additional PP1C-binding site comprising the last 22 residues of the bovine protein was characterized (Beullens et al., 2000). A polybasic inhibitory sequence, immediately N-terminal the RVXF motif a ($^{191}$RPKRKRKNSR$^{200}$) is required for the inhibition of PP1C (Monique Beullens et al., 1999). This inhibitory sequence is not present in N-terminal-truncated shortest isoform γ (or Ard1 protein). All three proteins can bind to RNA and have $Mg^{2+}$-dependent ssRNase activity, cleaving its substrate at specific sites in A + U-rich regions. NIPP1 has an essential role in pre-mRNA splicing that seems unrelated to its ability to bind to RNA o PP1C (Monique Beullens & Bollen, 2002).

NIPP1 plays a role in the nuclear localization of PP1C and is involved in nuclear functions, including transcription and splicing regulation (Van Dessel et al., 2015).

The FHA domain binds substrates, such as the CDC5L and SAP155 splicing factors, only when they are phosphorylated at threonine in a Thr-Pro dipeptide-motif, prompting their dephosphorylation by the associated PP1C, (Kumeta et al., 2008; Minnebo et al., 2013). PP1C–NIPP1 acts as transcriptional repressor since is involved in the PP1-dependent dephosphorylation of Thr-Pro-motifs of the histone methyltransferase EZH2, the catalytic subunit of the Polycomb Repressive Complex 2 (PRC2). Association of EZH2 with NIPP1 occurs by generation of a docking site in EZH2, after its phosphorylation on Thr416 by CDKs, for the FHA domain of NIPP1 (Minnebo et al., 2013). Thus, NIPP1 can inhibits PP1C activity, and this fact enables de phosphorylation of EZH2 on other Thr-Pro motifs promoting its binding to chromatin. EZH2 catalyzes the trimethylation of H3 in the Lys 27, regulating cell proliferation and differentiation by silencing Polycomb target genes (Minnebo et al., 2013).

Very recent work M. Bollen's laboratory determined that NIPP1 confers a resistance to chemically induced skin and hepatocellular carcinogenesis that was concomitant with the high expression levels of DNA-repair factors (Verbinnen et al., 2020a). NIPP1 has also been identified as a key regulator of epidermal homeostasis (Verbinnen et al., 2020).

### 5.4.4 PPP1R10 (PNUTS)

This protein, also called Phosphatase 1 NUclear Targeting Subunit (PNUTS) is present in vertebrates and ubiquitously expressed in human tissues and, together with NIPP1, the most abundant PP1C binding protein in the nucleus. First identified in humans as p99 (Kreivi et al., 1997), it is a 940-aminoacids protein involved in RNA splicing. PNUTS is an intrinsically disordered protein that binds to PP1C through a region comprised between residues 394 and 433. This region contains the $^{396}$RKTVTW$^{401}$ sequence, where the consensus Phe of the RVXF motif is replaced by Trp, closely followed by a hydrophobic **ΦΦ** motif ($^{410}$YF$^{411}$) that bind the hydrophobic groove in PP1 (Choy et al., 2014). Interestingly, a the third PP1C interaction site was defined in PNUTS from Phe-413 through Asn-424, centered on Arg420, which binds the PP1 C-terminal groove (Fig. 5). This Arg motif is conserved in spinophilin, but not in NIPP-1 (Choy et al., 2014).

Phosphorylation of PNUTS by PKA negatively regulates its interaction with PP1C (Kim et al., 2003). It is worth noting the elevated quantity of Gly residues (40%) in the long predicted low complexity region (from 644 to 905) that conform seven RGG motifs, postulated as responsible for the binding to RNA. In fact, it has been demonstrated that is PNUTS that anchors PP1C to specific RNA-associated complexes in the nucleus (Kim et al., 2003).

This is a ubiquitous expressed protein that plays a role in multiple nuclear functions, including cell cycle progression, DNA repair and apoptosis.

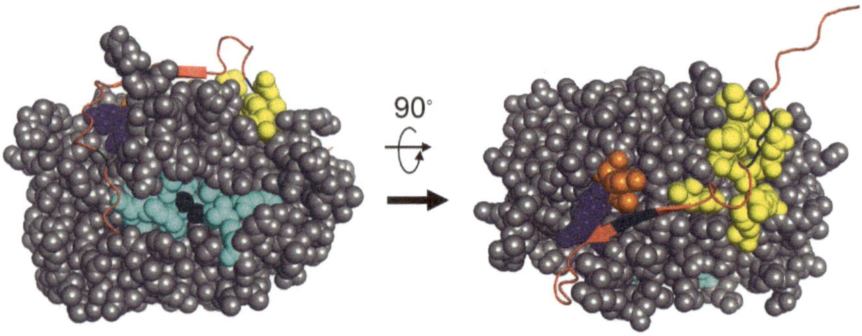


**Fig. 5** Interactions between PP1C and PNUTS. The structures correspond to PDB 4MOY coordinates for human PP1Cα and rat PNUTS (residues 390–419) (Choy et al., 2014). The PP1C catalytic site is shown in cyan (metals in black), RVXF-docking site in yellow, residues interacting with the ϕϕ-motif in orange, and residues interacting with the Arg-motif in deep blue. The corresponding PNUTS motifs are shown in lack on an otherwise red cartoon structure.

PNUTS is involved in chromosome modification in anaphase (Landsverk, Kirkhus, Bollen, Küntziger, & Collas, 2005; Lee, You, Dobrota, & Skalnik, 2010). It has been found associated with the RNA polymerase II (RNA Pol II) and, *in vitro*, PNUTS:PP1C dephosphorylates the Ser-5 of the C-terminal domain of RNA Pol II (Ciurciu et al., 2013; Jerebtsova et al., 2011). In consequence, PNUTS plays crucial role in regulation of RNA Pol II-dependent transcription. Other possible PNUTS substrates have been identified, such as p110Rb, p54 and MDM2.

As NIPP1, PNUTS is involved in pre-mRNA splicing, chromatin decondensation in DNA damage response, and maintenance of telomere stability, as well as modulation of p110Rb and PTEN among other functions (Landsverk et al., 2010, 2005; Zhu, Fisher, Bessho, & Peng, 2017).

## 5.5 The myosin phosphatase targeting subunit family

This family of mostly vertebrate Myosin Phosphatase Targeting Protein regulatory subunits is composed by PPP1R12A, PPP1R12B, PPP1R12C, PPP1R16A and PPP1R16B (Table 1 and Fig. 6A) and they share several common structural motifs required to target $PP1C_{\beta/\delta}$ to specific localization and/or substrates. A set of conserved motifs as the RVXF and the MYPhoNE in addition to a series of 29 to 33-residues ankyrin repeats are common among the members of this family. It is known that these proteins

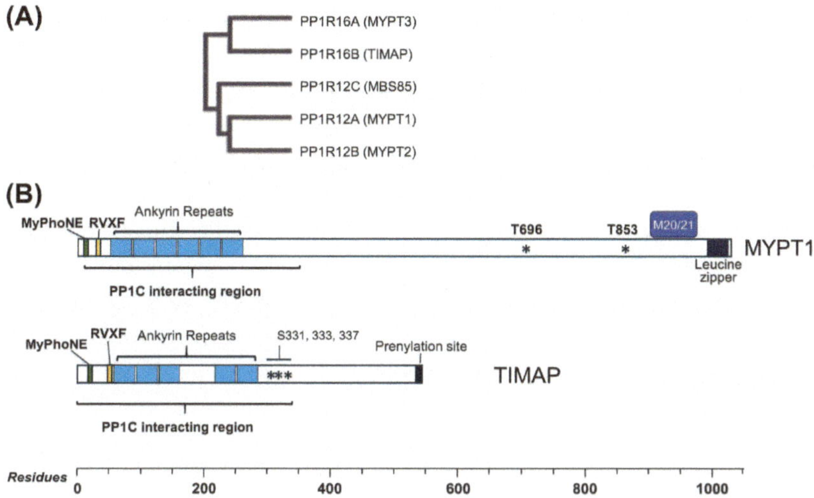


**Fig. 6** (A) Phylogenetic relationship between PP1R12 and PP1R16 members. (B) Cartoon depicting the most relevant features of a representative of the PP1R12 family (MYPT1) and the PP1R16 family (TIMAP). Motifs have been extracted from the literature and by Scansite analysis.

are also regulated by phosphorylation by several protein kinases leading to inhibition of the myosin phosphatase activity (Grassie, Moffat, Walsh, & MacDonald, 2011).

### 5.5.1 PPP1R12A (MYPT1)

PPP1R12A is known as the myosin-binding subunit of myosin phosphatase or myosin phosphatase target subunit 1 (MYPT1) and was first identified as involved in regulation of cell morphology and motility. This is a vertebrate-specific protein that, when interacts with PP1C (exclusively with the PP1Cβ/δ isoform) and with the 20/21 kDa microtubules binding protein (M20/21), conforms the myosin phosphatase (MP) trimeric holoenzyme (Scotto-Lavino et al., 2010; Xia, Stull, & Kamm, 2005). Although the M20/21 subunit binds to the C-terminal region of MYPT1 (Fig. 6B), it does not have effect on PP1C activity. MP regulates the phosphorylation state of the myosin light chain and, in consequence, regulates the function of smooth muscle contractility (Ito, Nakano, Erdodi, & Hartshorne, 2004).

Four isoforms (from A to D) have been identified in human, consequence of splicing variants, ranging from the isoform A (1030 residues) to the shortest D (974 amino acids). These proteins are predicted to have several ankyrin repeats in their N-terminal region (from residues 38 to

296), except isoform B, that does not present the first repeat and part of the second since it lacks the 87 most N-terminal residues. The ankyrin repeats domain consists of 7 or 8 ankyrin repeats (although some prediction software detects only six), where each repeat folds in a conserved β-hairpin/loop-helix-loop-helix structure that promotes protein-protein interaction [see (Grassie et al., 2011) for a review]. Three coiled-coil domains are predicted in all four isoforms, being the closest to the C-terminal end of the protein shorter in isoform C which lacks, therefore, a cGMP-dependent protein kinase interacting domain (Sharma et al., 2008; Surks & Mendelsohn, 2003). PP1R12 is a protein in which the low complexity regions are abundant. A RVXF motif, essential for PP1C binding, has been identified by plasmon resonance near the N-terminal end ($^{35}$KVKF$^{38}$). Once the KVKF motif interacts with PP1C, a most N-terminal $^{10}$RNEQLKRW$^{17}$ MyPhoNE motif become available to also interact with PP1C. Binding of MyPhoNE to PP1C have a steric effect, placing the N-terminus of MYPT1 near the catalytic site of PP1C, where it may interact with myosin substrate (Terrak et al., 2004). Interaction of the RVXF and MyPhoNE motifs (that are absent in isoform B) with PP1C facilitates the binding of two more regions: i) from 40 to 296 (the ankyrin repeats) that specifically binds to a region of the unstructured C-terminal tail of PP1Cβ/δ (Scotto-Lavino et al., 2010), and ii) from 304 to 511 (Tóth et al., 2000). This model implies that they are multiple and hierarchical interaction between PP1C and MYPT1. The MyPhoNE motif is common to other members of the PP1CR12 family.

MYPT1 is involved in regulation of cell morphology and motility. The MP controls the phosphorylation state of the 20 kDa myosin light chain, thus regulating smooth muscle contractility. Although several phosphorylation sites have been identified in MYPT1 it has been demonstrated (reviewed in (Grassie et al., 2011)) that phosphorylation of two threonine residues (Thr-696 and Thr-853) located in the C-terminal region has an inhibitory effect on MP activity (Murányi et al., 2005). It has been hypothesized that phosphorylation of these threonine residues triggers structural changes in MYPT1 that allows the interaction of a phosphorylated inhibitory peptide (C-kinase-dependent Phosphatase Inhibitor or CPI-17) with the substrate binding groove and the catalytic center of PP1C (Kiss, Erdődi, & Lontay, 2019). Phosphorylation of Thr-696 and Thr-853 was first identified as a dependent on RhoA and Rho-kinase but other protein kinases could also participate in these events, such as the Zipper-Interacting Protein Kinase (ZIPK), the Integrin-Linked Kinase (ILK), myotonic dystrophy kinase, Raf-1, or p21-activated protein kinase (Ito, Nakano, Erdodi, & Hartshorne, 2004; Murányi et al., 2005).

In addition to the well-stablished cellular role as regulator of cell morphology and motility, new unrelated functions for MYPT1a s component of the MP have been unveiled during the last few years. Thus, MYPT1 plays a role in cell cycle, development, cell migration, adhesion, gene expression regulation, neurotransmitter release, tumorigenesis among other [see (Kiss, Erdődi, & Lontay, 2019) for an extensive and recent review]. For example, recent evidences indicate that MYPT1 suppresses tumor angiogenesis and inhibits gastric cancer proliferation through inhibition of RhoA phosphorylation (Liang et al., 2018; Wang & Sun, 2018). In fact MYPT1 has been found down-regulated in gastric cancer cells (Wang & Sun, 2018). Several loss-of-function variants of MYPT1 have been also identified that are associated with human diseases (Hughes et al., 2020).

### *5.5.2 PPP1R12B (MYPT2, PP1bp55)*

In humans, MYPT2 is present as several isoforms with important differences in structure, ranging from 1043 to 186 residues (isoform h and g, respectively). The longest protein has a domain distribution similar to the MYPT1 proteins, with several ankyrin repeats in its N-terminal end and two coiled-coil regions, one predicted between the ankyrin repeats, and the second and longest, located at the C-terminus, lies, in fact, in the cAMP-dependent protein kinase binding domain, characteristic of the PPP1R12 family. Only the two largest isoforms (h and a) share both N— and C-terminal domains. Two more isoforms are C-terminal truncated versions of the protein (isoforms e and f), but retain the complete set of ankyrin repeats, and three transcripts encode for shortest isoforms, containing the few last exons (isoforms d, c and g). The phosphorylatable Thr-696 of MYPT1 is conserved in PP1C12B and PP1C12C but Thr-853 is only conserved in PPP1R12B.

MYPT2 is not limited to vertebrates since related sequences have been found in insects and worm. Fruit fly and worm proteins are most similar in primary structure (mainly in its N-terminal region) to isoforms f or e, but its domain organization is analog to the longest human isoforms a or h. Several isoforms (c, d and g) lack the N-terminal PP1C-binding sites. Interestingly, the structure of the ortholog protein in fishes is similar to that of isoforms c, d and g, keeping the C-terminal coiled-coil domain that binds cGMP-dependent protein kinase but lacking the PP1C-binding sites (Dippold & Fisher, 2014).

In contrast to the widely expressed MYPT1, MYPT2 is more specific of cardiac and skeletal muscle and brain. However, both proteins share several features. MYPT2 also conform a trimeric MP by binding to M20/21 and to PP1Cβ/δ, increasing the phosphatase activity toward phosphorylated myosin. Ankyrin repeats are detected in the N-terminal two-thirds region (between residues 57 to 306) of MYPT2 (Fujioka et al., 1998). Its C-terminal domain (956–982), present in isoforms h, a, d, c and g, has leucine-zipper motifs that binds to the active form of RhoA. MYPT2 is also a target for the Rho-kinase (Okamoto et al., 2006). The longest isoforms h and a conserve the two inhibitory phosphorylation sites for the Rho kinase). The first site, equivalent to Thr-696 of MYPT1, corresponds to Thr-707 in MYPT2 isoform h and to Thr-646 in isoform a. The second site (Thr-853 in MYPT1) is Thr-869 in the longest isoform h, and is also present in isoforms a, d and c.

### *5.5.3 PPP1R12C (MBS85)*

MBS85 is involved in assembly of the actin cytoskeleton. Ubiquitously expressed in human tissues, it was identified as a substrate of myotonic dystrophy kinase-related Cdc42 binding kinase-α (MRCKα) that mediates Cdc42-dependent actin reorganization (Tan, Ng, Lim, & Leung, 2001). Four splicing forms are detected, with minimal changes giving rise to deletion of up to two residues. Isoform a is the largest form, with 782 residues, and isoform b, lacking Gln-410 and Ala-609, is the shortest. The $^{45}$RTVRF$^{49}$ sequence conforms the RVXF motif for PP1C interaction. All isoforms maintain the typical domains of this family, with several predicted ankyrin repeats in the N-terminal half of the protein and the cGMP-dependent kinase interacting domain spanning along the last 100 residues of the C-terminal end.

## 5.6 The PPP1R16 family

This family is composed by PPP1R16A (MYPT3) and PPP1R16B (also known as TGF-β Inhibited Membrane Associated Protein, TIMAP). Both proteins contain a RVXF motif ($^{66}$QVLF$^{69}$ in MYPT3 and $^{63}$QVLF$^{66}$ in TIMAP) and are present not only in vertebrates, but orthologous proteins are encoded on the genome of cnidarians and anthropods. Identified isoforms of MYPT3 range from 582 to 519 residues and contain several ankyrin repeats on the N-terminal portion of the proteins. Four variants of TIMAP, consequence of alternative splicing, have been identified, ranging from 571 to 368 amino acids, being ankyrin repeats also predicted in the N-terminal region of the protein. The shorter isoform X2 lacks the first segment of ankyrin repeats, whereas the longest isoform X1 possesses a different N-terminal sequence. TIMAP form a complex with PP1Cβ/δ in endothelial cells, where is highly expressed.

MYPT3 and TIMAP may bind to the membrane through their respective prenylation consensus CAAX box site, where C is the prenylated Cys, AA were initially defined as aliphatic amino acids and X is the C-terminal residue of the protein (Casey, 1992) ($^{564}$CRIS$^{567}$ in TIMAP and $^{525}$CLLM$^{528}$ in MYPT3) (Boratkó & Csortos, 2017). ROK phosphorylation sites found in the members of the PP1C12 family (MYPT1, MYPT2 and MBS85) are missing in the PPP1R16 proteins. However, three serine residues in the central region of MYPT3 (Ser-340, Ser-341 and Ser-353), identified as *in vivo* targets for PKA, are absent in the PPP1R12 members. PKA phosphorylation would have the opposite effect to that of other know kinases that phosphorylate MYPT family members, since PKA-phosphorylated MYPT3 would act as a PP1c activator (Yong, Tan, Lim, & Leung, 2006). It has been proposed a regulatory mechanism in which phosphorylation of the serine residues of the central domain of MYPT3 by PKA disrupts the interaction between the phosphorylated central region and the ankyrin repeats, resulting in activation of PP1C (Yong et al., 2006). Although the MYPT3 phosphorylation sites for PKA are conserved in TIMAP and it has been demonstrated that TIMAP is a substrate form PKA, the proposed regulatory mechanism is different from that described for MYPT3. PKA phosphorylation of TIMAP Ser-337 (equivalent to Ser-340 of MYPT3) is necessary to prime the phosphorylation of Ser-333 by GSK3β (not conserved in MYPT3). This phosphorylation controls PP1C activity by reducing the association between TIMAP and PP1C, thus enhancing PP1C activity. Additionally, the more active (P)-TIMAP: PP1C complex would auto-dephosphorylate TIMAP at these residues, thus acting as a negative feed-back loop to fine tune PP1 activity (Li et al., 2007).

TIMAP is also phosphorylated by PKCα in Ser-331, a residue that is not conserved in MYPT3. Phosphorylation by PKCα is a mechanism for endothelial barrier regulation, by inhibiting dephosphorylation of phospho-ERM (ezrin-radixin-moesin) proteins (Boratkó & Csortos, 2017a). Dephosphorylation of ERM proteins by PP1C:TIMAP causes its activation as cross-linkers between integral membrane proteins and the underlying actin filaments, having effects cell adhesion, motility and microvilli dynamics (Boratkó & Csortos, 2017). TIMAP might be involved in different cellular processes that are regulated by TGF-β, such as cell proliferation/apoptosis or cell adhesion [see (Boratkó & Csortos, 2017) for a recent review].

MYPT3 seems to target the myosin light chain phosphatase to membranes of endothelial cells, where it could be a regulator of the actin cytoskeleton dynamics. It is broadly expressed being lymph node, kidney and brain the tissues with higher levels of expression (Grassie et al., 2011).

## 5.7 The PPP1R13 (ASPP) family

ASPP proteins where first identified in fruit fly as interactors with PP1C (Bennett, Lyulcheva, & Alphey, 2006). The mammalian Apoptosis-Stimulating Protein of p53/Ankyrin repeat-, SH3-domain, and Pro-rich region containing proteins (ASPP) family comprises ASPP1, ASPP2 and iASPP (inhibitor of ASPP), having ASPP1 and 2 a consensus RVXF motif ($^{927}$RVRF$^{930}$ in isoform 1 of ASPP2) that differs from the $^{622}$RARL$^{625}$ found in the equivalent position of iASPP. This sequence precedes the ankyrin repeats that, in ASPP2, is predicted to span from 936 to 1029. A C-terminal SH3 domain is present in all three proteins. The N-terminus of ASPP1 and ASPP2 share a Ras-associated and a long coiled-coil domain (from residues 10 to 91 and from 129 to 330, respectively, in ASPP2 isoform 1), that are missing in iASPP. In fact, it has been proposed that iASPP is an inhibitor of the p53-dependent apoptotic process whereas ASPP1 and 2 are inductors of the same process [see (Sullivan & Lu, n.d.) for a complete review]. Very recently it has been determined the mechanism by which the SH3 domain of ASPP proteins discriminates between the different C-terminal tail of the PP1C isoforms (Bertran et al., 2019).

### *5.7.1 PPP1R13A or ASPP2*

All three identified isoforms of human PPP1R13A (ASPP2) differ only in their N-terminus, consequence of alternative splicing, giving rise to proteins ranging from 1134 (isoform 1) to 1005 residues (isoform 2). A RA (Ras Association) domain, similar to the Ras-GTP-binding domain found in the Raf kinase, is located at the N-terminal region of the isoform 1 (8–91), a segment that is previous to the initial methionine of the N-terminal-truncated isoforms 2 and X1 that, therefore, lack the mentioned domain.

ASPP2 is ubiquitously expressed and localizes to the perinuclear region of the cytoplasm and tight junctions. It controls cell proliferation/apoptosis, cell-cell contact remodeling and gene transcription through interaction with other regulatory molecules such as members of the p53 family (Helps, Barker, Elledge, & Cohen, 1995; Sottocornola et al., 2010). The possible role of PP1 in these p53-related functions has not been determined, although p53BP2, a fragment of ASPP2, was identified as an inhibitor of PP1C and it was hypothesized to be a regulator of p53 (Helps et al., 1995). ASPP proteins associated with PP1Cα and PP1Cβδ facilitate the interaction and dephosphorylation of proteins (such as Hec1) involved in the correct chromosome segregation and kinetochore-microtubule attachments (Zhang et al., 2015).

### *5.7.2 PPP1R13B or ASPP1*

Alternative splicing of the ASPP1 transcript gives to several isoforms of ASPP1, from the largest isoform X1 (containing 1193 residues) to the shortest X5 (873 residues). Except for isoform X5, these proteins have a Ras-Association domain near or at the N-terminal end. Ankyrin repeats region and a SH3 domain are located in all isoforms at the C-terminal end. ASPP1 is ubiquitously expressed but higher levels are found in testis and thyroid. ASPP1 induces the binding of p53-family proteins to promoters of proapoptotic genes, increasing their transcription to promote to apoptosis (Samuels-Lev et al., 2001).

### *5.7.3 PPP1R13-like or iASPP*

PPP1R13-Like (L) or iASPP is the shortest member of the family, since lacks about 300 residues of the N-terminal region of the longest isoforms of ASPP proteins. Therefore, iASPP only conserves the ankyrin repeats region and the SH3 domain at the C-terminal end. iASPP is the most evolutionary conserved member of this family, as it present in invertebrates (Protostomia and Eleutherozoa) and has been characterized in the worm *C. elegans* (Bergamaschi et al., 2003). Its structure resembles that of 53BP2, a 528 residues-C-terminal fragment of ASPP2 that has been described as its dominant negative fragment since competes with ASPP proteins for the binding to p53. This implies that iASPP inhibits the ability of ASPP proteins to stimulate the apoptotic function of p53 (Samuels-Lev et al., 2001).

Very recently the structures of PP1C bound to iASPP have been resolved revealing that the iASPP and PP1C interact through four different contact points that define a discontinuous interface (Zhou et al., 2019). Such interactions are mediated by the iASPP SH3 domain that binds to the PXXPXR motif of the C-terminal tail of PP1C, and by the $^{609}$SVLR$^{612}$ and $^{622}$RARL$^{625}$ regions, somewhat similar to the RVXF and SILK motifs, respectively (Bollen, Peti, Ragusa, & Beullens, 2010). In addition, conformational changes induced by the described interactions would position the ankyrin regions in a way that the first ankyrin repeat would define a fourth interaction interface. It has been proposed that plasticity of this interface may be key to determining the iASPP-mediated recruitment of p53 to PP1C prior to p53 dephosphorylation (Zhou et al., 2019). The consequence of the *in vitro* interaction of PP1Cα with iASPP is the increase of the phosphatase activity on phosphorylated Ser-15 of p53. This dephosphorylation relies on the integrity of the multiple contacts with PP1Cα (Zhou et al., 2019).

An *in vivo* role for iASPP, in which PP1C is required, is the cytokinetic complete abscission of the intercellular bridge that connects the mitotic cells. In this process iASPP targets PP1C to the 55 kDa centrosomal protein (CEP55), a cytokinetic abscission regulator that needs to be dephosphorylate in Ser-346 in late mitosis (K. Gao et al., 2018). In contrast to other members of the family, iASPP binds to PP1C through a non-canonical motif ($^{812}$RNYF$^{815}$) located at the end of its SH3 domain. In this motif Phe-815, not conserved in ASPP1 and 2 is essential for the iASPP-PP1C interaction (Llanos et al., 2011).

## 5.8 The PPP1R14 family

This is a family of small inhibitory subunits composed by PPP1R14A (CPI-17), PPP1R14B (PHI), PPP1R14C (KEPI) and PPP1R14D (GBPI-1) (Table 1) and are activated as inhibitors upon PKC phosphorylation (Liu, Gong, & Uhl, 2005). Members of this family are more broadly distributed that other myosin phosphatase regulatory subunits since ortholog proteins are encoded in fly and worm genomes (Dippold & Fisher, 2014). They are unrelated in sequence to other small PP1C inhibitors. A common feature of this family is that all members contain a central PP1 inhibitory domain (PP1 Holoenzyme Inhibitory or PHIN domain) in which a phosphorylatable Thr residue (Thr-38 in CPI-17) is located near the N-terminus of this domain (Fig. 7). Phosphorylation of this Thr is required to convert the protein into a Myosin Light Chain Phosphatase (MLCP) inhibitor (Eto, Senba, Morita, & Yazawa, 1997). Since contractility of vascular smooth muscle is hormonally controlled through the regulation of the MLCP activity, phosphorylation of the Thr residue is one of the regulatory mechanisms of this process. Defects in the PP1C14 family of proteins have been linked to a diversity of diseases, including cardiovascular disease and cancer. Further details about structure and functions of the members of this family are detailed in (Eto, 2009).

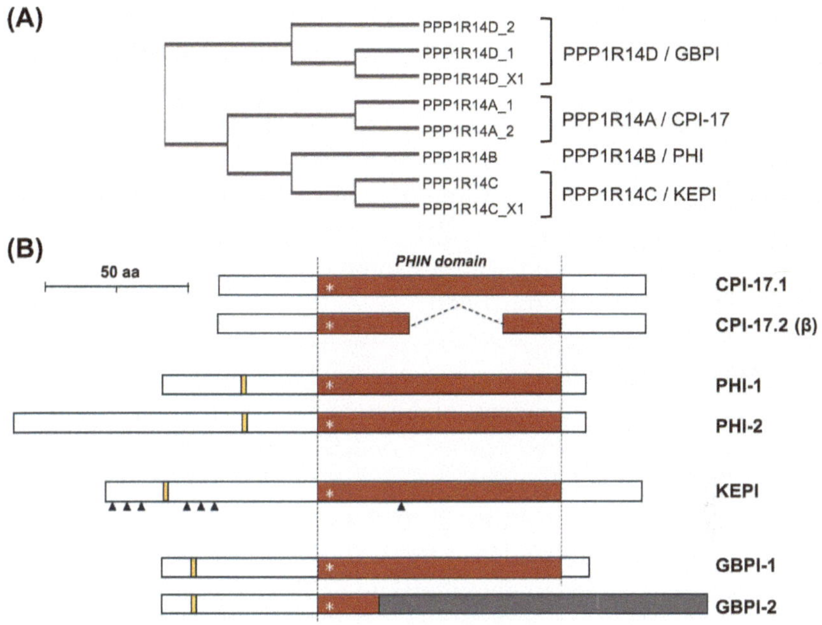


**Fig. 7** (A) Phylogenetic relationship between the different members of the PPP1R14A-D family. (B) Cartoon depicting the most relevant features of PPP1R14 proteins. Red boxes denote the PP1 Holoenzyme Inhibitory (PHIN) domain. Yellow boxes correspond to the RVXF motif. Triangles identify the diverse myristylation motifs found at the N-terminus in KEPI. Asterisks indicate the conserved phosphorylatable Thr residues present in all variants at the beginning of the PHIN domain.

### 5.8.1 *PPP1R14A (CPI-17)*

This 147-aminoacid protein was first identified as highly expressed in smooth muscle tissues such as aorta and bladder, with lower levels of expression in cardiac and skeletal muscle (Eto et al., 1997). A distinctive characteristic of CPI-17 is that, unlike other members of the family, it does not contain a RVXF motif in its N-terminal tail domain (Fig. 7).

PKC, Rho-dependent kinases or ILK phosphorylate Thr-38 of CPI-17 in response to agonist stimulation. This phosphorylation induces a conformational change in one of the four helices that creates new interactions with the catalytic site of PP1C, increasing its inhibitory capacity and thus acting as molecular switch for regulation of the smooth muscle contraction (Eto, 2009; Ohki et al., 2001). Two isoforms caused by alternative splicing have been identified, being isoform 2 (β) twenty-seven residues shorter since it lacks an acidic region comprised between position 68 and 94 of isoform 1 (Fig. 7). This shorter region is still predicted to behave as an inhibitory domain.

### 5.8.2 PPP1R14B (PHI)

Initially identified as Phospholipase C β neighboring gene (PNG or PHI-1) (Lagercrantz, Carson, Larsson, Nordenskjöld, & Weber, 1996), PPP1R14B is a 147-aminoacid protein ubiquitously expressed, with the highest expression levels detected in skin, appendix, placenta, colon and esophagus. In humans no alternative splicing forms have been described, but an alternative translational start site produces PHI-2, a protein of 203-residues with a longer N-terminal tail and whose expression is limited to muscle tissue (Eto, Karginov, & Brautigan, 1999).

### 5.8.3 PPP1R14C (KEPI)

Identified as a PKC-dependent inhibitor of PP1C (and hence named Kinase-Enhanced protein Phosphatase I or KEPI), this protein is phosphorylated by PKC in the KVT$^{73}$VK context (the Thr and neighbor residues are conserved in members of the family) and this phosphorylation is sufficient to convert this protein into a PP1C inhibitor (Liu et al., 2002). Other protein kinases described to phosphorylate KEPI are PKA (in RKLS$^{158}$, also conserved in PHI and in isoform 1 of GBPI), and CK-II (at S$^{43}$SRE, only found in one splicing variant of KEPI, and S$^{121}$DEE, also present in PHI, and the longest isoform of CPI-17). Several myristoylation sites are predicted in the N-terminal region. KEPI is expressed in heart, skeletal muscle and brain, where was found up-regulated upon morphine treatment of mice (Liu et al., 2002). KEPI affects neuronal activity by regulating multiple cellular processes and has been categorized as a potential tumor suppressor since it has been found down-regulated, on breast cancer and melanoma cell lines (Bonazzi, Irwin, & Hayward, 2009; Wenzel et al., 2007). A protective role of KEPI, promoting TNFα-driven inflammation in the host response again SASR-CoV infection, has been also identified by using mice models (McDermott et al., 2016).

### 5.8.4 PPP1R14D (GBPI)

Because of the tissues where this inhibitor is highly expressed, it has been named Gut and Brain Phosphatase Inhibitor (GBPI). mRNA of isoform 1 and 2 encode for proteins of 145 (GBPI-1) and 200 amino-acids (GBPI-2), respectively. GBPI-1 contains a complete PHIN domain, whereas in the longest version this domain is incomplete, and the amino acid sequence differs from GBPI-1 from residue 86 onwards (Fig. 7). GBPI-1 is phosphorylated by CK-II, PKA and PKC. Expression of mouse GBPI-2 was higher in testis (Liu et al., 2004). Expression of the PPP1R14D gene has been found altered in several conditions. It is overexpressed in brain of rats diabetic condition (Karthik & Ravikumar, 2011) and this gene is hypomethylated and down-regulated in human pancreatic adenocarcinoma (Zhang et al., 2020).

## 5.9 The PPP1R15 family (eIF2$\alpha$ phosphatases)

Composed of two members, GADD34 (PPP1R15A) and CReP (PPP1R15B), with predicted PP1C-binding region in their C-terminal region. These proteins form a complex with PP1C through a RVXF motif ($^{565}$KVRF$^{568}$ in GADD34 and $^{640}$KVTF$^{643}$ in CReP) within the PP1 binding domain. Both constitutive CReP—PP1C and inducible GADD34-PP1C complexes, are involved in the recruitment of PP1C in order to keep dephosphorylated eIF2$\alpha$ in the absence of stress, maintaining the cell in an active translation status. Dephosphorylation of eIF2$\alpha$ counteracts the effect of the GCN2, PKR, PERK and HRI kinases in response to amino acid starvation, viral dsRNA, ER stress and iron deficiency, respectively (Dalton, Healey, Irving, & Marciniak, 2012). Interestingly, a herpes simplex viral gene codes for an ortholog protein that drives mammalian PP1C to specifically dephosphorylate eIF2$\alpha$, escaping the inhibitory effect of PKR activation in response to the presence of viral dsRNA in the cell. As a result, the infected cell will keep protein translation active (Chou & Roizman, 1994; Jousse et al., 2003). Although orthologues of this family of proteins are absent in yeast cells, the yeast eIF2$\gamma$ subunit, encoded by *SUI4*, contains a RVXF motif in its N-terminal extension ($^{48}$KVAF$^{51}$) that recruits Glc7, the *S. cerevisiae* PP1C (Rojas, Gingras, & Dever, 2014), and perhaps also the PP1C-like phosphatase Ppz1 (Calafí et al., 2020), to dephosphorylate yeast eIF2$\alpha$.

### *5.9.1 PPP1R15A (GADD34)*

The human DNA-damage-induced transcript 34 (GADD34) is a 674-aminoacid protein whose mRNA levels increase upon stressful conditions. Broadly expressed at high levels in bone marrow, it has been defined as a part of a negative feedback regulatory loop that promotes translation recovery at the end of stress-induced responses (Ma & Hendershot, 2003). GADD34 is one of the stress-response proteins that are translated when eIF2$\alpha$ is phosphorylated, as a consequence of the activation of the ATF4 transcription factor, and facilitates the dephosphorylation of the initiation factor to restore normal protein synthesis (Brush, Weiser, & Shenolikar, 2003).

GADD34 is an intrinsically disordered protein with an ER-targeting helix and four central PEST repeats. It binds to PP1C via RVXF ($^{555}$KVRF$^{558}$) and ΦΦ motifs ($^{564}$VH$^{565}$), located at the C-terminal region. GADD34 recruits, in an independent manner, PP1C and eIF2α, for dephosphorylation of the initiation factor at the stress recovery phase (Choy et al., 2015).

GADD34 orthologous proteins are present in multicellular organisms. It should be noted the existence of short viral proteins that mimic the function or potentiate the expression of GADD34 to force PP1C to counterbalance the effect of the eIF2α kinases in response to the viral infection (Fung, Liao, & Liu, 2016). For example, the protein 7 of the porcine transmissible gastroenteritis cononavirus (TGEV) is a 78-residues peptide that, *in vivo*, binds to PP1C through its RVXF motif ($^{59}$RVIF$^{62}$). This "heterologous" protein 7-PP1C complex is required avoid mRNA and rRNA degradation as well as to maintain eIF2α active, counteracting thus the host antiviral translational shutoff response (Cruz et al., 2011).

### 5.9.2 PPP1R15B (CReP)

The Constitutive Repressor of eIF2α Phosphorylation (CReP) is a 713-aminoacid protein. A shorter isoform, originated by alternative splicing (X1, 670 aminoacids), that differs in the last 30 C-terminal residues, and hence, lacks the RVXF motif in the PP1C-binding domain, has been also detected. Unlike GADD34, expression of CReP is constitutive (Jousse et al., 2003) and PP1C-CReP is the responsible for keeping eIF2α dephosphorylated under basal condition. Stress-triggered translation inhibition decreases translation of the short-lived CReP, which contributes to increase phosphorylation levels of eIF2α (Jousse et al., 2003). Decreased binding of PP1C to eIF2α and impairing of eIF2α dephosphorylation has severe development and metabolic consequences. The change of the conserved Arg-658 by Cys in the PP1C-binding domain near the C-terminus, is responsible of a human syndrome that includes, among other manifestations, β-cell apoptosis and microcephaly (Abdulkarim et al., 2015).

## 5.10 PPP1R43 (AKAP1, AKAP149)

This is one of the A-Kinase Anchoring Protein that binds to regulatory subunits of PKA and anchors them to the mitochondrion. It is a 903-aminoacids protein with a predicted transmembrane region in the N-terminus (7–26) that localizes the protein to the mitochondrial and ER/nuclear envelope. Also, a KH (K homology RNA-binding) domain (residues 606–676) and a TUDOR domain (residues 756–815), present in several RNA-binding proteins, are predicted and it has been demonstrated that the KH domain of AKAP149 bind RNA *in vivo* and is required for the self-association of AKAP149 (Rogne et al., 2006). AKAP149 is present in vertebrates and broadly expressed in human tissues, with higher levels of mRNA detected in testis. AKAP149 was initially identified as required for the regulation of nuclear and chromosome dynamics at the end of mitosis. AKAP149 recruits PP1C to the nuclear envelope prior to assembly of B-type lamins and, in adipocytes, may anchor PKA and PP1C to the lipid droplet (Bridges, MacDonald, Wadzinski, & Moorhead, 2006; Steen, Martins, Taskén, & Collas, 2000).

*In vitro* binding of AKAP149 to PP1C is inhibited by phosphorylation of serine residues flanking the putative RVXF motif ($^{154}$GVLF$^{157}$). On the contrary, PKC phosphorylation of AKAP149 *in vitro* promotes the dissociation of PP1C (Küntziger, Rogne, Folstad, & Collas, 2006). Rogne and collaborators (Rogne, Stokka, Taskén, Collas, & Küntziger, 2008) determined that the primary *in vivo* binding site for PP1C is a second RVXF motif ($^{627}$RYVSF$^{631}$) located within the KH domain which explain the observation that binding of PP1C and RNA to AKAP149 is mutually exclusive. Phosphorylation of Ser-630, included in this C-terminal RYVSF sequence, might modulate the interaction since it decreases the binding to PP1C (Rogne et al., 2008). Expression of versions of AKAP149 mutated in its KH domain that are deficient in the binding to RNA leads to a phenotype of mitochondrial aggregation.

## 5.11 PPP1R55 (KNL1)

The kinetochore scaffold 1 (KNL1) is a 2342-aminoacid long protein. It is present in vertebrates, and participates in a multiprotein assembly that is necessary for the creation of kinetochore-microtubule attachments and chromosome segregation, coordinating the spindle assembly checkpoint (SAC) (Bajaj, Bollen, Peti, & Page, 2018). PP1-driven dephosphorylation of KNL1 silences the SAC and enable anaphase onset (Nijenhuis, Vallardi, Teixeira, Kops, & Saurin, 2014). High levels of expression have been detected in human testis. KNL1 binds PP1C via multiple SLiMs connected by a disordered loop. A SILK motif is critical for this interaction and its phosphorylation destabilize PP1C binding. Microtubule and PP1C binding to KNL1 is mutually exclusive since both microtubule- and PP1C -binding sites overlap (Bajaj et al., 2018).

Although lacking sequence similarity with human KLN1, the yeast Spc105/Knl1 protein contains a RVXF motif that brings PP1C (Glc7) to kinetochores. Kinetochores-localized Glc7 is required for the chromosome biorientation that precedes SAC silencing but is not necessary for stabilizing the kinetochore-microtubule attachments (Rosenberg, Cross, & Funabiki, 2011; Roy, Verma, Sim, Fontan, & Joglekar, 2019).

## 5.12 Regulatory subunits involved in mitotic exit (RepoMan and Ki-67)

These two nuclear proteins bind specifically to the PP1Cγ isoform during mitosis (Trinkle-Mulcahy et al., 2006). Their RVXF and ΦΦ motifs are located in the most conserved region among these two proteins, in addition to a particular PP1C interaction region identified in these proteins and named KiR-SLIM (Kumar et al., 2016). Both, PP1Cγ-RepoMan and PP1Cγ-Ki-67 complexes regulate histone dephosphorylation, chromatin remodeling and nuclear envelope reassembly (De Castro et al., 2017; Qian, Beullens, Lesage, & Bollen, 2013). Formation of both complexes during mitosis is dynamically regulated by the Aurora B kinase (Kumar et al., 2016).

### *5.12.1 PPP1R81 (RepoMan)*

This vertebrate-specific protein, also called RepoMan (Recruits PP1 onto Mitotic chromatin at anaphase), binds to PP1C to form an essential complex that coordinates chromatin reorganization and nuclear envelope re-assembly during the mitotic exit (anaphase-telophase transition). It is highly expressed in human testis and multiple isoforms have been described consequence of alternative splicing, being isoform 1 the longest (1023 amino acids). What makes RepoMan interesting is that it specifically binds and regulates PP1Cγ, that is the only isoform recruited to the chromatin at anaphase (Trinkle-Mulcahy et al., 2006; Vagnarelli & Earnshaw, 2012).

### *5.12.2 PPP1R105 (Ki-67)*

Ki-67 is a large nuclear protein (3256 aminoacids) required for cell proliferation that is highly expressed in human bone marrow and lymph node. Ki-67 contains a Forkhead associated domain near its N-terminus (residues 26–76), the PP1-binding region (residues 502–561), similar to RepoMan in sequence, near its RVXF motif, and also are predicted a set of 16 of the so-called KI67/chmadrin repeats, spanning from residues 1002 to 2929.

Ki-67 forms a steric and electrostatic charge barrier, that functions as a DNA surfactant, to prevent individual chromosomes from collapsing into chromatin mass when nuclear envelope is disassembled (Cuylen et al.,

2016). Expression levels of Ki-67 are used as a predictor of prognostic of many types of cancer (Byun, Lee, Hong, & Lee, 2019; Imaizumi et al., 2020).

## 6. Conclusion

In this chapter we have shown that PP1 activity is provided by a very limited number of highly conserved catalytic subunits in multicellular organisms (and only one in some unicellular eukaryotes). As other Ser/Thr phosphatases, PP1 has not a defined substrate specificity. Thus, the selection of the suitable protein substrates in the right cellular localization and time is accomplished by interaction with one or more additional regulatory proteins as part of the PP1 holoenzyme.

The number of characterized PP1 regulatory proteins largely exceed that found for other Ser/Thr phosphatases in both unicellular and multicellular organisms. In general, PP1-regulatory subunits have one of these two functions: i) they act as inhibitors of PP1C activity or ii) mediate the close interaction between PP1C and other molecules, thus favoring the direct interaction with the substrate (myosin, eIF2$\alpha$,..) or bringing PP1C to the cellular localization where the substrate is found (nucleus, membrane, or even the glycogen granule). In both cases, regulatory subunits can be, in turn, regulated by additional protein phospho/dephosphorylation mechanisms. In some cases, regulation involves the formation of trimeric complexes by binding two regulatory subunits to different regions of PP1C. However, in other cases, two subunits compete for the same binding regions with PP1C, therefore, their interaction is mutually exclusive.

The landscape of possible PP1 regulatory components is likely not fully explored. Current techniques to detect protein-protein interactions, such as the Split-BioID approach (De Munter et al., 2017), have been proposed to identify novel members of the already broad family of PP1C-binding proteins. These proteins, candidates to act as regulatory subunits, need to be further characterized to obtain the status of PP1-regulatory subunits. The identification of novel PP1C interactors has also expanded the number of motifs involved in such interaction.

The feasibility to analyze genome variations is allowing the identification of an increasing number of mutations in PP1C regulatory subunits as responsible of human syndromes and diseases that, in some cases, have been documented in this chapter. Alteration in the normal phosphorylation pattern is a major source of functional disturbance leading to disease. Thus, it is not surprising that protein kinase inhibitors have become one of the targets in which pharmaceutical companies have focused major efforts. In contrast the notion of protein phosphatases, and in particular of PP1C, as a likely drug target has been ignored for a long time. However, this conception has changed in the last few years (Ferreira, Beullens, Bollen, & Van Eynde, 2019; Thornton et al., 2019; Vagnarelli & Alessi, 2018) and there are growing evidences that PP1C regulatory subunits could be used as potential therapeutic targets to treat specific diseases.

## References

Abdulkarim, B., Nicolino, M., Igoillo-Esteve, M., Daures, M., Romero, S., Philippi, A., et al. (2015). A missense mutation in PPP1R15B causes a syndrome including diabetes, short stature, and microcephaly. *Diabetes, 64*(11), 3951–3962. http://dx.doi.org/10.2337/db15-0477.

Armstrong, C. G., Browne, G. J., Cohen, P., & Cohen, P. T. (1997). PPP1R6, a novel member of the family of glycogen-targetting subunits of protein phosphatase 1. *FEBS Letters, 418*(1–2), 210–214. http://dx.doi.org/10.1016/s0014-5793(97)01385-9.

Avanes, A., Lenz, G., & Momand, J. (2019). Darpp-32 and t-Darpp protein products of PPP1R1B: Old dogs with new tricks. *Biochemical Pharmacology, 160*, 71–79. http://dx.doi.org/10.1016/j.bcp.2018.12.008.

Ayllón, V., Cayla, X., García, A., Fleischer, A., & Rebollo, A. (2002). The anti-apoptotic molecules Bcl-xL and Bcl-w target protein phosphatase 1α to Bad. *European Journal of Immunology, 32*(7), 1847–1855. http://dx.doi.org/10.1002/1521-4141(200207)32:7<1847::AID-IMMU1847>3.0.CO;2-7.

Bajaj, R., Bollen, M., Peti, W., & Page, R. (2018). KNL1 binding to PP1 and microtubules is mutually exclusive. *Structure, 26*(10), 1327–1336. http://dx.doi.org/10.1016/j.str.2018.06.013. e4.

Bennett, D., Lyulcheva, E., & Alphey, L. (2006). Towards a comprehensive analysis of the protein phosphatase 1 interactome in Drosophila. *Journal of Molecular Biology, 364*(2), 196–212. http://dx.doi.org/10.1016/j.jmb.2006.08.094.

Bergamaschi, D., Samuels, Y., O'Neil, N. J., Trigiante, G., Crook, T., Hsieh, J. K., et al. (2003). iASPP oncoprotein is a key inhibitor of p53 conserved from worm to human. *Nature Genetics, 33*(2), 162–167. http://dx.doi.org/10.1038/ng1070.

Bertolotti, A. (2018). The split protein phosphatase system. *Biochemical Journal, 475*, 3707–3723. http://dx.doi.org/10.1042/BCJ20170726. Portland Press Ltd.

Bertran, M. T., Mouilleron, S., Zhou, Y., Bajaj, R., Uliana, F., Kumar, G. S., et al. (2019). ASPP proteins discriminate between PP1 catalytic subunits through their SH3 domain and the PP1 C-tail. *Nature Communications, 10*(1). http://dx.doi.org/10.1038/s41467-019-08686-0.

Beullens, M., & Bollen, M. (2002). The protein phosphatase-1 regulator NIPP1 is also a splicing factor involved in a late step of spliceosome assembly. *Journal of Biological Chemistry, 277*(22), 19855–19860. http://dx.doi.org/10.1074/jbc.M200847200.

Beullens, M., Van Eynde, A., Stalmans, W., & Bollen, M. (1992). The isolation of novel inhibitory polypeptides of protein phosphatase 1 from bovine thymus nuclei. *Journal of Biological Chemistry, 267*(23), 16538–16544.

Beullens, M., Van Eynde, A., Vulsteke, V., Connor, J., Shenolikar, S., Stalmans, W., et al. (1999). Molecular determinants of nuclear protein phosphatase-1 regulation by

NIPP-1. *Journal of Biological Chemistry, 274*(20), 14053–14061. http://dx.doi.org/10.1074/jbc.274.20.14053.

Beullens, M., Vulsteke, V., Van Eynde, A., Jagiello, I., Stalmans, W., & Bollen, M. (2000). The C-terminus of NIPP1 (nuclear inhibitor of protein phosphatase-1) contains a novel binding site for protein phosphatase-1 that is controlled by tyrosine phosphorylation and RNA binding. *Biochemical Journal, 352*(3), 651–658. http://dx.doi.org/10.1042/0264-6021:3520651.

Bharucha, J. P., Larson, J. R., Gao, L., Daves, L. K., & Tatchell, K. (2008). Ypi1, a positive regulator of nuclear protein phosphatase type 1 activity in *Saccharomyces cerevisiae. Molecular Biology of the Cell, 19*(3), 1032–1045. http://dx.doi.org/10.1091/mbc.E07-05-0499.

Bollen, M. (2001). Combinatorial control of protein phosphatase-1. *Trends in Biochemical Sciences, 26*(0968–0004), 426–431.

Bollen, M., Peti, W., Ragusa, M. J., & Beullens, M. (2010). The extended PP1 toolkit: Designed to create specificity. *Trends in Biochemical Sciences, 35*, 450–458. http://dx.doi.org/10.1016/j.tibs.2010.03.002.

Bonazzi, V. F., Irwin, D., & Hayward, N. K. (2009). Identification of candidate tumor suppressor genes inactivated by promoter methylation in melanoma. *Genes Chromosomes and Cancer, 48*(1), 10–21. http://dx.doi.org/10.1002/gcc.20615.

Boratkó, A., & Csortos, C. (2017). PKC mediated phosphorylation of TIMAP regulates PP1c activity and endothelial barrier function. *Biochimica et Biophysica Acta (BBA) - Molecular Cell Research, 1864*(2), 431–439. http://dx.doi.org/10.1016/j.bbamcr.2016.12.001.

Boratkó, A., & Csortos, C. (2017). TIMAP, the versatile protein phosphatase 1 regulator in endothelial cells. *IUBMB Life, 69*, 918–928. http://dx.doi.org/10.1002/iub.1695. Blackwell Publishing Ltd.

Bridges, D., MacDonald, J. A., Wadzinski, B., & Moorhead, G. B. G. (2006). Identification and characterization of D-AKAP1 as a major adipocyte PKA and PP1 binding protein. *Biochemical and Biophysical Research Communications, 346*(1), 351–357. http://dx.doi.org/10.1016/j.bbrc.2006.05.138.

Brush, M. H., Weiser, D. C., & Shenolikar, S. (2003). Growth arrest and DNA damage-inducible protein GADD34 targets protein phosphatase 1α to the endoplasmic reticulum and promotes dephosphorylation of the α subunit of eukaryotic translation initiation factor 2. *Molecular and Cellular Biology, 23*(4), 1292–1303. http://dx.doi.org/10.1128/mcb.23.4.1292-1303.2003.

Byun, S. S., Lee, M., Hong, S. K., & Lee, H. (2019). Elevated Ki-67 (MIB-1) expression as an independent predictor for unfavorable pathologic outcomes and biochemical recurrence after radical prostatectomy in patients with localized prostate cancer: A propensity score matched study. *PLoS One, 14*(11). http://dx.doi.org/10.1371/journal.pone.0224671.

Calafí, C., López-Malo, M., Velázquez, D., Zhang, C., Fernández-Fernández, J., Rodríguez-Galán, O., et al. (2020). Overexpression of budding yeast protein phosphatase Ppz1 impairs translation. *Biochimica et Biophysica Acta (BBA) - Molecular Cell Research, 1867*(8), 118727. http://dx.doi.org/10.1016/j.bbamcr.2020.118727.

Carmody, L. C., Baucum, A. J., Bass, M. A., & Colbran, R. J. (2008). Selective targeting of the γ1 isoform of protein phosphatase 1 to F-actin in intact cells requires multiple domains in spinophilin and neurabin. *The FASEB Journal, 22*(6), 1660–1671. http://dx.doi.org/10.1096/fj.07-092841.

Casey, P. J. (1992). Biochemistry of protein prenylation. *Journal of Lipid Research, 33*(12), 1731–1740. Retrieved from http://www.ncbi.nlm.nih.gov/pubmed/1479283.

Ceulemans, H., & Bollen, M. (2004). Functional diversity of protein phosphatase-1, a cellular economizer and reset button. *Physiological Reviews, 84*(1), 1–39. http://dx.doi.org/10.1152/physrev.00013.2003.

Ceulemans, H., De Maeyer, M., Stalmans, W., & Bollen, M. (1999). A capping domain for LRR protein interaction modules. *FEBS Letters, 456*(3), 349–351. http://dx.doi.org/10.1016/S0014-5793(99)00965-5.

Ceulemans, H., Stalmans, W., & Bollen, M. (2002). Regulator-driven functional diversification of protein phosphatase-1 in eukaryotic evolution. *BioEssays, 24*, 371–381. http://dx.doi.org/10.1002/bies.10069.

Ceulemans, H., Van Eynde, A., Pérez-Callejón, E., Beullens, M., Stalmans, W., & Bollen, M. (1999). Structure and splice products of the human gene encoding sds22, a putative mitotic regulator of protein phosphatase-1. *European Journal of Biochemistry, 262*(1), 36–42. http://dx.doi.org/10.1046/j.1432-1327.1999.00344.x.

Ceulemans, H., Vulsteke, V., De Maeyer, M., Tatchell, K., Stalmans, W., & Bollen, M. (2002). Binding of the concave surface of the Sds22 superhelix to the $\alpha4/\alpha5/\alpha6$-triangle of protein phosphatase-1. *Journal of Biological Chemistry, 277*(49), 47331–47337. http://dx.doi.org/10.1074/jbc.M206838200.

Chen, E., Choy, M. S., Petrényi, K., Kónya, Z., Erdődi, F., Dombrádi, V., et al. (2016). Molecular insights into the fungus-specific serine/threonine protein phosphatase Z1 in Candida albicans. *mBio,* 7(4). http://dx.doi.org/10.1128/mBio.00872-16. e00872-16.

Cheng, Y. L., & Chen, R. H. (2015). Assembly and quality control of the protein phosphatase 1 holoenzyme involves the Cdc48-Shp1 chaperone. *Journal of Cell Science, 128*(6), 1180–1192. http://dx.doi.org/10.1242/jcs.165159.

Chen, Y. H., Hansen, L., Chen, M. X., Bjørbaek, C., Vestergaard, H., Hansen, T., et al. (1994). Sequence of the human glycogen-associated regulatory subunit of type 1 protein phosphatase and analysis of its coding region and mRNA level in muscle from patients with NIDDM. *Diabetes, 43*(10), 1234–1241. http://dx.doi.org/10.2337/diabetes.43.10.1234.

Chia, P. H., Patel, M. R., & Shen, K. (2012). NAB-1 instructs synapse assembly by linking adhesion molecules and F-actin to active zone proteins. *Nature Neuroscience, 15*(2), 234–242. http://dx.doi.org/10.1038/nn.2991.

Chou, J., & Roizman, B. (1994). Herpes simplex virus 1 $\gamma$134.5 gene function, which blocks the host response to infection, maps in the homologous domain of the genes expressed during growth arrest and DNA damage. *Proceedings of the National Academy of Sciences of the United States of America, 91*(12), 5247–5251. http://dx.doi.org/10.1073/pnas.91.12.5247.

Choy, M. S., Hieke, M., Kumar, G. S., Lewis, G. R., Gonzalez-DeWhitt, K. R., Kessler, R. P., et al. (2014). Understanding the antagonism of retinoblastoma protein dephosphorylation by PNUTS provides insights into the PP1 regulatory code. *Proceedings of the National Academy of Sciences of the United States of America, 111*(11), 4097–4102. http://dx.doi.org/10.1073/pnas.1317395111.

Choy, M. S., Moon, T. M., Ravindran, R., Bray, J. A., Robinson, L. C., Archuleta, T. L., et al. (2019). SDS22 selectively recognizes and traps metal-deficient inactive PP1. *Proceedings of the National Academy of Sciences of the United States of America, 116*(41), 20472–20481. http://dx.doi.org/10.1073/pnas.1908718116.

Choy, M. S., Yusoff, P., Lee, I. C., Newton, J. C., Goh, C. W., Page, R., et al. (2015). Structural and functional analysis of the GADD34:PP1 eIF2$\alpha$ phosphatase. *Cell Reports, 11*(12), 1885–1891. http://dx.doi.org/10.1016/j.celrep.2015.05.043.

Ciurciu, A., Duncalf, L., Jonchere, V., Lansdale, N., Vasieva, O., Glenday, P., et al. (2013). PNUTS/PP1 regulates RNAPII-mediated gene expression and is necessary for developmental growth. *PLoS Genetics, 9*(10). http://dx.doi.org/10.1371/journal.pgen.1003885.

Connor, J. H., Kleeman, T., Barik, S., Honkanen, R. E., & Shenolikar, S. (1999). Importance of the beta12-beta13 loop in protein phosphatase-1 catalytic subunit for inhibition by toxins and mammalian protein inhibitors. *Journal of Biological Chemistry, 274*(32), 22366–22372. http://dx.doi.org/10.1074/jbc.274.32.22366.

Connor, J. H., Quan, H. N., Ramaswamy, N. T., Zhang, L., Barik, S., Zheng, J., et al. (1998). Inhibitor-1 interaction domain that mediates the inhibition of protein phosphatase-1. *Journal of Biological Chemistry, 273*(42), 27716–27724. http://dx.doi.org/10.1074/jbc.273.42.27716.

Connor, J. H., Weiser, D. C., Li, S., Hallenbeck, J. M., & Shenolikar, S. (2001). Growth arrest and DNA damage-inducible protein GADD34 assembles a novel signaling complex containing protein phosphatase 1 and inhibitor 1. *Molecular and Cellular Biology, 21*(20), 6841–6850. http://dx.doi.org/10.1128/mcb.21.20.6841-6850.2001.

Cruz, J. L. G., Sola, I., Becares, M., Alberca, B., Plana, J., Enjuanes, L., et al. (2011). Coronavirus gene 7 counteracts host defenses and modulates virus virulence. *PLoS Pathogens,* 7(6). http://dx.doi.org/10.1371/journal.ppat.1002090.

Cuylen, S., Blaukopf, C., Politi, A. Z., Muller-Reichert, T., Neumann, B., Poser, I., et al. (2016). Ki-67 acts as a biological surfactant to disperse mitotic chromosomes. *Nature, 535*(7611), 308–312. http://dx.doi.org/10.1038/nature18610.

Dalton, L. E., Healey, E., Irving, J., & Marciniak, S. J. (2012). Phosphoproteins in stress-induced disease. *Progress in Molecular Biology and Translational Science, 106*, 189–221. http://dx.doi.org/10.1016/B978-0-12-396456-4.00003-1. Elsevier B.V.

De Castro, I. J., Budzak, J., Di Giacinto, M. L., Ligammari, L., Gokhan, E., Spanos, C., et al. (2017). Repo-Man/PP1 regulates heterochromatin formation in interphase. *Nature Communications, 8*. http://dx.doi.org/10.1038/ncomms14048.

De Munter, S., Görnemann, J., Derua, R., Lesage, B., Qian, J., Heroes, E., et al. (2017). Split-BioID: A proximity biotinylation assay for dimerization-dependent protein interactions. *FEBS Letters, 591*, 415–424. http://dx.doi.org/10.1002/1873-3468.12548. Wiley Blackwell.

Dent, P., Campbell, D. G., Caudwell, F. B., & Cohen, P. (1990). Identification of three in vivo phosphorylation sites on the glycogen-binding subunit of protein phosphatase 1 from rabbit skeletal muscle, and their response to adrenaline. *FEBS Letters, 259*(2), 281–285. http://dx.doi.org/10.1016/0014-5793(90)80027-G.

Dey, S., Brothag, C., & Vijayaraghavan, S. (2019). Signaling enzymes required for sperm maturation and fertilization in mammals. *Frontiers in Cell and Developmental Biology,* 7(341), 1–15. http://dx.doi.org/10.3389/fcell.2019.00341. Frontiers Media S.A.

Dippold, R. P., & Fisher, S. A. (2014). A bioinformatic and computational study of myosin phosphatase subunit diversity. *American Journal of Physiology - Regulatory, Integrative and Comparative Physiology, 307*(3), R256–R270. http://dx.doi.org/10.1152/ajpregu.00145.2014.

Doherty, M. J., Moorhead, G., Morrice, N., Cohen, P., & Cohen, P. T. (1995). Amino acid sequence and expression of the hepatic glycogen-binding (GL)-subunit of protein phosphatase-1. *FEBS Letters, 375*(3), 294–298. http://dx.doi.org/10.1016/0014-5793(95)01184-g.

Doostparast Torshizi, A., Duan, J., & Wang, K. (2018). Transcriptional network analysis on brains reveals a potential regulatory role of PPP1R3F in autism spectrum disorders. *BMC Research Notes, 11*(1), 489. http://dx.doi.org/10.1186/s13104-018-3594-0.

Dunn, J. S., Mlynarski, W. M., Pezzolesi, M. G., Borowiec, M., Powers, C., Krolewski, A. S., et al. (2006). Examination of PPP1R3B as a candidate gene for the type 2 diabetes and MODY loci on chromosome 8p23. *Annals of Human Genetics, 70*(5), 587–593. http://dx.doi.org/10.1111/j.1469-1809.2005.00248.x.

Edler, M. C., Salek, A. B., Watkins, D. S., Kaur, H., Morris, C. W., Yamamoto, B. K., et al. (2018). Mechanisms regulating the association of protein phosphatase 1 with spinophilin and neurabin. *ACS Chemical Neuroscience, 9*(11), 2701–2712. http://dx.doi.org/10.1021/acschemneuro.8b00144.

Egloff, M. P., Johnson, D. F., Moorhead, G., Cohen, P. T., Cohen, P., & Barford, D. (1997). Structural basis for the recognition of regulatory subunits by the catalytic subunit of protein phosphatase 1. *The EMBO Journal, 16*(8), 1876–1887. http://dx.doi.org/10.1093/emboj/16.8.1876.

Eiteneuer, A., Seiler, J., Weith, M., Beullens, M., Lesage, B., Krenn, V., et al. (2014). Inhibitor-3 ensures bipolar mitotic spindle attachment by limiting association of SDS 22 with kinetochore-bound protein phosphatase-1. *The EMBO Journal, 33*(22), 2704–2720. http://dx.doi.org/10.15252/embj.201489054.

El-Rifai, W., Smith, M. F., Li, G., Beckler, A., Carl, V. S., Montgomery, E., et al. (2002). Gastric cancers overexpress DARPP-32 and a novel isoform, t-DARPP. *Cancer Research, 62*(14), 4061–4064.

Endo, S., Zhou, X., Connor, J., Wang, B., & Shenolikar, S. (1996). Multiple structural elements define the specificity of recombinant human inhibitor-1 as a protein phosphatase-1 inhibitor. *Biochemistry, 35*(16), 5220–5228. http://dx.doi.org/10.1021/bi952940f.

Esteves, S. L. C., Domingues, S. C., Da Cruz E Silva, O. A. B., Fardilha, M., & Da Cruz E Silva, E. F. (2012). Protein phosphatase 1α interacting proteins in the human brain. *OMICS: A Journal of Integrative Biology, 16*, 3–17. http://dx.doi.org/10.1089/omi.2011.0041.

Eto, M. (2009). Regulation of cellular protein phosphatase-1 (PP1) by phosphorylation of the CPI-17 family, C-kinase-activated PP1 Inhibitors. *Journal of Biological Chemistry, 284*, 35273–35277. http://dx.doi.org/10.1074/jbc.R109.059972.

Eto, M., Karginov, A., & Brautigan, D. L. (1999). A novel phosphoprotein inhibitor of protein type-1 phosphatase holoenzymes. *Biochemistry, 38*(51), 16952–16957. http://dx.doi.org/10.1021/bi992030o.

Eto, M., Senba, S., Morita, F., & Yazawa, M. (1997). Molecular cloning of a novel phosphorylation-dependent inhibitory protein of protein phosphatase-1 (CPI17) in smooth muscle: Its specific localization in smooth muscle. *FEBS Letters, 410*(2–3), 356–360. http://dx.doi.org/10.1016/S0014-5793(97)00657-1.

Ferreira, M., Beullens, M., Bollen, M., & Van Eynde, A. (2019). Functions and therapeutic potential of protein phosphatase 1: Insights from mouse genetics. *Biochimica et Biophysica Acta (BBA) - Molecular Cell Research, 1866*(1), 16–30. http://dx.doi.org/10.1016/j.bbamcr.2018.07.019.

Fong, N. M., Jensen, T. C., Shah, A. S., Parekh, N. N., Saltiel, A. R., & Brady, M. J. (2000). Identification of binding sites on protein targeting to glycogen for enzymes glycogen metabolism. *Journal of Biological Chemistry, 275*(45), 35034–35039. http://dx.doi.org/10.1074/jbc.M005541200.

Fujiki, H., & Suganuma, M. (2009). Carcinogenic aspects of protein phosphatase 1 and 2A inhibitors. *Progress in Molecular and Subcellular Biology, 46*, 221–254. http://dx.doi.org/10.1007/978-3-540-87895-7_8.

Fujioka, M., Takahashi, N., Odai, H., Araki, S., Ichikawa, K., Feng, J., et al. (1998). A new isoform of human myosin phosphatase targeting/regulatory subunit (MYPT2): cDNA cloning, tissue expression, and chromosomal mapping. *Genomics, 49*(1), 59–68. http://dx.doi.org/10.1006/geno.1998.5222.

Fung, T. S., Liao, Y., & Liu, D. X. (2016). Regulation of stress responses and translational control by coronavirus. *Viruses, 8*(7), 184. http://dx.doi.org/10.3390/v8070184. MDPI AG.

Futter, M., Uematsu, K., Bullock, S. A., Kim, Y., Hemmings, H. C., Nishi, A., et al. (2005). Phosphorylation of spinophilin by ERK and cyclin-dependent PK 5 (Cdk5). *Proceedings of the National Academy of Sciences of the United States of America, 102*(9), 3489–3494. http://dx.doi.org/10.1073/pnas.0409802102.

Gao, J., Hu, X.-D., Yang, H., & Xia, H. (2018). Distinct roles of protein phosphatase 1 bound on neurabin and spinophilin and its regulation in AMPA receptor trafficking and LTD induction. *Molecular Neurobiology, 55*(9), 7179–7186. http://dx.doi.org/10.1007/s12035-018-0886-2.

Gao, K., Zhang, Y., Shi, Q., Zhang, J., Zhang, L., Sun, H., et al. (2018). IASPP-PP1 complex is required for cytokinetic abscission by controlling CEP55 dephosphorylation article. *Cell Death and Disease, 9*(5). http://dx.doi.org/10.1038/s41419-018-0561-6.

García-Gimeno, M. A., Muñoz, I., Ariño, J., & Sanz, P. (2003). Molecular characterization of Ypi1, a novel *Saccharomyces cerevisiae* type 1 protein phosphatase inhibitor. *Journal of Biological Chemistry, 278*(48), 47744–47752. http://dx.doi.org/10.1074/jbc.M306157200.

Garcia, A., Cayla, X., Caudron, B., Deveaud, É., Roncal, F., & Rebollo, A. (2004). New insights in protein phosphorylation: A signature for protein phosphatase 1 interacting proteins. *Comptes Rendus Biologies, 327*(2), 93–97. http://dx.doi.org/10.1016/j.crvi.2004.01.001.

Godet, A. N., Guergnon, J., Maire, V., Croset, A., & Garcia, A. (2010). The combinatorial PP1-binding consensus motif (R/K)x (0,1)v/IxFxx(R/K)x(R/K) is a new apoptotic signature. *PloS One, 5*(4). http://dx.doi.org/10.1371/journal.pone.0009981.

Goldberg, J., Huang, H. B., Kwon, Y. G., Greengard, P., Nairn, A. C., & Kuriyan, J. (1995). Three-dimensional structure of the catalytic subunit of protein serine/threonine phosphatase-1. *Nature, 376*(6543), 745–753. http://dx.doi.org/10.1038/376745a0.

Grassie, M. E., Moffat, L. D., Walsh, M. P., & MacDonald, J. A. (2011). The myosin phosphatase targeting protein (MYPT) family: A regulated mechanism for achieving substrate specificity of the catalytic subunit of protein phosphatase type 1δ. *Archives of Biochemistry and Biophysics, 510*(2), 147–159. http://dx.doi.org/10.1016/j.abb.2011.01.018.

Grossman, S. D., Futter, M., Snyder, G. L., Allen, P. B., Nairn, A. C., Greengard, P., et al. (2004). Spinophilin is phosphorylated by Ca2+/calmodulin-dependent protein kinase II resulting in regulation of its binding to F-actin. *Journal of Neurochemistry, 90*(2), 317–324. http://dx.doi.org/10.1111/j.1471-4159.2004.02491.x.

Grusche, F. A., Hidalgo, C., Fletcher, G., Sung, H. H., Sahai, E., & Thompson, B. J. (2009). Sds22, a PPI phosphatase regulatory subunit, regulates epithelial cell polarity and shape [Sds22 in epithelial morphology]. *BMC Developmental Biology, 9*(1). http://dx.doi.org/10.1186/1471-213X-9-14.

Guerrero, R., Vernia, S., Sanz, R., Abreu-Rodríguez, I., Almaraz, C., García-Hoyos, M., et al. (2011). A PTG variant contributes to a Milder Phenotype in Lafora disease. *PloS One, 6*(6), e21294. http://dx.doi.org/10.1371/journal.pone.0021294.

Haghighi, K., Pritchard, T. J., Liu, G.-S., Singh, V. P., Bidwell, P., Lam, C. K., et al. (2015). Human G109E-inhibitor-1 impairs cardiac function and promotes arrhythmias. *Journal of Molecular and Cellular Cardiology, 89*, 349–359. http://dx.doi.org/10.1016/j.yjmcc.2015.10.004.

Hansen, L., Hansen, T., Vestergaard, H., Bjørbæk, C., Echwald, S. M., Clausen, J. O., et al. (1995). A widespread amino acid polymorphism at codon 905 of the glycogen-associated regulatory subunit of protein phosphatase-1 is associated with insulin resistance and hypersecretion of insulin. *Human Molecular Genetics, 4*(8), 1313–1320. http://dx.doi.org/10.1093/hmg/4.8.1313.

Hayashida, Y., Goi, T., Hirono, Y., Katayama, K., Urano, T., & Yamaguchi, A. (2005). PPP1R3 gene (protein phosphatase 1) alterations in colorectal cancer and its relationship to metastasis. *Oncology Reports, 13*(6), 1223–1227. http://dx.doi.org/10.3892/or.13.6.1223.

Helps, N. R., Barker, H. M., Elledge, S. J., & Cohen, P. T. W. (1995). Protein phosphatase 1 interacts with p53BP2, a protein which binds to the tumour suppressor p53. *FEBS Letters, 377*(3), 295–300. http://dx.doi.org/10.1016/0014-5793(95)01347-4.

Helps, N. R., Street, A. J., Elledge, S. J., & Cohen, P. T. (1994). Cloning of the complete coding region for human protein phosphatase inhibitor 2 using the two hybrid system and expression of inhibitor 2 in *E. coli*. *FEBS Letters, 340*(1–2), 93–98. http://dx.doi.org/10.1016/0014-5793(94)80179-7.

Hendrickx, A., Beullens, M., Ceulemans, H., Den Abt, T., Van Eynde, A., Nicolaescu, E., et al. (2009). Docking motif-guided mapping of the interactome of protein phosphatase-1. *Chemistry and Biology, 16*(4), 365–371. http://dx.doi.org/10.1016/j.chembiol.2009.02.012.

Heroes, E., Van der Hoeven, G., Choy, M. S., Garcia, J. del P., Ferreira, M., Nys, M., et al. (2019). Structure-guided exploration of SDS22 interactions with protein phosphatase PP1 and the splicing factor BCLAF1. *Structure, 27*(3), 507–518. http://dx.doi.org/10.1016/j.str.2018.12.002. e5.

Hofmann, K., & Bucher, P. (1995). The FHA domain: A putative nuclear signalling domain found in protein kinases and transcription factors. *Trends in Biochemical Sciences, 20*, 347–349. http://dx.doi.org/10.1016/S0968-0004(00)89072-6.

Hsieh-Wilson, L. C., Allen, P. B., Watanabe, T., Nairn, A. C., & Greengard, P. (1999). Characterization of the neuronal targeting protein spinophilin and its interactions with protein phosphatase-1. *Biochemistry, 38*(14), 4365–4373. http://dx.doi.org/10.1021/bi982900m.

Huang, F. L., & Glinsmann, W. (1976). A second heat-stable protein inhibitor of phosphorylase phosphatase from rabbit muscle. *FEBS Letters, 62*(3), 326–329. http://dx.doi.org/10.1016/0014-5793(76)80086-5.

Huang, F. L., & Glinsmann, W. H. (1976). Separation and characterization of two phosphorylase phosphatase inhibitors from rabbit skeletal muscle. *European Journal of Biochemistry, 70*(2), 419–426. http://dx.doi.org/10.1111/j.1432-1033.1976.tb11032.x.

Huang, W., Jin, Y., Yuan, Y., Bai, C., Wu, Y., Zhu, H., et al. (2014). Validation and target gene screening of hsa-miR-205 in lung squamous cell carcinoma. *Chinese Medical Journal, 127*(2), 272–278. Retrieved from http://www.ncbi.nlm.nih.gov/pubmed/24438615.

Huang, H. S., & Lee, E. Y. C. (2008). Protein phosphatase-1 inhibitor-3 is an in vivo target of caspase-3 and participates in the apoptotic response. *Journal of Biological Chemistry, 283*(26), 18135–18146. http://dx.doi.org/10.1074/jbc.M709735200.

Huang, H. S., Pozarowski, P., Gao, Y., Darzynkiewicz, Z., & Lee, E. Y. C. (2005). Protein phosphatase-1 inhibitor-3 is co-localized to the nucleoli and centrosomes with PP1γ1 and PP1α, respectively. *Archives of Biochemistry and Biophysics, 443*(1–2), 33–44. http://dx.doi.org/10.1016/j.abb.2005.08.021.

Hughes, J. J., Alkhunaizi, E., Kruszka, P., Pyle, L. C., Grange, D. K., Berger, S. I., et al. (2020). Loss-of-Function variants in PPP1R12A: From isolated sex reversal to holoprosencephaly spectrum and urogenital malformations. *The American Journal of Human Genetics, 106*(1), 121–128. http://dx.doi.org/10.1016/j.ajhg.2019.12.004.

Hurley, T. D., Yang, J., Zhang, L., Goodwin, K. D., Zou, Q., Cortese, M., et al. (2007). Structural basis for regulation of protein phosphatase 1 by inhibitor-2. *Journal of Biological Chemistry, 282*(39), 28874–28883. http://dx.doi.org/10.1074/jbc.M703472200.

Imaizumi, K., Suzuki, T., Kojima, M., Shimomura, M., Sakuyama, N., Tsukada, Y., et al. (2020). Ki67 expression and localization of T cells after neoadjuvant therapies as reliable predictive markers in rectal cancer. *Cancer Science, 111*(1), 23–35. http://dx.doi.org/10.1111/cas.14223.

Ingebritsen, T. S., & Cohen, P. (1983). The protein phosphatases involved in cellular regulation. 1. Classification and substrate specificities. *European Journal of Biochemistry, 132*(2), 255–261. http://dx.doi.org/10.1111/j.1432-1033.1983.tb07357.x.

Ito, M., Nakano, T., Erdodi, F., & Hartshorne, D. J. (2004). Myosin phosphatase: Structure, regulation and function. *Molecular and Cellular Biochemistry, 259*, 197–209. http://dx.doi.org/10.1023/B:MCBI.0000021373.14288.00.

Jerebtsova, M., Klotchenko, S. A., Artamonova, T. O., Ammosova, T., Washington, K., Egorov, V. V., et al. (2011). Mass spectrometry and biochemical analysis of RNA polymerase II: Targeting by protein phosphatase-1. *Molecular and Cellular Biochemistry, 347*(1–2), 79–87. http://dx.doi.org/10.1007/s11010-010-0614-3.

Ji, X., Wang, S., Tang, H., Zhang, Y., Zhou, F., Zhang, L., et al. (2019). PPP1R3C mediates metformin-inhibited hepatic gluconeogenesis. *Metabolism Clinical and Experimental, 98*, 62–75. http://dx.doi.org/10.1016/j.metabol.2019.06.002.

Joshi, R. N., Binai, N. A., Marabita, F., Sui, Z., Altman, A., Heck, A. J. R., et al. (2017). Phosphoproteomics reveals regulatory T cell-mediated DEF6 dephosphorylation that affects cytokine expression in human conventional T cells. *Frontiers in Immunology, 8*(SEP), 1163. http://dx.doi.org/10.3389/fimmu.2017.01163.

Joshi, R. N., Fernandes, S. J., Shang, M. M., Kiani, N. A., Gomez-Cabrero, D., Tegnér, J., et al. (2019). Phosphatase inhibitor PPP1R11 modulates resistance of human T cells toward Treg-mediated suppression of cytokine expression. *Journal of Leukocyte Biology, 106*(2), 413–430. http://dx.doi.org/10.1002/JLB.2A0618-228R.

Jousse, C., Oyadomari, S., Novoa, I., Lu, P., Zhang, Y., Harding, H. P., et al. (2003). Inhibition of a constitutive translation initiation factor 2α phosphatase, CReP, promotes survival of stressed cells. *Journal of Cell Biology, 163*(4), 767–775. http://dx.doi.org/10.1083/jcb.200308075.

Karthik, D., & Ravikumar, S. (2011). Characterization of the brain proteome of rats with diabetes mellitus through two-dimensional electrophoresis and mass spectrometry. *Brain Research, 1371*, 171–179. http://dx.doi.org/10.1016/j.brainres.2010.11.066.

Kelsall, I. R., Voss, M., Munro, S., Cuthbertson, D. J. R., & Cohen, P. T. W. (2011). R3F, a novel membrane-associated glycogen targeting subunit of protein phosphatase 1 regulates glycogen synthase in astrocytoma cells in response to glucose and extracellular signals. *Journal of Neurochemistry, 118*(4), 596–610. http://dx.doi.org/10.1111/j.1471-4159.2011.07345.x.

Kim, Y. M., Watanabe, T., Allen, P. B., Kim, Y. M., Lee, S. J., Greengard, P., et al. (2003). PNUTS, a protein Phosphatase 1 (PP1) NUclear Targeting Subunit: Characterization of its PP1 and RNA-binding domains and regulation by phosphorylation. *Journal of Biological Chemistry, 278*(16), 13819–13828. http://dx.doi.org/10.1074/jbc.M209621200.

Kiss, A., Erdődi, F., & Lontay, B. (2019). Myosin phosphatase: Unexpected functions of a long-known enzyme. *Biochimica et Biophysica Acta (BBA) - Molecular Cell Research, 1866*, 2–15. http://dx.doi.org/10.1016/j.bbamcr.2018.07.023. Elsevier B.V.

Kita, A., Matsunaga, S., Takai, A., Kataiwa, H., Wakimoto, T., Fusetani, N., et al. (2002). Crystal structure of the complex between calyculin A and the catalytic subunit of protein phosphatase 1. *Structure, 10*(5), 715–724. http://dx.doi.org/10.1016/S0969-2126(02)00764-5.

Kobe, B., & Kajava, A. V. (2001). The leucine-rich repeat as a protein recognition motif. *Current Opinion in Structural Biology, 11*, 725–732. http://dx.doi.org/10.1016/S0959-440X(01)00266-4.

Konopaske, G. T., Subburaju, S., Coyle, J. T., & Benes, F. M. (2015). Altered prefrontal cortical MARCKS and PPP1R9A mRNA expression in schizophrenia and bipolar disorder. *Schizophrenia Research, 164*(1–3), 100–108. http://dx.doi.org/10.1016/j.schres.2015.02.005.

Kreivi, J. P., Trinkle-Mulcahy, L., Lyon, C. E., Morrice, N. A., Cohen, P., & Lamond, A. I. (1997). Purification and characterisation of p99, a nuclear modulator of protein phosphatase 1 activity. *FEBS Letters, 420*(1), 57–62. http://dx.doi.org/10.1016/S0014-5793(97)01485-3.

Kumar, G. S., Gokhan, E., De Munter, S., Bollen, M., Vagnarelli, P., Peti, W., et al. (2016). The Ki-67 and RepoMan mitotic phosphatases assemble via an identical, yet novel mechanism. *ELife, 5*(AUGUST). http://dx.doi.org/10.7554/eLife.16539.

Kumeta, H., Ogura, K., Adachi, S., Fujioka, Y., Tanuma, K., Kikuchi, K., et al. (2008). The NMR structure of the NIPP1 FHA domain. *Journal of Biomolecular NMR, 40*(3), 219–224. http://dx.doi.org/10.1007/s10858-008-9222-x.

Küntziger, T., Rogne, M., Folstad, R. L. S., & Collas, P. (2006). Association of PP1 with its regulatory subunit AKAP149 is regulated by serine phosphorylation flanking the RVXF motif of AKAP149. *Biochemistry, 45*(18), 5868–5877. http://dx.doi.org/10.1021/bi060066s.

Lagercrantz, J., Carson, E., Larsson, C., Nordenskjöld, M., & Weber, G. (1996). Isolation and characterization of a novel gene close to the human phosphoinositide-specific phospholipase C β3 gene on chromosomal region 11q13. *Genomics, 31*(3), 380–384. http://dx.doi.org/10.1006/geno.1996.0063.

Landsverk, H. B., Kirkhus, M., Bollen, M., Küntziger, T., & Collas, P. (2005). PNUTS enhances in vitro chromosome decondensation in a PP1-dependent manner. *Biochemical Journal, 390*(3), 709–717. http://dx.doi.org/10.1042/BJ20050678.

Landsverk, H. B., Mora-Bermúdez, F., Landsverk, O. J. B., Hasvold, G., Naderi, S., Bakke, O., et al. (2010). The protein phosphatase 1 regulator PNUTS is a new component of the DNA damage response. *EMBO Reports, 11*(11), 868–875. http://dx.doi.org/10.1038/embor.2010.134.

Lee, S. K., Moon, J. W., Lee, Y. W., Lee, J. O., Kim, S. J., Kim, N., et al. (2015). The effect of high glucose levels on the hypermethylation of protein phosphatase 1 regulatory subunit 3C (PPP1R3C) gene in colorectal cancer. *Journal of Genetics, 94*(1), 75–85. http://dx.doi.org/10.1007/s12041-015-0492-2.

Lee, J. H., You, J., Dobrota, E., & Skalnik, D. G. (2010). Identification and characterization of a novel human PP1 phosphatase complex. *Journal of Biological Chemistry, 285*(32), 24466–24476. http://dx.doi.org/10.1074/jbc.M110.109801.

Lesage, B., Beullens, M., Pedelini, L., Garcia-Gimeno, M. A., Waelkens, E., Sanz, P., et al. (2007). A complex of catalytically inactive protein phosphatase-1 sandwiched between Sds22 and inhibitor-3. *Biochemistry, 46*(31), 8909–8919. http://dx.doi.org/10.1021/bi7003119.

Letunic, I., & Bork, P. (2018). 20 Years of the SMART protein domain annotation resource. *Nucleic Acids Research, 46*(D1), D493–D496. Retrieved from https://academic.oup.com/nar/article/46/D1/D493/4429069.

Liang, Y., Zhuo, Y., Lin, Z., Jiang, F., Dai, Q., Lu, J., et al. (2018). Decreased expression of MYPT1 contributes to tumor angiogenesis and poor patient prognosis in human prostate cancer. *Current Molecular Medicine, 18*. http://dx.doi.org/10.2174/1566524018666180705111342.

Li, J., Ivansson, E., Klevebring, D., Tobin, N. P., Lindström, L. S., Holm, J., et al. (2017). Molecular differences between screen-detected and interval breast cancers are largely explained by PAM50 subtypes. *Clinical Cancer Research, 23*(10), 2584–2592. http://dx.doi.org/10.1158/1078-0432.CCR-16-0967.

Li, L., Kozlowski, K., Wegner, B., Rashid, T., Yeung, T., Holmes, C., et al. (2007). Phosphorylation of TIMAP by glycogen synthase kinase-3β activates its associated protein phosphatase 1. *Journal of Biological Chemistry, 282*(35), 25960–25969. http://dx.doi.org/10.1074/jbc.M703532200.

Lin, T. H., Chen, Y. C., Chyan, C. L., Tsay, L. H., Tang, T. C., Jeng, H. H., et al. (2003). Phosphorylation by glycogen synthase kinase of inhibitor-2 does not change its structure in free state. *FEBS Letters, 554*(3), 253–256. http://dx.doi.org/10.1016/S0014-5793(03)01097-4.

Lipskaia, L., Bobe, R., Chen, J., Turnbull, I. C., Lopez, J. J., Merlet, E., et al. (2014). Synergistic role of protein phosphatase inhibitor 1 and sarco/endoplasmic reticulum $Ca^{2+}$ -ATPase in the acquisition of the contractile phenotype of arterial smooth muscle cells. *Circulation, 129*(7), 773–785. http://dx.doi.org/10.1161/CIRCULATIONAHA.113.002565.

Liu, J., & Brautigan, D. L. (2000). Glycogen synthase association with the striated muscle glycogen-targeting subunit of protein phosphatase-1: Synthase activation involves scaffolding regulated by β-adrenergic signaling. *Journal of Biological Chemistry, 275*(34), 26074–26081. http://dx.doi.org/10.1074/jbc.M003843200.

Liu, H., Krizek, J., & Bretscher, A. (1992). Construction of a GAL1-regulated yeast cDNA expression library and its application to the identification of genes whose overexpression causes lethality in yeast. *Genetics, 132*(0016–6731), 665–673.

Liu, D., Vleugel, M., Backer, C. B., Hori, T., Fukagawa, T., Cheeseman, I. M., et al. (2010). Regulated targeting of protein phosphatase 1 to the outer kinetochore by KNL1 opposes Aurora B kinase. *Journal of Cell Biology, 188*(6), 809–820. http://dx.doi.org/10.1083/jcb.201001006.

Liu, Q. R., Gong, J. P., & Uhl, G. R. (2005). Families of protein phosphatase 1 modulators activated by protein kinases A and C: Focus on brain. *Progress in Nucleic Acid Research and Molecular Biology, 79*, 371–404. http://dx.doi.org/10.1016/S0079-6603(04)79008-X. Academic Press.

Liu, Q. R., Zhang, P. W., Lin, Z., Li, Q. F., Woods, A. S., Troncoso, J., et al. (2004). GBPI, a novel gastrointestinal- and brain-specific PP1-inhibitory protein, is activated by PKC and inactivated by PKA. *Biochemical Journal, 377*(1), 171–181. http://dx.doi.org/10.1042/BJ20030128.

Liu, Q. R., Zhang, P. W., Zhen, Q., Walther, D., Wang, X. B., & Uhl, G. R. (2002). KEPI, a PKC-dependent protein phosphatase 1 inhibitor regulated by morphine. *Journal of Biological Chemistry, 277*(15), 13312–13320. http://dx.doi.org/10.1074/jbc.M107558200.

Llanos, S., Royer, C., Lu, M., Bergamaschi, D., Lee, W. H., & Lu, X. (2011). Inhibitory member of the apoptosis-stimulating proteins of the p53 family (iASPP) interacts with protein phosphatase 1 via a noncanonical binding motif. *Journal of Biological Chemistry, 286*(50), 43039–43044. http://dx.doi.org/10.1074/jbc.M111.270751.

Luo, W., Xu, C., Ayello, J., Dela Cruz, F., Rosenblum, J. M., Lessnick, S. L., et al. (2018). Protein phosphatase 1 regulatory subunit 1A in ewing sarcoma tumorigenesis and metastasis. *Oncogene, 37*(6), 798–809. http://dx.doi.org/10.1038/onc.2017.378.

Luo, X., Zhang, Y., Ruan, X., Jiang, X., Zhu, L., Wang, X., et al. (2011). Fasting-induced protein phosphatase 1 regulatory subunit contributes to postprandial blood glucose homeostasis via regulation of hepatic glycogenesis. *Diabetes, 60*(5), 1435–1445. http://dx.doi.org/10.2337/db10-1663.

MacKelvie, S. H., Andrews, P. D., & Stark, M. J. (1995). The *Saccharomyces cerevisiae* gene SDS22 encodes a potential regulator of the mitotic function of yeast type 1 protein phosphatase. *Molecular and Cellular Biology, 15*(7), 3777–3785. http://dx.doi.org/10.1128/mcb.15.7.3777.

MacMillan, L. B., Bass, M. A., Cheng, N., Howard, E. F., Tamura, M., Strack, S., et al. (1999). Brain actin-associated protein phosphatase 1 holoenzymes containing spinophilin, neurabin, and selected catalytic subunit isoforms. *Journal of Biological Chemistry, 274*(50), 35845–35854. http://dx.doi.org/10.1074/jbc.274.50.35845.

Ma, Y., & Hendershot, L. M. (2003). Delineation of a negative feedback regulatory loop that controls protein translation during endoplasmic reticulum stress. *Journal of Biological Chemistry, 278*(37), 34864–34873. http://dx.doi.org/10.1074/jbc.M301107200.

Marquina, M., González, A., Barreto, L., Gelis, S., Muñoz, I., Ruiz, A., et al. (2012). Modulation of yeast alkaline cation tolerance by ypi1 requires calcineurin. *Genetics, 190*(4), 1355–1364. http://dx.doi.org/10.1534/genetics.112.138370.

Marquina, M., Queralt, E., Casamayor, A., & Ariño, J. (2012). Lack of the Glc7 phosphatase regulatory subunit Ypi1 activates the morphogenetic checkpoint. *The International Journal of Biochemistry and Cell Biology, 44*(11), 1862–1871. http://dx.doi.org/10.1016/j.biocel.2012.06.029.

McDermott, J. E., Mitchell, H. D., Gralinski, L. E., Eisfeld, A. J., Josset, L., Bankhead, A., et al. (2016). The effect of inhibition of PP1 and TNFα signaling on pathogenesis of

SARS coronavirus. *BMC Systems Biology, 10*(1). http://dx.doi.org/10.1186/s12918-016-0336-6.

Mehta, M. B., Shewale, S. V., Sequeira, R. N., Millar, J. S., Hand, N. J., & Rader, D. J. (2017). Hepatic protein phosphatase 1 regulatory subunit 3B (Ppp1r3b) promotes hepatic glycogen synthesis and thereby regulates fasting energy homeostasis. *Journal of Biological Chemistry, 292*(25), 10444–10454. http://dx.doi.org/10.1074/jbc.M116.766329.

Meiselbach, H., Sticht, H., & Enz, R. (2006). Structural analysis of the protein phosphatase 1 docking motif: Molecular description of binding specificities identifies interacting proteins. *Chemistry and Biology, 13*(1), 49–59. http://dx.doi.org/10.1016/j.chembiol.2005.10.009.

Minnebo, N., Görnemann, J., O'Connell, N., Van Dessel, N., Derua, R., Vermunt, M. W., et al. (2013). NIPP1 maintains EZH2 phosphorylation and promoter occupancy at proliferation-related target genes. *Nucleic Acids Research, 41*(2), 842–854. http://dx.doi.org/10.1093/nar/gks1255.

Moorhead, G. B. G., De Wever, V., Templeton, G., & Kerk, D. (2009). Evolution of protein phosphatases in plants and animals. *Biochemical Journal, 417*, 401–409. http://dx.doi.org/10.1042/BJ20081986. Portland Press.

Munro, S., Ceulemans, H., Bollen, M., Diplexcito, J., & Cohen, P. T. W. (2005). A novel glycogen-targeting subunit of protein phosphatase 1 that is regulated by insulin and shows differential tissue distribution in humans and rodents. *FEBS Journal, 272*(6), 1478–1489. http://dx.doi.org/10.1111/j.1742-4658.2005.04585.x.

Murányi, A., Derkach, D., Erdodi, F., Kiss, A., Ito, M., & Hartshorne, D. J. (2005). Phosphorylation of Thr695 and Thr850 on the myosin phosphatase target subunit: Inhibitory effects and occurrence in A7r5 cells. *FEBS Letters, 579*(29), 6611–6615. http://dx.doi.org/10.1016/j.febslet.2005.10.055.

Niazi, R. K., Sun, J., Have, C. T., Hollensted, M., Linneberg, A., Pedersen, O., et al. (2019). Increased frequency of rare missense PPP1R3B variants among Danish patients with type 2 diabetes. *PLoS One, 14*(1). http://dx.doi.org/10.1371/journal.pone.0210114.

Nicolaou, P., Hajjar, R. J., & Kranias, E. G. (2009). Role of protein phosphatase-1 inhibitor-1 in cardiac physiology and pathophysiology. *Journal of Molecular and Cellular Cardiology, 47*(3), 365–371. http://dx.doi.org/10.1016/j.yjmcc.2009.05.010.

Nicolaou, P., Rodriguez, P., Ren, X., Zhou, X., Qian, J., Sadayappan, S., et al. (2009). Inducible expression of active protein phosphatase-1 inhibitor-1 enhances basal cardiac function and protects against ischemia/reperfusion injury. *Circulation Research, 104*(8), 1012–1020. http://dx.doi.org/10.1161/CIRCRESAHA.108.189811.

Nijenhuis, W., Vallardi, G., Teixeira, A., Kops, G. J. P. L., & Saurin, A. T. (2014). Negative feedback at kinetochores underlies a responsive spindle checkpoint signal. *Nature Cell Biology, 16*(12), 1257–1264. http://dx.doi.org/10.1038/ncb3065.

Ohki, S. Y., Eto, M., Kariya, E., Hayano, T., Hayashi, Y., Yazawa, M., et al. (2001). Solution NMR structure of the myosin phosphatase inhibitor protein CPI-17 shows phosphorylation-induced conformational changes responsible for activation. *Journal of Molecular Biology, 314*(4), 839–849. http://dx.doi.org/10.1006/jmbi.2001.5200.

Ohkura, H., & Yanagida, M. (1991). *S. pombe* gene sds22+ essential for a midmitotic transition encodes a leucine-rich repeat protein that positively modulates protein phosphatase-1. *Cell, 64*(1), 149–157. http://dx.doi.org/10.1016/0092-8674(91)90216-L.

Okamoto, R., Kato, T., Mizoguchi, A., Takahashi, N., Nakakuki, T., Mizutani, H., et al. (2006). Characterization and function of MYPT2, a target subunit of myosin phosphatase in heart. *Cellular Signalling, 18*(9), 1408–1416. http://dx.doi.org/10.1016/j.cellsig.2005.11.001.

Oligschlaeger, Y., Miglianico, M., Dahlmans, V., Rubio-Villena, C., Chanda, D., Garcia-Gimeno, M. A., et al. (2016). The interaction between AMPKβ2 and the PP1-targeting subunit R6 is dynamically regulated by intracellular glycogen content. *Biochemical Journal, 473*(7), 937–947. http://dx.doi.org/10.1042/BJ20151035.

O'Connell, N., Nichols, S. R., Heroes, E., Beullens, M., Bollen, M., Peti, W., et al. (2012). The molecular basis for substrate specificity of the nuclear NIPP1:PP1 holoenzyme. *Structure, 20*(10), 1746–1756. http://dx.doi.org/10.1016/j.str.2012.08.003.

Oliver, C. J., & Shenolikar, S. (1998). Physiologic importance of protein phosphatase inhibitors. *Frontiers in Bioscience, 3*, D961–D972. http://dx.doi.org/10.2741/a336.

Pathak, A., del Monte, F., Zhao, W., Schultz, J.-E., Lorenz, J. N., Bodi, I., et al. (2005). Enhancement of cardiac function and suppression of heart failure progression by inhibition of protein phosphatase 1. *Circulation Research, 96*(7), 756–766. http://dx.doi.org/10.1161/01.RES.0000161256.85833.fa.

Paul, D., Bargale, A. B., Rapole, S., Shetty, P. K., & Santra, M. K. (2019). Protein phosphatase 1 regulatory subunit SDS22 inhibits breast cancer cell tumorigenesis by functioning as a negative regulator of the AKT signaling pathway. *Neoplasia, 21*(1), 30–40. http://dx.doi.org/10.1016/j.neo.2018.10.009.

Pedelini, L., Marquina, M., Ariño, J., Casamayor, A., Sanz, L., Bollen, M., et al. (2007). YPI1 and SDS22 proteins regulate the nuclear localization and function of yeast type 1 phosphatase Glc7. *Journal of Biological Chemistry, 282*(5), 3282–3292. http://dx.doi.org/10.1074/jbc.M607171200.

Peggie, M. W., MacKelvie, S. H., Bloecher, A., Knatko, E. V., Tatchell, K., & Stark, M. J. R. (2002). Essential functions of Sds22p in chromosome stability and nuclear localization of PP1. *Journal of Cell Science, 115*(1), 195–206.

Peti, W., Nairn, A. C., & Page, R. (2013). Structural basis for protein phosphatase 1 regulation and specificity. *FEBS Journal, 280*, 596–611. http://dx.doi.org/10.1111/j.1742-4658.2012.08509.x.

Piton, A., Gauthier, J., Hamdan, F. F., Lafrenière, R. G., Yang, Y., Henrion, E., et al. (2011). Systematic resequencing of X-chromosome synaptic genes in autism spectrum disorder and schizophrenia. *Molecular Psychiatry, 16*(8), 867–880. http://dx.doi.org/10.1038/mp.2010.54.

Plattner, F., & Bibb, J. A. (2012). Serine and threonine phosphorylation. *Basic Neurochemistry*, 467–492. http://dx.doi.org/10.1016/B978-0-12-374947-5.00025-0. Elsevier.

Posch, M., Khoudoli, G. A., Swift, S., King, E. M., DeLuca, J. G., & Swedlow, J. R. (2010). Sds22 regulates aurora B activity and microtubule-kinetochore interactions at mitosis. *Journal of Cell Biology, 191*(1), 61–74. http://dx.doi.org/10.1083/jcb.200912046.

Pritchard, T. J., Kawase, Y., Haghighi, K., Anjak, A., Cai, W., Jiang, M., et al. (2013). Active inhibitor-1 maintains protein hyper-phosphorylation in aging hearts and halts remodeling in failing hearts. *PLoS One, 8*(12), e80717. http://dx.doi.org/10.1371/journal.pone.0080717.

Qian, J., Beullens, M., Lesage, B., & Bollen, M. (2013). Aurora B defines its own chromosomal targeting by opposing the recruitment of the phosphatase scaffold Repo-Man. *Current Biology, 23*(12), 1136–1143. http://dx.doi.org/10.1016/j.cub.2013.05.017.

Ragusa, M. J., Dancheck, B., Critton, D. A., Nairn, A. C., Page, R., & Peti, W. (2010). Spinophilin directs protein phosphatase 1 specificity by blocking substrate binding sites. *Nature Structural and Molecular Biology, 17*(4), 459–464. http://dx.doi.org/10.1038/nsmb.1786.

Reis, H. J., Rosa, D. V. F., Guimarães, M. M., Souza, B. R., Barros, A. G. A., Pimenta, F. J., et al. (2007). Is DARPP-32 a potential therapeutic target? *Expert Opinion on Therapeutic Targets, 11*, 1649–1661. http://dx.doi.org/10.1517/14728222.11.12.1649.

Renouf, S., Beullens, M., Wera, S., Van Eynde, A., Sikela, J., Stalmans, W., et al. (1995). Molecular cloning of a human polypeptide related to yeast sds22, a regulator of protein phosphatase-1. *FEBS Letters, 375*(1–2), 75–78. http://dx.doi.org/10.1016/0014-5793(95)01180-M.

Rodríguez-Sanoja, R., Oviedo, N., & Sánchez, S. (2005). Microbial starch-binding domain. *Current Opinion in Microbiology, 8*, 260–267. http://dx.doi.org/10.1016/j.mib.2005.04.013.

Rogne, M., Landsverk, H. B., Van Eynde, A., Beullens, M., Bollen, M., Collas, P., et al. (2006). The KH-Tudor domain of A-kinase anchoring protein 149 mediates RNA-dependent self-association. *Biochemistry, 45*(50), 14980–14989. http://dx.doi.org/10.1021/bi061418y.

Rogne, M., Stokka, A. J., Taskén, K., Collas, P., & Küntziger, T. (2008). Mutually exclusive binding of PP1 and RNA to AKAP149 affects the mitochondrial network. *Human Molecular Genetics, 18*(5), 978–987. http://dx.doi.org/10.1093/hmg/ddn425.

Rojas, M., Gingras, A. C., & Dever, T. E. (2014). Protein phosphatase PP1/GLC7 interaction domain in yeast eIF2γ bypasses targeting subunit requirement for eIF2α dephosphorylation. *Proceedings of the National Academy of Sciences of the United States of America, 111*(14). http://dx.doi.org/10.1073/pnas.1400129111.

Rosenberg, J. S., Cross, F. R., & Funabiki, H. (2011). KNL1/Spc105 recruits PP1 to silence the spindle assembly checkpoint. *Current Biology, 21*(11), 942–947. http://dx.doi.org/10.1016/j.cub.2011.04.011.

Roy, B., Verma, V., Sim, J., Fontan, A., & Joglekar, A. P. (2019). Delineating the contribution of Spc105-bound PP1 to spindle checkpoint silencing and kinetochore microtubule attachment regulation. *The Journal of Cell Biology, 218*(12), 3926–3942. http://dx.doi.org/10.1083/jcb.201810172.

Rubio-Villena, C., Garcia-Gimeno, M. A., & Sanz, P. (2013). Glycogenic activity of R6, a protein phosphatase 1 regulatory subunit, is modulated by the laforin-malin complex. *The International Journal of Biochemistry and Cell Biology, 45*(7), 1479–1488. http://dx.doi.org/10.1016/j.biocel.2013.04.019.

Rubio-Villena, C., Sanz, P., & Garcia-Gimeno, M. A. (2015). Structure-function analysis of PPP1R3D, a protein phosphatase 1 targeting subunit, reveals a binding motif for 14-3-3 proteins which regulates its glycogenic properties. *PLoS One, 10*(6), e0131476. http://dx.doi.org/10.1371/journal.pone.0131476.

Sacco, F., Perfetto, L., Castagnoli, L., & Cesareni, G. (2012). The human phosphatase interactome: An intricate family portrait. *FEBS Letters, 586*, 2732–2739. http://dx.doi.org/10.1016/j.febslet.2012.05.008.

Samuels-Lev, Y., O'Connor, D. J., Bergamaschi, D., Trigiante, G., Hsieh, J. K., Zhong, S., et al. (2001). ASPP proteins specifically stimulate the apoptotic function of p53. *Molecular Cell, 8*(4), 781–794. http://dx.doi.org/10.1016/S1097-2765(01)00367-7.

Sánchez-Pozos, K., Ortíz-López, M. G., Peña-Espinoza, B. I., de los Ángeles Granados-Silvestre, M., Jiménez-Jacinto, V., Verleyen, J., et al. (2018). Whole-exome sequencing in maya indigenous families: Variant in PPP1R3A is associated with type 2 diabetes. *Molecular Genetics and Genomics, 293*(5), 1205–1216. http://dx.doi.org/10.1007/s00438-018-1453-2.

Scotto-Lavino, E., Garcia-Diaz, M., Du, G., & Frohman, M. A. (2010). Basis for the isoform-specific interaction of myosin phosphatase subunits protein phosphatase 1c β and myosin phosphatase targeting subunit 1. *Journal of Biological Chemistry, 285*(9), 6419–6424. http://dx.doi.org/10.1074/jbc.M109.074773.

Sharma, K., D'Souza, R. C. J., Tyanova, S., Schaab, C., Wiśniewski, J. R., Cox, J., et al. (2014). Ultradeep human phosphoproteome reveals a distinct regulatory nature of Tyr and Ser/Thr-based signaling. *Cell Reports, 8*(5), 1583–1594. http://dx.doi.org/10.1016/j.celrep.2014.07.036.

Sharma, A. K., Zhou, G. P., Kupferman, J., Surks, H. K., Christensen, E. N., Chou, J. J., et al. (2008). Probing the interaction between the coiled coil leucine zipper of cGMP-dependent protein kinase Iα and the C terminus of the myosin binding subunit of the myosin light chain phosphatase. *Journal of Biological Chemistry, 283*(47), 32860–32869. http://dx.doi.org/10.1074/jbc.M804916200.

Shi, Y. (2009). Serine/threonine phosphatases: Mechanism through structure. *Cell, 139*, 468–484. http://dx.doi.org/10.1016/j.cell.2009.10.006, 1097-4172 (Electronic).

Smith, R. D., & Walker, J. C. (1993). Expression of multiple type 1 phosphoprotein phosphatases in *Arabidopsis thaliana*. *Plant Molecular Biology, 21*(2), 307–316. http://dx.doi.org/10.1007/BF00019946.

Somaiah, N., Beird, H. C., Barbo, A., Song, J., Mills Shaw, K. R., Wang, W. L., et al. (2018). Targeted next generation sequencing of well-differentiated/dedifferentiated liposarcoma reveals novel gene amplifications and mutations. *Oncotarget, 9*(28), 19891–19899. http://dx.doi.org/10.18632/oncotarget.24924.

Sottocornola, R., Royer, C., Vives, V., Tordella, L., Zhong, S., Wang, Y., et al. (2010). ASPP2 binds par-3 and controls the polarity and proliferation of neural progenitors during CNS development. *Developmental Cell, 19*(1), 126–137. http://dx.doi.org/10.1016/j.devcel.2010.06.003.

Steen, R. L., Martins, S. B., Taskén, K., & Collas, P. (2000). Recruitment of protein phosphatase 1 to the nuclear envelope by A-kinase anchoring protein AKAP149 is a prerequisite for nuclear lamina assembly. *Journal of Cell Biology, 150*(6), 1251–1261. http://dx.doi.org/10.1083/jcb.150.6.1251.

Stender, S., Smagris, E., Lauridsen, B. K., Kofoed, K. F., Nordestgaard, B. G., Tybjærg-Hansen, A., et al. (2018). Relationship between genetic variation at PPP1R3B and levels of liver glycogen and triglyceride. *Hepatology, 67*(6), 2182–2195. http://dx.doi.org/10.1002/hep.29751.

Sullivan, A., & Lu, X. (January 29, 2007). Aspp: A new family of oncogenes and tumour suppressor genes. *British Journal of Cancer, 96*, 196–200. http://dx.doi.org/10.1038/sj.bjc.6603525.

Surks, H. K., & Mendelsohn, M. E. (2003). Dimerization of cGMP-dependent protein kinase 1alpha and the myosin-binding subunit of myosin phosphatase: Role of leucine zipper domains. *Cellular Signalling, 15*(10), 937–944. http://dx.doi.org/10.1016/s0898-6568(03)00057-3.

Takane, K., Midorikawa, Y., Yagi, K., Sakai, A., Aburatani, H., Takayama, T., et al. (2014). Aberrant promoter methylation of PPP1R3C and EFHD1 in plasma of colorectal cancer patients. *Cancer Medicine, 3*(5), 1235–1245. http://dx.doi.org/10.1002/cam4.273.

Tang, P. M., Bondor, J. A., Swiderek, K. M., & DePaoli-Roach, A. A. (1991). Molecular cloning and expression of the regulatory (RG1) subunit of the glycogen-associated protein phosphatase. *Journal of Biological Chemistry, 266*(24), 15782–15789. Retrieved from http://www.ncbi.nlm.nih.gov/pubmed/1651919.

Tang, X., Hui, Z., Cui, X., Garg, R., Kastan, M. B., & Xu, B. (2008). A novel ATM-dependent pathway regulates protein phosphatase 1 in response to DNA damage. *Molecular and Cellular Biology, 28*(8), 2559–2566. http://dx.doi.org/10.1128/mcb.01711-07.

Tan, I., Ng, C. H., Lim, L., & Leung, T. (2001). Phosphorylation of a novel myosin binding subunit of protein phosphatase 1 reveals a conserved mechanism in the regulation of actin cytoskeleton. *Journal of Biological Chemistry, 276*(24), 21209–21216. http://dx.doi.org/10.1074/jbc.M102615200.

Terrak, M., Kerff, F., Langseimo, K., Tao, T., & Dominguez, R. (2004). Structural basis of protein phosphatase 1 regulation. *Nature, 429*(6993), 780–784. http://dx.doi.org/10.1038/nature02582.

Terry-Lorenzo, R. T., Elliot, E., Weiser, D. C., Prickett, T. D., Brautigan, D. L., & Shenolikar, S. (2002). Neurabins recruit protein phosphatase-1 and inhibitor-2 to the actin cytoskeleton. *Journal of Biological Chemistry, 277*(48), 46535–46543. http://dx.doi.org/10.1074/jbc.M206960200.

Thornton, B. P., Johns, A., Al-Shidhani, R., Álvarez-Carretero, S., Storer, I. S. R., Bromley, M. J., et al. (2019). Identification of functional and druggable sites in Aspergillus fumigatus essential phosphatases by virtual screening. *International Journal of Molecular Sciences, 20*(18). http://dx.doi.org/10.3390/ijms20184636.

Tóth, A., Kiss, E., Herberg, F. W., Gergely, P., Hartshorne, D. J., & Erdödi, F. (2000). Study of the subunit interactions in myosin phosphatase by surface plasmon resonance. *European Journal of Biochemistry, 267*(6), 1687–1697. http://dx.doi.org/10.1046/j.1432-1327.2000.01158.x.

Trinkle-Mulcahy, L., Andersen, J., Yun, W. L., Moorhead, G., Mann, M., & Lamond, A. I. (2006). Repo-Man recruits PP1γ to chromatin and is essential for cell viability. *Journal of Cell Biology, 172*(5), 679–692. http://dx.doi.org/10.1083/jcb.200508154.

Turnbull, J., Epp, J. R., Goldsmith, D., Zhao, X., Pencea, N., Wang, P., et al. (2014). PTG protein depletion rescues malin-deficient Lafora disease in mouse. *Annals of Neurology, 75*(3), 442–446. http://dx.doi.org/10.1002/ana.24104.

Vagnarelli, P., & Alessi, D. R. (2018). PP1 phosphatase complexes: Undruggable No longer. *Cell, 174*, 1049–1051. http://dx.doi.org/10.1016/j.cell.2018.08.007. Cell Press.

Vagnarelli, P., & Earnshaw, W. C. (2012). Repo-Man-PP1. *Nucleus, 3*(2), 138–142. http://dx.doi.org/10.4161/nucl.19267.

Van Dessel, N., Boens, S., Lesage, B., Winkler, C., Görnemann, J., Van Eynde, A., et al. (2015). Protein phosphatase PP1-NIPP1 activates mesenchymal genes in HeLa cells. *FEBS Letters, 589*(12), 1314–1321. http://dx.doi.org/10.1016/j.febslet.2015.04.017.

Van Eynde, A., Wera, S., Beullens, M., Torrekens, S., Van Leuven, F., Stalmans, W., et al. (1995). Molecular cloning of NIPP-1, a nuclear inhibitor of protein phosphatase- 1, reveals homology with polypeptides involved in RNA processing. *Journal of Biological Chemistry, 270*(47), 28068–28074. http://dx.doi.org/10.1074/jbc.270.47.28068.

Verbinnen, I., Boens, S., Ferreira, M., Szekér, K., Van Wijk, L., Van Eynde, A., et al. (2020). Enhanced DNA-repair capacity and resistance to chemically induced carcinogenesis upon deletion of the phosphatase regulator NIPP1. *Oncogenesis, 9*(3), 30. http://dx.doi.org/10.1038/s41389-020-0214-3.

Verbinnen, I., Ferreira, M., & Bollen, M. (2017). Biogenesis and activity regulation of protein phosphatase 1. *Biochemical Society Transactions, 45*, 89–99. http://dx.doi.org/10.1042/BST20160154. Portland Press Ltd.

Verbinnen, I., Jonkhout, M., Liakath-Ali, K., Szekér, K., Ferreira, M., Boens, S., et al. (2020). Phosphatase regulator NIPP1 restrains chemokine-driven skin inflammation. *Journal of Investigative Dermatology*. Published online ahead of print, S0022-202X(20)30035-X. https://dx.doi.org/10.1016/j.jid.2020.01.008.

Wakula, P., Beullens, M., Ceulemans, H., Stalmans, W., & Bollen, M. (2003). Degeneracy and function of the ubiquitous RVXF motif that mediates binding to protein phosphatase-1. *Journal of Biological Chemistry, 278*(21), 18817–18823. http://dx.doi.org/10.1074/jbc.M300175200.

Walaas, S. I., Hemmings, H. C., Greengard, P., & Nairn, A. C. (2011). Beyond the dopamine receptor: Regulation and roles of serine/threonine protein phosphatases. *Frontiers in Neuroanatomy, 5*, 50. http://dx.doi.org/10.3389/fnana.2011.00050.

Walker, K. S., Watt, P. W., & Cohen, P. (2000). Phosphorylation of the skeletal muscle glycogen-targetting subunit of protein phosphatase 1 in response to adrenaline in vivo. *FEBS Letters, 466*(1), 121–124. http://dx.doi.org/10.1016/s0014-5793(99)01771-8.

Wang, M., & Cohen, S. N. (1994). ard-1: A human gene that reverses the effects of temperature-sensitive and deletion mutations in the *Escherichia coli* rne gene and encodes an activity producing RNase E-like cleavages. *Proceedings of the National Academy of Sciences of the United States of America, 91*(22), 10591–10595. http://dx.doi.org/10.1073/pnas.91.22.10591.

Wang, F., & Sun, Y. (2018). Overexpression of myosin phosphatase target subunit 1 (MYPT1) inhibits tumor progression and metastasis of gastric cancer. *Medical Science Monitor, 24*, 2508–2517. http://dx.doi.org/10.12659/MSM.906852.

Weiser, D. C., Sikes, S., Li, S., & Shenolikar, S. (2004). The inhibitor-1 C terminus facilitates hormonal regulation of cellular protein phosphatase-1: Functional implications for inhibitor-1 isoforms. *Journal of Biological Chemistry, 279*(47), 48904–48914. http://dx.doi.org/10.1074/jbc.M404416200.

Weith, M., Seiler, J., van den Boom, J., Kracht, M., Hülsmann, J., Primorac, I., et al. (2018). Ubiquitin-independent disassembly by a p97 AAA-ATPase complex drives PP1 holoenzyme formation. *Molecular Cell, 72*(4), 766–777.e6. http://dx.doi.org/10.1016/j.molcel.2018.09.020.

Wenzel, K., Daskalow, K., Herse, F., Seitz, S., Zacharias, U., Schenk, J. A., et al. (2007). Expression of the protein phosphatase 1 inhibitor KEPI is downregulated in breast cancer cell lines and tissues and involved in the regulation of the tumor suppressor EGR1 via the MEK-ERK pathway. *Biological Chemistry, 388*(5), 489–495. http://dx.doi.org/10.1515/BC.2007.062.

Worby, C. A., Gentry, M. S., & Dixon, J. E. (2008). Malin decreases glycogen accumulation by promoting the degradation of protein targeting to glycogen (PTG). *Journal of Biological Chemistry, 283*(7), 4069–4076. http://dx.doi.org/10.1074/jbc.M708712200.

Wu, J. Q., Guo, J. Y., Tang, W., Yang, C. S., Freel, C. D., Chen, C., et al. (2009). PP1-mediated dephosphorylation of phosphoproteins at mitotic exit is controlled by inhibitor-1 and PP1 phosphorylation. *Nature Cell Biology, 11*(5), 644–651. http://dx.doi.org/10.1038/ncb1871.

Wu, J., Liu, J., Thompson, I., Oliver, C. J., Shenolikar, S., & Brautigan, D. L. (1998). A conserved domain for glycogen binding in protein phosphatase-1 targeting subunits. *FEBS Letters, 439*(1–2), 185–191. http://dx.doi.org/10.1016/s0014-5793(98)01371-4.

Xia, D., Stull, J. T., & Kamm, K. E. (2005). Myosin phosphatase targeting subunit 1 affects cell migration by regulating myosin phosphorylation and actin assembly. *Experimental Cell Research, 304*(2), 506–517. http://dx.doi.org/10.1016/j.yexcr.2004.11.025.

Yadav, L., Tamene, F., Göös, H., van Drogen, A., Katainen, R., Aebersold, R., et al. (2017). Systematic analysis of human protein phosphatase interactions and dynamics. *Cell Systems, 4*(4), 430–444. http://dx.doi.org/10.1016/j.cels.2017.02.011. e5.

Yang, H., Hou, H., Pahng, A., Gu, H., Nairn, A. C., Tang, Y. P., et al. (2015). Protein phosphatase-1 inhibitor-2 is a novel memory suppressor. *Journal of Neuroscience, 35*(45), 15082–15087. http://dx.doi.org/10.1523/JNEUROSCI.1865-15.2015.

Yao, G., Wan, J., Mu, C., Liu, Q., Wang, Y., & Sang, J. (2016). Sds22 participates in Glc7 mediated Rad53 dephosphorylation in MMS-induced DNA damage in Candida albicans. *Fungal Genetics and Biology, 93*, 50–61. http://dx.doi.org/10.1016/j.fgb.2016.06.003.

Yong, J., Tan, I., Lim, L., & Leung, T. (2006). Phosphorylation of myosin phosphatase targeting subunit 3 (MYPT3) and regulation of protein phosphatase 1 by protein kinase A. *Journal of Biological Chemistry, 281*(42), 31202–31211. http://dx.doi.org/10.1074/jbc.M607287200.

Yu, J., Deng, T., & Xiang, S. (2018). Structural basis for protein phosphatase 1 recruitment by glycogen-targeting subunits. *FEBS Journal, 285*(24), 4646–4659. http://dx.doi.org/10.1111/febs.14699.

Zeng, Q., Huang, Y., Zeng, L., Huang, Y., Cai, D., & Zhang, H. (2012). IPP5, a novel inhibitor of protein phosphatase 1, suppresses tumor growth and progression of cervical carcinoma cells by inducing G2/M arrest. *Cancer Genetics, 205*(9), 442–452. http://dx.doi.org/10.1016/j.cancergen.2012.06.002.

Zeng, Q.-Y., Huang, Y., Zeng, L.-J., Huang, M., Huang, Y.-Q., & Zhu, Q.-F. (2014). Sensitization of cervical carcinoma cells to paclitaxel by an IPP5 active mutant. *Asian Pacific Journal of Cancer Prevention: Asian Pacific Journal of Cancer Prevention, 15*(19), 8337–8343. http://dx.doi.org/10.7314/apjcp.2014.15.19.8337.

Zeng, Q., Huang, Y., Zeng, L., Lan, X., Huang, Y., He, S., et al. (2009). Effect of IPP5, a novel inhibitor of PP1, on apoptosis and the underlying mechanisms involved. *Biotechnology and Applied Biochemistry, 54*(4), 231–238. http://dx.doi.org/10.1042/BA20090168.

Zhang, Y., Gan, W., Tian, C., Li, H., Lin, X., & Chen, Y. (2013). Association of PPP1R3B polymorphisms with blood lipid and C-reactive protein levels in a Chinese population. *Journal of Diabetes, 5*(3), 275–281. http://dx.doi.org/10.1111/1753-0407.12028.

Zhang, S., Guha, S., & Volkert, F. C. (1995). The Saccharomyces SHP1 gene, which encodes a regulator of phosphoprotein phosphatase 1 with differential effects on glycogen metabolism, meiotic differentiation, and mitotic cell cycle progression. *Molecular and Cellular Biology, 15*(0270–7306), 2037–2050.

Zhang, L., Qi, Z., Gao, Y., & Lee, E. Y. C. (2008). Identification of the interaction sites of Inhibitor-3 for protein phosphatase-1. *Biochemical and Biophysical Research Communications, 377*(2), 710–713. http://dx.doi.org/10.1016/j.bbrc.2008.10.062.

Zhang, W., Shang, S., Yang, Y., Lu, P., Wang, T., Cui, X., et al. (2020). Identification of DNA methylation-driven genes by integrative analysis of DNA methylation and transcriptome data in pancreatic adenocarcinoma. *Experimental and Therapeutic Medicine, 19*(4). http://dx.doi.org/10.3892/etm.2020.8554.

Zhang, P., Zhang, Y., Gao, K., Wang, Y., Jin, X., Wei, Y., et al. (2015). ASPP1/2-PP1 complexes are required for chromosome segregation and kinetochore-microtubule attachments. *Oncotarget, 6*(39), 41550–41565. http://dx.doi.org/10.18632/oncotarget.6355.

Zhao, S., & Lee, E. Y. C. (1997). A protein phosphatase-1-binding motif identified by the panning of a random peptide display library. *Journal of Biological Chemistry, 272*(45), 28368–28372. http://dx.doi.org/10.1074/jbc.272.45.28368.

Zhou, Y., Millott, R., Kim, H. J., Peng, S., Edwards, R. A., Skene-Arnold, T., et al. (2019). Flexible tethering of ASPP proteins facilitates PP-1c catalysis. *Structure, 27*(10), 1485–1496.e4. http://dx.doi.org/10.1016/j.str.2019.07.012.

Zhu, S., Fisher, L. A., Bessho, T., & Peng, A. (2017). Protein phosphatase 1 and phosphatase 1 nuclear targeting subunit-dependent regulation of DNA-dependent protein kinase and non-homologous end joining. *Nucleic Acids Research, 45*(18), 10583–10594. http://dx.doi.org/10.1093/nar/gkx686.

CHAPTER NINE

# Cytochromes P450 (P450s): A review of the class system with a focus on prokaryotic P450s

**James D. Finnigan[a], Carl Young[a], Darren J. Cook[a], Simon J. Charnock[a,]* and Gary W. Black[b,]***

[a]Prozomix Limited, Haltwhistle, Northumberland, United Kingdom
[b]Hub for Biotechnology in the Built Environment, Faculty of Health and Life Sciences, Northumbria University, Newcastle upon Tyne, United Kingdom
*Corresponding authors: E-mail: gary.black@northumbria.ac.uk; simon.charnock@prozomix.com

## Contents



*Advances in Protein Chemistry and Structural Biology*, Volume 122
ISSN 1876-1623
https://doi.org/10.1016/bs.apcsb.2020.06.005

## Abstract

Cytochromes P450 (P450s) are a large superfamily of heme-containing monooxygenases. P450s are found in all Kingdoms of life and exhibit incredible diversity, both at sequence level and also on a biochemical basis. In the majority of cases, P450s can be assigned into one of ten classes based on their associated redox partners, domain architecture and cellular localization. Prokaryotic P450s now represent a large diverse collection of annotated/known enzymes, of which many have great potential biocatalytic potential. The self-sufficient P450 classes (Class VII/VIII) have been explored significantly over the past decade, with many annotated and biochemically characterized members. It is clear that the prokaryotic P450 world is expanding rapidly, as the number of published genomes and metagenome studies increases, and more P450 families are identified and annotated (CYP families).

**Keywords:** Biotechnology applications of P450s; Cytochromes P450; P450s; Classes of P450; Heme-containing monooxygenases.

## 1. P450s

Cytochromes P450 (P450s) are one of the largest and most diverse superfamily of enzymes (Sezutsu, Goff, & Feyereisen, 2013). The first experimental evidence of P450s surfaced in the mid-1950s with the work of Axelrod (1955) investigating the metabolism of xenobiotics in rabbit liver microsomes. However, it was not until 1958, that P450s were formally identified by Klingenberg where he observed the presence of a carbon monoxide binding pigment, with a large, broad absorbance band with a peak at 450 nm (Klingenberg, 1958). This maximum absorbance is the source of the name P450, i.e. Pigment-450 (Hildebrandt, Remmer, & Estabrook, 1968). In 1964, Omura and Sato concluded that P450s were heme *b*-containing enzymes, and therefore possess protoporphyrin IX as their prosthetic heme factor (Lewis & Wiseman, 2005; Omura & Sato, 1964).

P450s are ubiquitous throughout Nature, but are not present in some *Archaea* or *Eubacteria* (Kelly & Kelly, 2013), and also most notably, *Escherichia coli* (Urlacher & Girhard, 2012). The human genome contains 57 P450s, many of which are associated with phase I drug metabolism (Zanger & Schwab, 2013). These hold acute importance in research involving drug interactions and variability in drug metabolism across individuals (Danielson, 2002; Guengerich, Wu, & Bartleson, 2005).

The diversity of the P450 superfamily is demonstrated by the various reactions they catalyze, and also by the collection of chemically disparate substrates P450s act upon (Danielson, 2002). P450s catalyze oxidation reactions, most commonly hydroxylation, but also others including *N*– and *O*-dealkylations, sulfoxidations and epoxidations (Robin et al., 2011).

There are over 340,000 putative P450s in the open-database with 41,048 formally annotated across 2,252 defined families, with over 590 of these assigned to prokaryotic P450s (Nelson, 2017). In general, P450s require the delivery of electrons for catalysis, typically via redox partner proteins such as flavodoxin reductases and flavodoxins (Hannemann, Bichet, Ewen, & Bernhardt, 2007). In respect to the prokaryotic P450s, a large amount of current knowledge relates to research into two classical P450s, P450cam and P450BM3. P450cam was the first isolated prokaryotic P450, identified from the Gram-negative bacterium *Pseudomonas putida* and its biological role was characterized as the hydroxylation of the 5-methylene carbon of camphor (Keller, Wuthrich, & Debrunner, 1972). In 1985, Poulos et al. successfully solved the first P450 superfamily crystal structure, that of P450cam (Poulos, Finzel, Gunsalus, Wagner, & Kraut, 1985). P450BM3 was first discovered in 1974 by Fulco et al. and was isolated from the bacterium *Bacillus megaterium*. The group also characterized the P450 as a medium-to long-chain fatty acid hydroxylase (Miura & Fulco, 1974). As the P450 was the third P450 to be isolated from *B. megaterium* it was assigned the name P450BM3 (Whitehouse, Bell, & Wong, 2012). P450BM3 was the first identified self-sufficient P450, with the P450 heme domain fused to a C-terminal reductase domain, used to shuffle electrons needed for catalysis, containing two flavin groups; flavin adenine dinucleotide (FAD) and flavin mononucleotide (FMN) at an equimolar ratio (Narhi & Fulco, 1986).

## 2. Physiological roles of P450s

The number and diversity of P450s within an organism varies immensely, depending on physiological requirements often linked with environmental factors. In comparison to mammals and prokaryotes, some plants have a notable high number of P450s (Table 1).

### 2.1 Prokaryotic P450s

Prokaryotic P450s represent a large proportion of P450 diversity and activities (Nelson, 2017). They are non-membrane bound, cytosolic enzymes evolved toward different roles such as the hydroxylation of fatty acids and xenobiotic metabolism, for example herbicide degradation (Lamb &

**Table 1** Table of example organisms with their total number of putative P450 genes.

| Organism | Number of P450s | References |
|---|---|---|
| *Aspergillus nidulans* | 111 | Kelly, Krasevec, Mullins and Nelson (2009) |
| *Fusarium oxysporum* | 169 | Durairaj, Hur and Yun (2016) |
| *Homo sapien* | 57 | Lewis (2004) |
| *Sorghum bicolor* | 372 | Mizutani (2012) |

Waterman, 2013). The prokaryotic P450s are well studied, partially due to their soluble nature, allowing easy manipulation *in vitro*, in contrast to their eukaryotic counterparts, which are predominantly membrane bound. This ease-of-use likely contributed to the early structure of P450cam (Poulos et al., 1985). The prokaryotic P450s span over 605 families, compared to 277 plant and 805 fungal families (Nelson, 2017).

In comparison to eukaryotes, P450 encoding genes are usually not abundant within individual prokaryotes, with some bacteria, notably *E. coli*, containing no P450 genes (Urlacher & Girhard, 2012). However, some prokaryotes have a large collection of P450s, such as *Streptomyces coelicolor* A3(2)'s 33 P450 genes (Lamb et al., 2003).

## 2.2 Fungal P450s

In common with the prokaryotes, the fungal Kingdom is extremely large and diverse with over one million estimated species from all temperature zones of the world (Choi & Kim, 2017). There are currently 7,925 fungal P450s with assigned CYP names spread across 805 fungal CYP families (Nelson, 2017). Only two of these families are possessed by most fungal species, CYP51 and CYP61, both of which are critical for ergosterol biosynthesis, a major component of the fungal plasma membrane (Ichinose, 2012). The biodiversity of P450s presented by a fungus is associated with environmental factors (Chen et al., 2014). One example of a diverse fungal P450ome is that of the white rot fungus *Phanerochaete chrysosporium,* containing 126 P450s distributed across 12 CYP families (Syed & Yadav, 2012).

## 2.3 Mammalian P450s

Mammalian P450s are among the most well studied, in particular the human P450s with respect to medical research. The detoxification activities of the human P450s are well understood and are found localized in liver microsomes. Specifically, they are bound to the endoplasmic reticulum via an

N-terminal transmembrane anchor sequence (Williams, Cosme, Sridhar, Johnson, & Mcree, 2000). The human P450ome comprises 57 P450s distributed across 18 CYP families, with the majority belonging to families, CYP1-4 (Guengerich, Waterman, & Egli, 2016). One of the most well-studied human P450s is CYP3A4 which is responsible for approx. 50% of all P450-dependent drug metabolism (Sim & Ingelman-Sundberg, 2010).

## 2.4 Plant P450s

The plant P450s represent the most abundant P450s in Nature, but not necessarily the most diverse. Currently there are 16,219 annotated plant P450s spread across 277 families, in comparison to the 805 fungal families containing 7,925 members (Nelson, 2017). In many plant species, P450 genes account for around 1% of the total genes in their genome (Mizutani, 2012). In comparison to most prokaryotic and mammalian organisms the number of P450 genes is much greater, for example the genomes of *Arabidopsis thaliania* contains 245 and *Sorghum bicolor* contains 372 putative P450 genes (Mizutani, 2012; Nelson & Werck-Reichhart, 2011). The plant P450s can be split into either; the A-type P450s, which represent a group of plant-specific enzymes involved in the biosynthesis of natural products such as lignin and the non-A-type P450s which demonstrate activities such as fatty acid metabolism (Werck-Reichhart & Feyereisen, 2000).

## 2.5 Viral P450s

The first viral P450s were identified in *Acanthamoeba polyphaga mimivirus* (APMV) in 2009 (Lamb et al., 2009). The genome of this virus is larger than some bacterial genomes and contains a predicted 911 coding sequences. Viral P450s are extremely rare, as previous to these viral P450s being identified, 2,180 viral genomes had been annotated with no putative CYP genes found. *A. polyphaga* contains a total of two P450s, known as MIMI_L532 and MIMI_L808 (Lamb et al., 2009). A BLAST analysis of MIMI_L532 shows sequence homology to other viral P450s, but also 26% sequence identity to a P450 from *Nocardia* sp. CNY236. BLAST analysis of MIMI_L808 also shows sequence homology to other viral P450s and to eukaryotic P450s, such as the arthropod *Folsomia candida*. In 2019, a viral P450 was identified within the prokaryotic CYP102 family, this P450, CYP102L1 was identified in *Mycobacteriophage Adler*. Due to its high degree of similarity to its microbial CYP102 counterparts, it is suggested that this gene was directly obtained from the viral host (Lamb et al., 2019).

# 3. P450 nomenclature and classes

## 3.1 CYP nomenclature

Due to the ubiquity of P450s, a nomenclature system was developed. The CYP nomenclature system was first proposed in 1987 by Nebert and Gonzalez (Nebert & Gonzalez, 1987) and revised in both 1989 and 1991 (Nebert et al., 1989, 1991). In 1995, the Cytochrome P450 homepage was created as a means of presenting information on P450 annotation and nomenclature (Nelson, 2009). The nomenclature system divides the P450 superfamily into CYP families, based solely on amino acid sequence identity. P450 sequences sharing more than 40% sequence identity are clustered into the same family and sequences that share more than 55% sequence identity into the same sub-family (Gricman, Vogel, & Pleiss, 2014; Nelson, 2006). All P450 names are preceded by 'CYP' followed by the family number, and then by sub-family letter and finally by the P450 number within the sub-family. Some notable examples are CYP101A1, which represents P450cam, and CYP102A1, which represents P450BM3. All P450 annotation is undertaken by the Committee of Standardized Cytochrome P450 Nomenclature (Nelson, 2006). Although P450s in each CYP family share more than 40% sequence homology, they often differ in their substrate specificity and physiological roles (Moody & Loveridge, 2014). In 2018, approximately 55,000 P450s have been annotated with a CYP name (Nelson & Nebert, 2018).

## 3.2 Classes of P450

Historically the P450 superfamily was split into two classes, Class I and II. The cytosolic Class I P450s and the membrane-bound eukaryotic Class II P450s. This was expanded to Class III and IV upon the discovery of the self-sufficient families CYP102A and CYP116B respectively (Roberts, Grogan, Greter, Flitsch, & Turner, 2002). However, as the number of characterized P450s with novel electron transport mechanisms expanded further and more diverse catalysts were discovered, a new ten Class system was proposed. Most putative P450s biocatalysts can be placed into one of the ten distinct classes, based on associated redox partners, domain architecture and cellular localization (Fig. 1) (Hannemann et al., 2007).

### *3.2.1 Class I*

Class I systems (Fig. 1, Ia and Ib) encompass most bacterial P450 systems along with the mitochondrial P450 systems. These P450s utilize a three

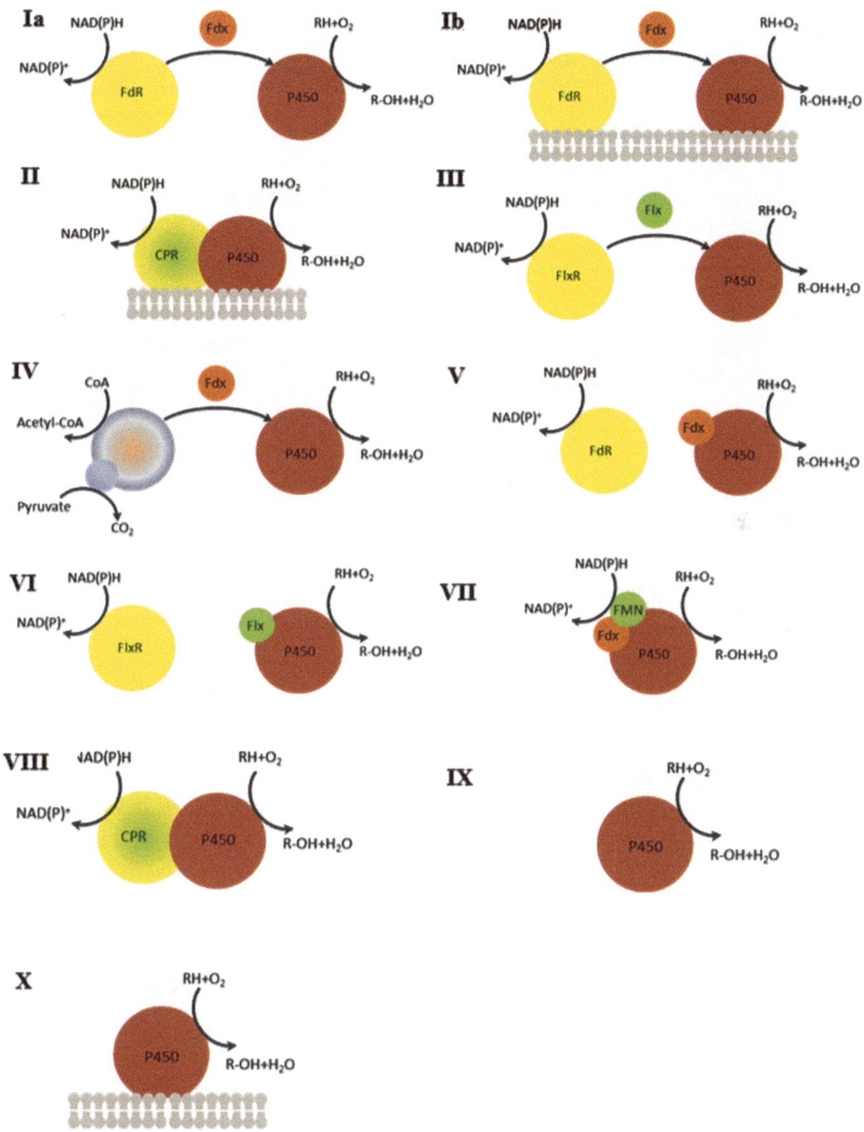


**Fig. 1** Graphical representation of the domain architectures of the 10 P450 classes. Domains highlighted on figure.

component system, all of which are expressed from separate genes but co-function for P450 activity. They comprise a FAD-dependent ferredoxin reductase (FdxR), a ferredoxin (Fdx), containing an iron-sulfur cluster (either, 1Fe—0S, 2Fe—2S, 3Fe—4S, 4Fe—3S or 4Fe—4S) and the P450 heme. They all use NAD(P)H as an electron donor. Class I P450s can be split into two subclasses Ia and Ib, Ia being cytosolic (prokaryotic) and Ib being bound to the mitochondrial membrane (eukaryotic) (Hannemann et al., 2007). In prokaryotes, all three components are non-membrane-bound, but only the ferredoxin component in eukaryotic systems is non-membrane-bound, shuttling electrons between its mitochondrial membrane-bound reductase and P450 domains.

The prokaryotic Class Ia P450 systems are primarily involved in the metabolism of exogenous compounds for use as carbon sources. For example, their involvement in the metabolism of fatty acids and xenobiotics (Volokhan, Sletta, Ellingsen, & Zotchev, 2006). The prokaryotic Class I systems account for the vast majority of all prokaryotic P450s. P450cam (CYP101A1) is a model example of a bacterial Class I P450, first discovered in 1972 and was the first P450 to have its structure determined by x-ray crystallography (Poulos et al., 1985). P450cam utilizes a ferredoxin [2Fe—2S], termed putidaredoxin and a FAD-dependent ferredoxin reductase which is referred to as putidaredoxin reductase. The natural role of P450cam is the 5-exo hydroxylation of camphor, allowing the bacteria to utilize camphor as its sole carbon source (Poulos, Finzel, & Howard, 1987). Although being one of the best studied Class I P450s, P450cam's CYP family only contains a few further characterized P450s; CYP101B1, with characterized activity toward a number of substrates including norisoprenoids (Hall & Bell, 2014). CYP101C1, with characterized activity toward α-/β-ionone (Ma et al., 2011). CYP101J2, CYP101J3 and CYP101J4 with characterized activity toward 1,8-Cineole (Unterweger et al., 2016).

In terms of whole CYP family studies of prokaryotic P450s, the CYP105 is the most reported. The CYP105 family contains a large array of characterized P450s, catalyzing many diverse reactions (Moody & Loveridge, 2014). Table 2 highlights some examples of these characterized CYP105 P450s along with other characterized Class I P450s.

In mammals, the mitochondrial Class Ib systems play a critical role in the biosynthesis of the cholesterol-derived steroid hormones and vitamin D. One example of this is the adrenal steroid hydroxylase P450scc (CYP11A1). The natural role of this enzyme is the conversion of cholesterol into pregnenolone by the three step cleavage of the cholesterol side-chain (Chien, Rosal, & Chung, 2017). The adrenal steroid hydroxylases rely on an electron transport chain containing an NAD(P)H dependent ferredoxin reductase, referred to as adrenodoxin reductase and a [2Fe—2S] ferredoxin, denoted as a adrenodoxin (Beilke et al., 2002).

**Table 2** Examples of characterized Class I P450s.

| CYP name | Organism | Example substrate(s) | References |
|---|---|---|---|
| CYP101A1 | *Pseudomonas putida* | Camphor | Poulos et al. (1987) |
| CYP101B1 | *Novosphingobium aromaticivorans* | Norisoprenoids | Hall and Bell (2014) |
| CYP101C1 | *Novosphingobium aromaticivorans* | α-/β-ionone | Ma et al. (2011) |
| CYP101D1 | *Novosphingobium aromaticivorans* | Camphor | Batabyal and Poulos (2013) |
| CYP101J2-4 | *Sphingobium yanoikuyae* B2 | 1,8-Cineole | Unterweger et al. (2016) |
| CYP102B1 | *Streptomyces coelicolor* | Arachidonic acid | Lamb et al. (2010) |
| CYP105A1 | *Streptomyces griseolus* | Vitamin D3 | Moody and Loveridge (2014) |
| CYP105A3 | *Streptomyces carbophilus* | Mevastatin | Watanabe, Nara and Serizawa (1995) |
| CYP105D1 | *Streptomyces griseus* | Papaverine | Shen, Zhao, Liu and Liu (2019) |
| CYP105D5 | *Streptomyces coelicolor* | Fatty acids | Chun, Shimada, Waterman and Guengerich (2006) |
| CYP105D7 | *Streptomyces avermitilis* | Daidzein | Pandey et al. (2010) |
| | | Diclofenac | Xu et al. (2015) |
| CYP105F2 | *Streptomyces peucetius* | Oleandomycin | Shrestha, Oh, Liou and Sohng (2008) |
| CYP105L1 | *Streptomyces fradiae* | Tylosin | Mclean, Leys and Munro (2015a) |
| CYP107A1 | *Saccharopolyspora erythraea* | 6-Deoxyerythronolide B | Shafiee and Hutchinson (1987) |
| CYP107D1 | *Streptomyces antibioticus* | Testosterone | Agematu et al. (2006) |
| | | Secondary bile acids | Grobe, Wszolek, Brundiek, Fekete and Bornscheuer (2020) |
| CYP107F1 | *Streptomyces griseus* | 1,3,6,8-tetrahydroxynaphthalene | Funa, Funabashi, Ohnishi and Horinouchi (2005) |
| CYP108A1 | *Pseudomonas aeruginosa* | α-terpineol | Hasemann, Ravichandran, Peterson and Deisenhofer (1994) |
| CYP108B1 | *Pseudomonas nitroreducens* TPJM | α-terpineol | Wong et al. (2018) |
| CYP108D1 | *Novosphingobium aromaticivorans* | α-terpineol | Bell et al. (2012a) |

*(Continued)*

**Table 2** Examples of characterized Class I P450s.—cont'd

| CYP name | Organism | Example substrate(s) | References |
|---|---|---|---|
| CYP109B1 | *Bacillus subtilis* | α-/β-ionone | Girhard, Klaus, Khatri, Bernhardt and Urlacher (2010) |
| CYP113A1 | *Saccharopolyspora erythraea* | Erythromycin D | Savino, Sciara, Miele, Kendrew and Vallone (2008) |
| CYP116A1 | *Rhodococcus erythropolis* | S-Ethyl dipropylcarbamothioate | Nagy et al. (1995) |
| CYP120A1 | *Synechocystis* sp. PCC 6803 | Retinoic acid | Alder, Bigler, Werck-Reichhart and AL-Babili (2009) |
| CYP121A1 | *Mycobacterium tuberculosis* | Dicyclotyrosine | Kishk et al. (2019) |
| CYP124A1 | *Mycobacterium tuberculosis* | Methyl-branched lipids | Johnston, Kells, Podust and Ortiz de Montellano (2009) |
| CYP125A1 | *Mycobacterium tuberculosis* | Cholesterol | Ouellet et al. (2010) |
| CYP147F1 | *Streptomyces peucetius* | Long chain fatty acids | Bhattarai, Liou and Oh (2013) |
| CYP147G1 | *Mycobacterium marinum* | Fatty acids | Child, Rossi and Bell (2019) |
| CYP153A1 | *Acinetobacter* sp. EB104 | Alkanes | van Beilen et al., (2006) |
| CYP154A1 | *Streptomyces coelicolor* | Dipentaenone | Cheng, Lamb, Kelly, Lei and Guengerich (2010) |
| CYP154H1 | *Thermobifida fusca* | Styrene | Schallmey, den Besten, Teune, Kembaren and Janssen (2011) |
| CYP170A1 | *Streptomyces coelicolor* | epi-isozizaene | Zhao et al. (2008) |
| CYP175A1 | *Thermus thermophilus* | β-Carotene | Momoi, Hofmann, Schmid and Urlacher (2006) |
| CYP183A1 | *Streptomyces avermitilis* | Pentalenene | Quaderer, Omura, Ikeda and Cane (2006) |
| CYP191A1 | *Mycobacterium smegmatis* | Fatty acid hydroxylation | Jo et al. (2017) |
| CYP199A1 | *Bradyrhizobium japonicum* | 2-Naphthoic acid | Furuya and Kino (2009) |
| CYP199A2 | *Rhodopseudomonas palustris* | 4-Ethylbenzoic acid | Bell et al. (2008) |
| | | 2-Naphthoic acid | Furuya and Kino (2009) |
| CYP199A4 | *Rhodopseudomonas palustris* | 4-methoxybenzoic acid | Bell et al. (2012b) |
| CYP238A1 | *Pseudomonas putida* KT2440 | Nerolidol | Bell et al. (2013) |
| CYP268A2 | *Mycobacterium marinum* | Geranyl acetate | Child, Naumann, Bruning, & Bell, 2018 |
| CYP1232A24 | *Paenarthrobacter nitroguajacolicus* | 3,4-dimethoxyphenylacetic acid | Klenk et al. (2019) |
| CYP1232F1 | *Paenarthrobacter nitroguajacolicus* | 3,4-dimethoxyphenylacetic acid | Klenk et al. (2019) |

### 3.2.2 Class II

The most ubiquitous P450s found in eukaryotes belong to the Class II system (Fig. 1, II). Unlike the mitochondrial Class Ib systems, these P450s can perform vastly dissimilar catalytic activities, ranging from the metabolism of endogenous compounds such as fatty acids in mammals, to a role in cutin and lignin barrier synthesis in plants. Generally, the Class II P450s are localized to the membrane of the endoplasmic reticulum and employ a two component system, a P450 domain and a FAD and FMN containing NAD(P)H cytochrome P450 reductase domain (CPR). The Class II reductase has evolved as a fusion of two ancestral proteins, a [2Fe—2S] ferredoxin and a ferredoxin reductase, where ferredoxin binding and utility was lost in favor of FMN-binding during evolutionary events (Porter & Kasper, 1986). Although the Class II system is found most predominantly in eukaryotes, bacterial examples do exist, such as P450sca-2 from *Streptomyces carbophilus*. This enzyme is involved in the stereoselective conversion of mevastatin to pravastatin, and although industrially important, the true biocatalytic value of P450sca-2 and other Class II P450s is limited, due to their membrane-bound nature (Ba, Li, Zhang, Duan, & Lin, 2013).

### 3.2.3 Class III

The Class III system (Fig. 1, III) shares similarities with the Class I system, but also some substantial differences. Both classes share the same primary composition of the three component system, with the electrons being transferred from the initial electron donor NAD(P)H via a FAD-containing ferredoxin reductase and a second redox protein to the P450 domain. Whereas in Class I systems the secondary mediator redox protein is an iron-sulfur protein, Class III systems utilize an FMN-containing redox protein, referred to as a flavodoxin. The redox partners of the Class III system resemble those of the eukaryotic Class II system, except unlike the fused CPR domain of Class IIs, the redox partners of the Class III system are separate proteins. An example of a Class III system is P450cin (CYP176A1) from *Citrobacter braakii* which catalyzes the hydroxylation of cineole to (S)-6β-hydroxycineole (Hawkes, Adams, Burlingame, Ortiz de Montellano, & de Voss, 2002). Particular recombinant P450s demonstrate 'crossover' between the Class I and Class III redox systems. One example of this is the ability of P450BioI (CYP107H1), a typical Class I P450 domain from *Bacillus subtilis*, to be reduced by a Class III redox system, specifically an *E. coli* flavodoxin reductase (FdR) and citrodoxin, a flavodoxin from *C. braakii* (Hannemann et al., 2007; Stok & de Voss, 2000).

### 3.2.4 Class IV

Class IV systems (Fig. 1, IV), which are much less common in Nature, are found exclusively in extremophiles. The first discovered Class IV system was CYP119A1 from *Sulfolobus solfataricus*, which at 368 amino acids, is approximately 10% smaller than most P450s (Wright, Harris, Solow, White, & Kennelly, 1996). This system is unique in that it does not require NAD(P)H for the initial electron donation, instead utilizing pyruvic acid as the reducing equivalent for the 2-oxoacid-ferredoxin oxidoreductase. This then utilizes a thermostable ferredoxin to shuttle the electrons to the P450 domain (Puchkaev & Ortiz de Montellano, 2005).

### 3.2.5 Class V

Class V systems (Fig. 1, V) employ two components, a currently unknown NAD(P)H-dependent ferredoxin reductase and a P450-ferredoxin fusion protein (Hannemann et al., 2007). To date, only a single Class V P450 system has been characterized, a sterol 14α-demethylase from *Methylococcus capsulatus* (MCCYP51FX). This system contains the P450 domain fused to a ferredoxin [3Fe—4S] domain at the C-terminus, by means of an alanine-rich linker region (Jackson et al., 2002).

### 3.2.6 Class VI

The Class VI system (Fig. 1, VI) shares some similarities with the Class V system, as they function as two separate proteins. They comprise a P450 heme domain fused to the C-terminus of their native flavodoxin partner and a NAD(P)H dependent flavoprotein reductase. As with the Class V P450s, the Class VI P450s are extremely sparse in Nature. P450 XplA from *Rhodococcus rhodochrous* strain 11Y was the first characterized example of the Class VI system. Exhibiting great potential in the field of bioremediation, XplA mediates the microbial metabolism of hexahydro-1,3,5-trinitro-1,3,5-triazine (Royal Demolition Explosive, RDX), found in military explosives. P450 XplA has previously been expressed alongside its native reductase XplB within transgenic plants and shown to remove and metabolize saturated RDX from soil leachate. This therefore presents a potential sustainable bioremediation approach for decontamination of military sites (Rylott et al., 2011b). Approximately thirty homologs of P450 XplA have been identified (Seth-Smith, Edwards, Rosser, Rathbone, & Bruce, 2008), the majority belonging to *Rhodococcus* species and one characterized XplA homolog from *Gordonia* sp. KTR9 (Indest et al., 2010).

### 3.2.7 Class VII

The bacterial Class VII P450 system (Fig. 1, VII) demonstrates a completely novel architecture, encompassing a single fused gene comprising a N-terminal P450 domain, a phthalate dioxygenase reductase (PDOR) and the P450 domain. The phthalate reductase is not commonly associated with P450s, but is found within the Class VII system. They are an iron-sulfur flavoprotein which employs FMN to transfer electrons from the electron donor NADH to the electron acceptor ferredoxin [2Fe—2S] (Correll, Batie, Ballou, & Ludwig, 1992).

Class VII P450s are of great interest to the biocatalysis and drug metabolism communities, due to their promiscuous substrate specificity, along with their ability to catalyze various reactions such as hydroxylations, epoxidations and sulfoxidations (O'reilly et al., 2013). A further common activity performed by many Class VIIs, synonymous with many other P450s, is the *O*-dealkylation of 7-ethoxycoumarin to 7-hydroxycoumarin (Fig. 2) (Roberts et al., 2003).

Since the initial discovery of the first Class VII P450, CYP116B2 (P450RhF) in 2002 (Roberts et al., 2002), several further Class VII P450s also from the CYP116B subfamily, have been characterized (Roberts et al., 2002, 2003; Tavanti, Porter, Sabatini, Turner, & Flitsch, 2018). This P450, from *Rhodococcus sp.* strain NCIMB 9784, has been subject to extensive research and demonstrates a diverse selection of reactions, such as the *O*-dealkylation of 7-ethoxycoumarin and the hydroxylation of the anti-inflammatory compound, diclofenac (Klenk et al., 2017; Ueno et al., 2014). In 2006, Liu et al. cloned, expressed and characterized CYP116B3. This P450, isolated from *Rhodococcus ruber*, demonstrated hydroxylation activity toward polycyclic aromatic hydrocarbons such as acenaphthene, ethyl benzene, flourene, indene, naphthalene, toluene and *m*-xylene (Liu, Schmid, & Urlacher, 2006). The third characterized Class VII P450, P450SMO isolated from *Rhodococcus* sp. ECU006, was completed in 2010 by Zhang et al. This enzyme demonstrated high levels of sulfoxidation activity and stereoselective sulfoxidation on several sulfides (Zhang, Li, Yang, & Xu, 2010). Interestingly, although P450SMO had been

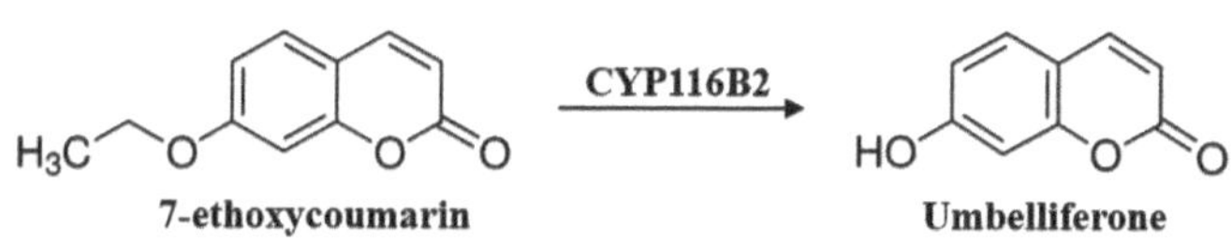


**Fig. 2** The CYP116B2 mediated *O*-dealkylation of 7-ethoxycoumarin to umbelliferone.

characterized in 2010, a CYP name was not formally assigned, P450SMO has now been annotated as CYP116B236. The first annotated member of the CYP116B subfamily, CYP116B1, was cloned, expressed and characterized in 2012 by Warman et al., and demonstrated NADPH-dependant oxidation of thiocarbamate herbicides such as *S*-ethyl dipropylthiocarbamate (EPTC) and *S*-propyl dipropylthiocarbamate (vernolate) (Warman et al., 2012).

Over the past several years, many further Class VII P450s have been described and characterized within the literature. P450LaMO was first described in 2014 by Yin et al. The enzyme was isolated from *Labrenzia aggregate* and incorrectly annotated as CYP116B4 by Yin et al. but was formally annotated as CYP116B39 by the Committee of Standardized Cytochrome P450 Nomenclature. P450LaMO catalyzes a wide range of reactions, such as the dealkylation of *p*-nitroanisole, hydroxylation of alkyl aromatic hydrocarbons, bicyclic hydrocarbons and terpenoids, and the epoxidation of olefin compounds (Yin et al., 2014). The first Class VII P450 to be characterized against its natural compounds was completed in 2015 by Minerdi et al. This enzyme, CYP116B5 isolated from *Acinetobacter radioresistens* S13, was characterized as an alkane oxidase and was shown to catalyze the oxidation of tetradecane to tetradecanol (Minerdi et al., 2015). In 2016, Li et al. developed a HT-colorimetric assay for probing the demethylation activity of Class VII enzymes. The study extensively characterized three Class VII P450s, CYP116B4 (P450RpMO), CYP116B52 (P450ArMO) and CYP116B146 ortholog (P450CtMO). These three enzymes, described and cloned previously (Yin et al., 2014) were characterized against a large, representative selection of methyl ethers. These substrate were used as probes in assays which ultimately end in the formation of a purple product following demethylation activity (Li et al., 2016). The Purpald™-P450 assays rely on the detection of the P450 reaction biproducts, such as formaldehyde and can detect the formation of as little as 1 nmol formaldehyde (Quesenberry & Lee, 1996). Purpald™-based assays have been used successfully in various P450 characterization projects, such as the characterization of the terminal hydroxylase activity of P450BM3 mutants (Meinhold, Peters, Hartwick, Hernandez, & Arnold, 2006; Peters, Meinhold, Glieder, & Arnold, 2003). All three enzymes, CYP116B4, CYP116B52 and CYP116B146 ortholog were characterized against methyl ethers of napthols, naphthalenones, monoterpenes and alkanes (Li et al., 2016).

In 2017, a panel of five thermostable Class VII P450s was identified, expressed and characterized by Tavanti et al.; CYP116B29 (P450-TB, from *Thermobispora bispora*), CYP116B46 (P450-TT, from *Tepidiphilus thermophiles*), CYP116B63 (P450-JT, from *Jhaorihella thermophile*), CYP116B64 (P450-AX, from *Albidovulum xiamenense*) and CYP116B65 (P450-AT, from *Amycolatopsis thermoflava*). These P450s, all isolated from thermostable organisms, were characterized against a panel of substrates and various activity fingerprints identified. These include the hydroxylation of diclofenac to 5-hydroxydiclofenac, demethylation of 7-methoxycoumarin to 7-hydroxycoumarin and the sulfoxidation of methyl phenyl sulfide to methyl phenyl sulfoxide (Tavanti, Porter, Sabatini, Turner, & Flitsch, 2018). In 2018, Tavanti *et* al published the first crystal structure of a Class VII P450; CYP116B46 (P450-TT) (Tavanti et al., 2018). A further Class VII P450 was identified and characterized by the same group in 2018. This P450, CYP116B62 (P450-Hal1) identified from the halophilic bacterium *Halomonas sp.* NCIMP 172, demonstrated a wide substrate scope and is the first Class VII P450 to have reported activity toward the saturated fatty acid, dodecanoic acid (Porter et al., 2018).

Furthermore, a number of Class VII P450s have been reported outside of the atypical CYP116B subfamily, with a number of natural Class VII P450s reported with P450 domains that are annotated within the CYP153A subfamily (Fiorentini et al., 2018; Nie et al., 2014).

### 3.2.8 Class VIII

Class VIII systems are similar to Class VII systems in that they comprise a single-fused 'self-sufficient' protein. Both classes also contain an N-terminal P450 domain but contain wholly different redox partners. Class VIII systems contain a P450 domain fused to a eukaryotic-like diflavin reductase partner, known as a cytochrome P450 reductase (CPR). This CPR domain contains an FMN-containing flavin followed by a FAD-dependent C-terminal reductase (Whitehouse et al., 2012). Throughout Nature, the Class VIII enzymes are found in both bacteria and fungi, but are much less ubiquitous than the Class I/II systems (Durairaj et al., 2016). However, their fused-single gene architecture and bacterial and fungal origins make them prime candidates for biotechnological applications. One key example of a Class VIII P450 system is that of P450BM3 from *Bacillus megaterium* (CYP102A1). Since its discovery in 1974 (Miura & Fulco, 1974), P450BM3 has been subject to extensive characterization and protein engineering studies in an attempt to change its natural substrate specificity and improve catalytic efficiency (Butler et al., 2013; Narhi & Fulco, 1986; Venkataraman et al., 2014). The natural role of P450BM3 is medium and long chain fatty acid hydroxylation, specifically at the ω-1, ω-2 and ω-3 positions in the presence of NAD(P)H and oxygen. It demonstrates a high conversion activity for many fatty acids, but specifically arachidonic acid, where it demonstrates the most rapid P450 catalyzed hydroxylation known (Kcat = 17,000 min-1) (Warman et al., 2005). A well-researched fungal example of a Class VIII system is that of P450foxy from *Fusarium oxysporum* (CYP505A1), which shares a number of properties with the bacterial Class VIII systems. These similarities are their high catalytic turnover, regio-specificity of the hydroxylation position and general secondary structure, although P450foxy is membrane bound, unlike its Class VIII bacterial counterparts (Nakayama et al., 1996).

Their self-sufficient nature, coupled with protein engineering studies allows researchers to turn these enzymes into versatile biocatalysts, overcoming both the need for external redox partners and limited substrate scope (Whitehouse et al., 2012). P450BM3 has been at the forefront of prokaryotic P450 research for decades and is the best characterized self-sufficient P450 (Whitehouse et al., 2012). The P450 was originally characterized as a medium-to long-chain fatty acid hydroxylase in the 1970s (Miura & Fulco, 1974). Many groups have undertaken engineering attempts by mutating the wild-type P450BM3, using rational and random mutagenic strategies, to allow the enzyme to catalyze many diverse reactions (Whitehouse et al., 2012).

Following the characterization of P450BM3 many further Class VIII P450 systems have been discovered and characterized. Examples of these characterized systems can be found in Table 3. Although the Class VIII P450s are regularly referred to as 'fatty acid hydroxylases', many other substrates are converted by wild type Class VIII P450s, such as the conversion of indole by CYP102G4 (Kim, Lee, Jung, & Kim, 2017).

### 3.2.9 Class IX

The Class IX system (Fig. 1, IX) comprises of nitric oxide reductases. Unlike the Class I-VIII P450 systems, these P450s do not rely on redox partner proteins for the delivery of electrons. An example of a Class IX system is P450nor from *F. oxysporum* (CYP55). This nitric oxide reductase represents a soluble eukaryotic P450 which obtains its electrons from NADH (Kelly & Kelly, 2013; Tomura, Obika, Fukamizu, & Shoun, 1994). P450nor and other Class IX systems play a role in nitrate respiration under anaerobic

**Table 3** Characterized Class VIII P450s and example substrates.

| CYP name | Organism | Substrates | References |
|---|---|---|---|
| CYP102A1 | *Bacillus megaterium* | • Saturated and unsaturated medium chain fatty acids | Whitehouse et al. (2012) |
| CYP102A2 | *Bacillus subtilis* | • Saturated and unsaturated medium chain fatty acids | Budde, Maurer, Schmid and Urlacher (2004); Gustafsson et al. (2004) |
| CYP102A3 | *Bacillus subtilis* | • Saturated and unsaturated medium chain fatty acids | Gustafsson et al. (2004); Lentz, Urlacher and Schmid (2004) |
| CYP102A4 | *Bacillus anthracis* | • Not currently characterized; pathogenic bacterial origin | Chowdhary, Alemseghed and Haines (2007) |
| CYP102A5 | *Bacillus cereus* | • Saturated and unsaturated medium chain fatty acids<br>• 4-Nitrophenol<br>• Coumarin<br>• Aniline<br>• Chlorzoxazone | Chowdhary et al. (2007); Pendyala, Chaganti, Thadikamala and Reddy Shetty (2013) |
| CYP102A6 | *Bradyrhizobium japonicum* | • Saturated medium chain fatty acids | Lamb et al. (2010) |
| CYP102A7 | *Bacillus licheniformis* | • 7-Ethoxycoumarin<br>• Cyclic and acyclic terpenes | Dietrich et al. (2008) |
| CYP102A16 | *Bacillus sp.* C3 | • Saturated medium chain fatty acids | Cook (2015); Li et al. (2012) |
| CYP102A18 | *Ktedonobacter racemifer* DSM44963 | • Saturated fatty acids | Munday, Maddigan, Young and Bell (2016) |
| CYP102A25 | *Bacillus marmarensis* | • Saturated and unsaturated medium chain fatty acids | Porter et al. (2017) |
| CYP102A26 | *Pontibacillus halophilus* | • Saturated and unsaturated medium chain fatty acids | Porter et al. (2017) |
| CYP102D1 | *Streptomyces avermitilis* | • Saturated and unsaturated medium chain fatty acids<br>• Daidzein<br>• Anthracene<br>• Coumarin | Choi et al. (2012) |

*(Continued)*

**Table 3** Characterized Class VIII P450s and example substrates.—cont'd

| CYP name | Organism | Substrates | References |
|---|---|---|---|
| CYP102G4 | *Streptomyces cattleya* | • Saturated and unsaturated medium chain fatty acids<br>• Indole<br>• Benzophenone<br>• Flavones | Kim, Lee, Jung, & Kim, 2017 |
| CYP102K1 | *Azorhizobium caulinodans* | • Saturated fatty acids<br>• dicarboxylic acid monoesters | Mertens (2015) |
| CYP102P1 | *Ktedonobacter racemifer* DSM44963 | • Saturated fatty acids<br>• Various benzoic acid substrates | Maddigan and Bell (2017) |

conditions, specifically catalyzing the final step in reducing nitric oxide (NO) to nitrous oxide ($N_2O$) (Takaya et al., 1999).

### 3.2.10 Class X

Similarly to Class IX system, Class X systems (Fig. 1, X) do not rely on redox partner proteins for electron delivery, as Class X systems utilize an independent intramolecular transfer system to achieve P450 catalyzed substrate conversion. While being identified as P450s, these Class X P450 systems encompass hydrogen peroxide lyases (HPL, CYP74 B/C), allene oxide synthases (AOS, CYP74A), prostacyclin synthases (PGIS), thromboxane synthases (TBAX) and divinyl ether synthases (DES, CYP74D). HPL, AOS and DES are crucial enzymes in the plant lipoxygenase pathway. All these enzymes belong to the CYP74 family and are all localized in the membranes of chloroplasts (Froehlich, Itoh, & Howe, 2001). It has been shown that CYP74 family P450s exploit the acyl hydroperoxide groups of their substrates as oxygen donors, to form new carbon-oxygen bonds (Hannemann et al., 2007).

### 3.2.11 Other P450 fusions

There are examples of other P450-redox partner fusions reported within the literature which have yet to be assigned into a P450 class. The majority of these can easily be identified through the various readily available bioinformatics tools, such as BLAST and CDART (Conserved Domain Architecture Retrieval Tool (Geer, Domrachev, Lipman, & Bryant, 2002)). In 2014, McLean et al. used these tools in the search for new P450 fusions and to identify those with potential fused electron transport partners. This study highlighted a number of new P450-fusions and one prime example is the BDOR-P450 fusion of potential self-sufficient nature. This fusion contains a benzoate dioxygenase reductase fused at the N− terminus of a P450 and has been assigned to the CYP1049 family (Mclean, Luciakova, Belcher, Tee, & Munro, 2015b).

Furthermore, there are many examples of natural P450 fusion proteins with genes that provide functions other than the shuttling of electrons. The fusion PpoA, a heme-peroxidase fused to a P450, is a well-studied example that demonstrates activity toward C16 and C18 mono- and poly-unsaturated fatty acids. An example of a two phase reaction catalyzed by the PpoA protein, is the conversion from linoleic acid to 5,8-dihydroxyoctadecadienoic acid (Brodhun, Gobel, Hornung, & Feussner, 2009).

Due to the benefits of self-sufficient P450s, numerous attempts have been made at producing non-natural fusions of non-self-sufficient P450s with the known fused redox partners of characterized self-sufficient P450s, with some notable success. The majority of the fusion attempts employed various protein linkers, which are typically small peptides between 5 and 50 residues. For example, in 2006, Dodhia et al. successfully fused three N-terminally modified human P450s to the reductase domain of P450BM3. They demonstrated their previously characterized activities using solely the P450BM3 reductase for electron transport (Dodhia, Fantuzzi, & Gilardi, 2006). In 2010, Sabbadin et al. developed a high-throughput method of non-natural chimera creation, linking P450 heme domains to the reductase partners of P450RhF (Sabbadin et al., 2010).

## 4. Biotechnological applications of P450s

P450s have an exciting but not yet fully harnessed role in industry, having vast potential in multiple areas such as the conversion of terpenes, alkanes and aromatic compounds, all of which are in increased demand within the biocatalysis community (Bernhardt, 2006; Moody & Taylor, 2016). The diversity and ubiquity, and thus vast potential of P450s as biocatalysts has incited widespread research over the past few decades (Cook, Finnigan, Cook, Black, & Charnock, 2016), transforming currently characterized P450s into more versatile and effective biocatalysts, such as in the case of P450BM3 and P450RhF (Robin et al., 2011). However, in general, enzyme stability and limited substrate specificity scope represent limitations to the widespread application of P450s as effective biocatalysts at commercial scale. As such, enzyme engineering has been widely employed with model enzymes in an attempt to circumvent these problems, utilizing both site-saturation mutagenesis and advanced directed evolution approaches (O'reilly, Kohler, Flitsch, & Turner, 2011). The use of a P450BM3 mutant has been described in a potential semi-biosynthetic route toward the artemisinin group of antimalarial drugs (Dietrich et al., 2009), where the authors propose a P450 promoted multi-step oxidation of amorpha-4,11-diene toward the key intermediate compound dihydroartemisinic acid (Fig. 3). They subsequently demonstrated the key regio- and stereoselective epoxidation of the alkene at the 11 position which can subsequently be converted to the carboxylic acid via reductive ring opening to the alcohol and further oxidation steps.

A variant of P450BM3 has been developed to accept non-natural substrates for epoxidation and C—H hydroxylation (Münzer et al., 2005). The authors demonstrated the overall hydroxylation of cyclopentane carboxylic acid, via a protected intermediate, in a whole-cell biotransformation to give the products with either R or S selectivity in high enantiomeric excess, albeit with a low overall yield, via the use of two different mutants. This type of P450 mediated biocatalysis was proposed as a means toward the synthesis of valuable compounds such as the anti-HIV drug carbovir using similar substrates (Fig. 4).

It has been shown that the use of biocatalysts such as P450s can replace historic synthetic methods in the creation of APIs. One example of such reaction is the conversion of the *Penicillium citrinum* metabolite compactin (mevastatin) to the hyperlipidemia drug pravastatin, by a stereoselective 6$\beta$-hydroxylation using CYP105A3 (P450sca-2) (Watanabe et al., 1995).

P450s are also utilized in other industries, such as the use of P450s within the flavor and fragrance industry, which can lead to a more cost-effective

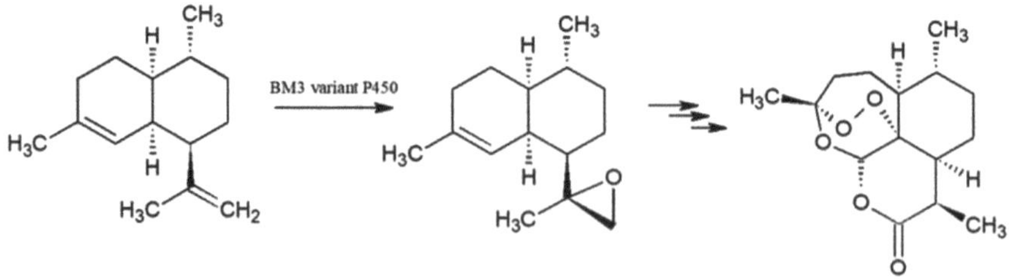


**Fig. 3** Regioselective epoxidation of amorpha-4,11-diene as a potential step toward artemisinin.

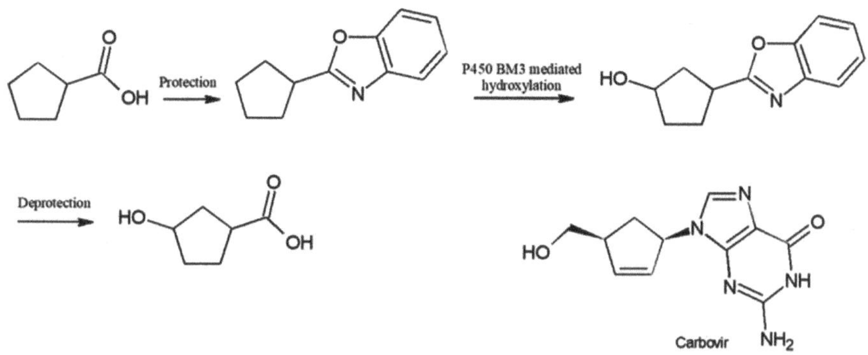


**Fig. 4** P450 mediated 4-hydroxylation of cylopentane carboxylic acid via an intermediate.

route of the production of high-value chemicals. (+)-Nootkatone is a prime example, originating from grapefruit, the chemical has many desirable properties in the flavor and fragrance sector. Valencene is a precursor of (+)-nootkatone and can be easily and cheaply sourced (Furusawa, Hashimoto, Noma, & Asakawa, 2005). In 2009, Girhard et al. screened 125 prokaroytic P450s for the regioselective oxidation of valencene to produce (+)-nootkatone, ultimately ending in the selection of CYP109B1 for further studies. Utilizing CYP109B1 in biotransformations yielded (+)-nootkatone as a product, however unwanted oxidation products made up 35% of the total products. Furthermore, the yield of (+)-nootkatone was increased to 97% following optimization using a biphasic reaction of water and 10% dodecane, with an overall productivity of 120 mg $L^{-1}$ (Girhard et al., 2009). Another example is the oxidation of abietic acid found in the resin of pine wood, the derivatives of which may be used in soaps and varnishes. In addition, such terpenoid compounds resemble the structure of many steroidal hormones and hydroxylation using P450s may offer routes to potentially biologically relevant molecules (Janocha et al., 2013).

Another industrial use of P450s is the detoxification of contaminated matter, which is an ever increasing issue in the modern day. The global industrial pollutants can be separated into three classes; polycyclic aromatic hydrocarbons (PAHs), polychlorinated biphenyls (PCBs) and polychlorinated dibenzo-p-dioxins (PCDDs) (Kumar, 2010). Through various P450 systems, from both prokaryotes and eukaryotes, these pollutants can be metabolized (Inouye, Shinkyo, Takita, Ohta, & Sakaki, 2002; Kumar, 2010). Activity of the petrochemical industry has also resulted in the production of significant hydrocarbon contamination. A number of microbial degradation routes have been identified and reviewed (Das & Chandran, 2011). Bacteria identified from contaminated soil, particularly pseudomonas, have been found to use n-alkanes as a carbon source (Das & Mukherjee, 2007), with metabolism occurring via iron containing alkane hydroxylases. Conversion of benzene and other aromatic hydrocarbons toward catechols, phenols, hydroquinones and benzoquinones are also known to occur via epoxidation catalyzed by P450 monooxygenases and dioxygenases. This pathway offers advantages over non-enzymatic reactions which typically require elevated temperatures and the use of metal based catalysts. P450 XplA (CYP177A1) offers novel bioremediative properties and has been extensively studied. XplA has been characterized as the mediator of the microbial degradation of a military explosive, hexahydro-1,3,5-trinitro-1,3,5-triazine also known as Royal demolition explosive (RDX) (Rylott et al., 2011a). RDX is the most commonly used military explosive, often found mixed with 2,4,6-trinitrotoluene (TNT). The use of these compounds on facilities such as military training grounds has led to the excessive build up and contamination of soil and groundwater. Chemical methods for degradation often involve the prolonged use of large quantities of caustic reagents such as calcium hydroxide. P450 XplA has been expressed within *Arabidopsis* and shown to metabolize RDX, most likely as a source of nitrogen (Rylott et al., 2011b).

## 5. Concluding remarks

The prokaryotic world is vastly unexplored, with unique and interesting organisms uncovered regularly and bringing with them new and exciting potential biotransformations (Whitman, Coleman, & Wiebe, 1998), of which P450s may play a role. In regards to diversity, these newly discovered organisms often reveal unexplored and novel sequences which are often exploited for biocatalytic routes as part of '(meta)genomic' studies (Kim et al., 2007; Zawodny et al., 2018). Along with the potential biotechnological applications, this sequence space is also expanding the known P450 world and this is reflected in the expanding known, annotated CYP families (Nelson, 2017).

## References

Agematu, H., Matsumoto, N., Fujii, Y., Kabumoto, H., Doi, S., Machida, K., et al. (2006). Hydroxylation of testosterone by bacterial cytochromes P450 using the *Escherichia coli* expression system. *Bioscience Biotechnology and Biochemistry, 70*, 307–311.

Alder, A., Bigler, P., Werck-Reichhart, D., & AL-Babili, S. (2009). In vitro characterization of Synechocystis CYP120A1 revealed the first nonanimal retinoic acid hydroxylase. *FEBS Journal, 276*, 5416–5431.

Axelrod, J. (1955). The enzymatic demethylation of ephedrine. *Journal of Pharmacology and Experimental Therapeutics, 114*, 430–438.

Ba, L., Li, P., Zhang, H., Duan, Y., & Lin, Z. (2013). Engineering of a hybrid biotransformation system for cytochrome P450sca-2 in *Escherichia coli*. *Biotechnology Journal, 8*, 785–793.

Batabyal, D., & Poulos, T. L. (2013). Crystal structures and functional characterization of wild-type CYP101D1 and its active site mutants. *Biochemistry, 52*, 8898–8906.

van Beilen, J. B., Funhoff, E. G., van Loon, A., Just, A., Kaysser, L., Bouza, M., et al. (2006). Cytochrome P450 alkane hydroxylases of the CYP153 family are common in alkane-degrading Eubacteria lacking integral membrane alkane hydroxylases. *Applied and Environmental Microbiology, 72*, 59–65.

Beilke, D., Weiss, R., Lohr, F., Pristovsek, P., Hannemann, F., Bernhardt, R., et al. (2002). A new electron transport mechanism in mitochondrial steroid hydroxylase systems based on structural changes upon the reduction of adrenodoxin. *Biochemistry, 41*, 7969–7978.

Bell, S. G., French, L., Rees, N. H., Cheng, S. S., Preston, G., & Wong, L. L. (2013). A phthalate family oxygenase reductase supports terpene alcohol oxidation by CYP238A1 from Pseudomonas putida KT2440. *Biotechnology and Applied Biochemistry, 60*, 9–17.

Bell, S. G., Xu, F., Forward, I., Bartlam, M., Rao, Z., & Wong, L. L. (2008). Crystal structure of CYP199A2, a para-substituted benzoic acid oxidizing cytochrome P450 from Rhodopseudomonas palustris. *Journal of Molecular Biology, 383*, 561–574.

Bell, S. G., Yang, W., Yorke, J. A., Zhou, W., Wang, H., Harmer, J., et al. (2012a). Structure and function of CYP108D1 from novosphingobium aromaticivorans DSM12444: An aromatic hydrocarbon-binding P450 enzyme. *Acta Crystallogr D Biol Crystallogr, 68*, 277–291.

Bell, S. G., Zhou, R., Yang, W., Tan, A. B., Gentleman, A. S., Wong, L. L., et al. (2012b). Investigation of the substrate range of CYP199A4: Modification of the partition between hydroxylation and desaturation activities by substrate and protein engineering. *Chemistry, 18*, 16677–16688.

Bernhardt, R. (2006). Cytochromes P450 as versatile biocatalysts. *Journal of Biotechnology, 124*, 128–145.

Bhattarai, S., Liou, K., & Oh, T. J. (2013). Hydroxylation of long chain fatty acids by CYP147F1, a new cytochrome P450 subfamily protein from Streptomyces peucetius. *Archives of Biochemistry and Biophysics, 539*, 63–69.

Brodhun, F., Gobel, C., Hornung, E., & Feussner, I. (2009). Identification of PpoA from Aspergillus nidulans as a fusion protein of a fatty acid heme dioxygenase/peroxidase and a cytochrome P450. *Journal of Biological Chemistry, 284*, 11792–11805.

Budde, M., Maurer, S. C., Schmid, R. D., & Urlacher, V. B. (2004). Cloning, expression and characterisation of CYP102A2, a self-sufficient P450 monooxygenase from Bacillus subtilis. *Applied Microbiology and Biotechnology, 66*, 180–186.

Butler, C. F., Peet, C., Mason, A. E., Voice, M. W., Leys, D., & Munro, A. W. (2013). Key mutations alter the cytochrome P450 BM3 conformational landscape and remove inherent substrate bias. *Journal of Biological Chemistry, 288*(35), 25387–25399. https://doi.org/10.1074/jbc.M113.479717.

Cheng, Q., Lamb, D. C., Kelly, S. L., Lei, L., & Guengerich, F. P. (2010). Cyclization of a cellular dipentaenone by Streptomyces coelicolor cytochrome P450 154A1 without oxidation/reduction. *Journal of the American Chemical Society, 132*, 15173–15175.

Chen, W., Lee, M. K., Jefcoate, C., Kim, S. C., Chen, F., & Yu, J. H. (2014). Fungal cytochrome p450 monooxygenases: Their distribution, structure, functions, family expansion, and evolutionary origin. *Genome Biol Evol, 6*, 1620–1634.

Chien, Y., Rosal, K., & Chung, B. C. (2017). Function of CYP11A1 in the mitochondria. *Molecular and Cellular Endocrinology, 441*, 55–61.

Child, S. A., Naumann, E. F., Bruning, J. B., & Bell, S. G. (2018). Structural and functional characterisation of the cytochrome P450 enzyme CYP268A2 from Mycobacterium marinum. *Biochemical Journal, 475*, 705–722.

Child, S. A., Rossi, V. P., & Bell, S. G. (2019). Selective ϖ-1 oxidation of fatty acids by CYP147G1 from Mycobacterium marinum. *Biochimica et Biophysica Acta (BBA) - General Subjects, 1863*, 408–417.

Choi, K. Y., Jung, E., Jung, D. H., Pandey, B. P., Yun, H., Park, H. Y., et al. (2012). Cloning, expression and characterization of CYP102D1, a self-sufficient P450 monooxygenase from Streptomyces avermitilis. *FEBS Journal, 279*, 1650–1662.

Choi, J., & Kim, S.-H. (2017). A genome tree of life for the fungi kingdom. *Proceedings of the National Academy of Sciences, 114*, 9391.

Chowdhary, P. K., Alemseghed, M., & Haines, D. C. (2007). Cloning, expression and characterization of a fast self-sufficient P450: CYP102A5 from Bacillus cereus. *Archives of Biochemistry and Biophysics, 468*, 32–43.

Chun, Y. J., Shimada, T., Waterman, M. R., & Guengerich, F. P. (2006). Understanding electron transport systems of Streptomyces cytochrome P450. *Biochemical Society Transactions, 34*, 1183–1185.

Cook, D. J. (2015). *Genome data mining towards Discovery of novel cytochromes P450 doctor of philosophy*. Northumbria University.

Cook, D. J., Finnigan, J. D., Cook, K., Black, G. W., & Charnock, S. J. (2016). Cytochromes P450: History, classes, catalytic mechanism, and industrial application. *Advances in Protein Chemistry and Structural Biology, 105*, 105–126.

Correll, C. C., Batie, C. J., Ballou, D. P., & Ludwig, M. L. (1992). Phthalate dioxygenase reductase: A modular structure for electron transfer from pyridine nucleotides to [2Fe-2S]. *Science, 258*, 1604–1610.

Danielson, P. B. (2002). The cytochrome P450 superfamily: Biochemistry, evolution and drug metabolism in humans. *Current Drug Metabolism, 3*, 561–597.

Das, N., & Chandran, P. (2011). Microbial degradation of petroleum hydrocarbon contaminants: An overview. *Biotechnology Research and Innovation, 2011*, 941810.

Das, K., & Mukherjee, A. K. (2007). Crude petroleum-oil biodegradation efficiency of Bacillus subtilis and *Pseudomonas aeruginosa* strains isolated from a petroleum-oil contaminated soil from North-East India. *Bioresource Technology, 98*, 1339–1345.

Dietrich, M., Eiben, S., Asta, C., Do, T. A., Pleiss, J., & Urlacher, V. B. (2008). Cloning, expression and characterisation of CYP102A7, a self-sufficient P450 monooxygenase from Bacillus licheniformis. *Applied Microbiology and Biotechnology, 79*, 931–940.

Dietrich, J. A., Yoshikuni, Y., Fisher, K. J., Woolard, F. X., Ockey, D., Mcphee, D. J., et al. (2009). A novel semi-biosynthetic route for artemisinin production using engineered substrate-promiscuous P450BM3. *ACS Chemical Biology, 4*, 261–267.

Dodhia, V. R., Fantuzzi, A., & Gilardi, G. (2006). Engineering human cytochrome P450 enzymes into catalytically self-sufficient chimeras using molecular Lego. *Journal of Biological Inorganic Chemistry, 11*, 903–916.

Durairaj, P., Hur, J.-S., & Yun, H. (2016). Versatile biocatalysis of fungal cytochrome P450 monooxygenases. *Microbial Cell Factories, 15*, 125.

Fiorentini, F., Hatzl, A. M., Schmidt, S., Savino, S., Glieder, A., & Mattevi, A. (2018). The extreme structural plasticity in the CYP153 subfamily of P450s directs development of designer hydroxylases. *Biochemistry, 57*, 6701–6714.

Froehlich, J. E., Itoh, A., & Howe, G. A. (2001). Tomato allene oxide synthase and fatty acid hydroperoxide lyase, two cytochrome P450s involved in oxylipin metabolism, are targeted to different membranes of chloroplast envelope. *Plant Physiology, 125*, 306–317.

Funa, N., Funabashi, M., Ohnishi, Y., & Horinouchi, S. (2005). Biosynthesis of hexahydroxyperylenequinone melanin via oxidative aryl coupling by cytochrome P-450 in Streptomyces griseus. *Journal of Bacteriology, 187*, 8149–8155.

Furusawa, M., Hashimoto, T., Noma, Y., & Asakawa, Y. (2005). Highly efficient production of nootkatone, the grapefruit aroma from valencene, by biotransformation. *Chemical and Pharmaceutical Bulletin, 53*, 1513–1514.

Furuya, T., & Kino, K. (2009). Discovery of 2-naphthoic acid monooxygenases by genome mining and their use as biocatalysts. *ChemSusChem, 2*, 645–649.

Geer, L. Y., Domrachev, M., Lipman, D. J., & Bryant, S. H. (2002). Cdart: Protein homology by domain architecture. *Genome Research, 12*, 1619–1623.

Girhard, M., Klaus, T., Khatri, Y., Bernhardt, R., & Urlacher, V. B. (2010). Characterization of the versatile monooxygenase CYP109B1 from Bacillus subtilis. *Applied Microbiology and Biotechnology, 87*, 595–607.

Girhard, M., Machida, K., Itoh, M., Schmid, R. D., Arisawa, A., & Urlacher, V. B. (2009). Regioselective biooxidation of (+)-valencene by recombinant *E. coli* expressing CYP109B1 from Bacillus subtilis in a two-liquid-phase system. *Microbial Cell Factories, 8*, 36.

Gricman, L., Vogel, C., & Pleiss, J. (2014). Conservation analysis of class-specific positions in cytochrome P450 monooxygenases: Functional and structural relevance. *Proteins, 82*, 491–504.

Grobe, S., Wszolek, A., Brundiek, H., Fekete, M., & Bornscheuer, U. T. (2020). Highly selective bile acid hydroxylation by the multifunctional bacterial P450 monooxygenase CYP107D1 (OleP). *Biotechnology Letters, 42*, 819–824.

Guengerich, F. P., Waterman, M. R., & Egli, M. (2016). Recent structural insights into cytochrome P450 function. *Trends in Pharmacological Sciences, 37*, 625–640.

Guengerich, F. P., Wu, Z. L., & Bartleson, C. J. (2005). Function of human cytochrome P450s: Characterization of the orphans. *Biochemical and Biophysical Research Communications, 338*, 465–469.

Gustafsson, M. C., Roitel, O., Marshall, K. R., Noble, M. A., Chapman, S. K., Pessegueiro, A., et al. (2004). Expression, purification, and characterization of Bacillus subtilis cytochromes P450 CYP102A2 and CYP102A3: Flavocytochrome homologues of P450 BM3 from Bacillus megaterium. *Biochemistry, 43*, 5474–5487.

Hall, E., & Bell, S. (2014). The efficient and selective biocatalytic oxidation of norisoprenoid and aromatic substrates by CYP101B1 from Novosphingobium aromaticivorans DSM12444. *RSC Advances, 5*.

Hannemann, F., Bichet, A., Ewen, K. M., & Bernhardt, R. (2007). Cytochrome P450 systems–biological variations of electron transport chains. *Biochimica et Biophysica Acta, 1770*, 330–344.

Hasemann, C. A., Ravichandran, K. G., Peterson, J. A., & Deisenhofer, J. (1994). Crystal structure and refinement of cytochrome P450terp at 2.3 A resolution. *Journal of Molecular Biology, 236*, 1169–1185.

Hawkes, D. B., Adams, G. W., Burlingame, A. L., Ortiz de Montellano, P. R., & de Voss, J. J. (2002). Cytochrome P450(cin) (CYP176A), isolation, expression, and characterization. *Journal of Biological Chemistry, 277*, 27725–27732.

Hildebrandt, A., Remmer, H., & Estabrook, R. W. (1968). Cytochrome P-450 of liver microsomes—one pigment or many. *Biochemical and Biophysical Research Communications, 30*, 607–612.

Ichinose, H. (2012). Molecular and functional diversity of fungal cytochrome P450s. *Biological and Pharmaceutical Bulletin, 35*, 833–837.

Indest, K. J., Jung, C. M., Chen, H. P., Hancock, D., Florizone, C., Eltis, L. D., et al. (2010). Functional characterization of pGKT2, a 182-kilobase plasmid containing the xplAB genes, which are involved in the degradation of hexahydro-1,3,5-trinitro-1,3,5-triazine by Gordonia sp. strain KTR9. *Applied and Environmental Microbiology, 76*, 6329—6337.
Inouye, K., Shinkyo, R., Takita, T., Ohta, M., & Sakaki, T. (2002). Metabolism of polychlorinated dibenzo-p-dioxins (PCDDs) by human cytochrome P450-dependent monooxygenase systems. *Journal of Agricultural and Food Chemistry, 50*, 5496—5502.
Jackson, C. J., Lamb, D. C., Marczylo, T. H., Warrilow, A. G., Manning, N. J., Lowe, D. J., et al. (2002). A novel sterol 14alpha-demethylase/ferredoxin fusion protein (MCCYP51FX) from Methylococcus capsulatus represents a new class of the cytochrome P450 superfamily. *Journal of Biological Chemistry, 277*, 46959—46965.
Janocha, S., Zapp, J., Hutter, M., Kleser, M., Bohlmann, J., & Bernhardt, R. (2013). Resin acid conversion with CYP105A1: An enzyme with potential for the production of pharmaceutically relevant diterpenoids. *ChemBioChem, 14*, 467—473.
Johnston, J. B., Kells, P. M., Podust, L. M., & Ortiz de Montellano, P. R. (2009). Biochemical and structural characterization of CYP124: A methyl-branched lipid omega-hydroxylase from *Mycobacterium tuberculosis*. *Proceedings of the National Academy of Sciences of the United States of America, 106*, 20687—20692.
Jo, H. Y., Park, S. H., Le, T. K., Ma, S. H., Kim, D., Ahn, T., et al. (2017). Peroxide-dependent oxidation reactions catalyzed by CYP191A1 from Mycobacterium smegmatis. *Biotechnology Letters, 39*, 1245—1252.
Keller, R. M., Wuthrich, K., & Debrunner, P. G. (1972). Proton magnetic resonance reveals high-spin iron (II) in ferrous cytochrome P450 cam from Pseudomonas putida. *Proceedings of the National Academy of Sciences of the United States of America, 69*, 2073—2075.
Kelly, S. L., & Kelly, D. E. (2013). Microbial cytochromes P450: Biodiversity and biotechnology. Where do cytochromes P450 come from, what do they do and what can they do for us? *Philosophical Transactions of the Royal Society of London B Biological Sciences, 368*, 20120476.
Kelly, D. E., Krasevec, N., Mullins, J., & Nelson, D. R. (2009). The CYPome (Cytochrome P450 complement) of Aspergillus nidulans. *Fungal Genetics and Biology, 46*(Suppl. 1), S53—S61.
Kim, B. S., Kim, S. Y., Park, J., Park, W., Hwang, K. Y., Yoon, Y. J., et al. (2007). Sequence-based screening for self-sufficient P450 monooxygenase from a metagenome library. *Journal of Applied Microbiology, 102*, 1392—1400.
Kim, J., Lee, P. G., Jung, E. O., & Kim, B. G. (2017). In vitro characterization of CYP102G4 from Streptomyces cattleya: A self-sufficient P450 naturally producing indigo. *Biochimica et Biophysica Acta, 1866*, 60—67.
Kishk, S. M., Mclean, K. J., Sood, S., Helal, M. A., Gomaa, M. S., Salama, I., et al. (2019). Synthesis and biological evaluation of novel cYY analogues targeting *Mycobacterium tuberculosis* CYP121A1. *Bioorganic and Medicinal Chemistry, 27*, 1546—1561.
Klenk, J. M., Fischer, M. P., Dubiel, P., Sharma, M., Rowlinson, B., Grogan, G., et al. (2019). Identification and characterization of cytochrome P450 1232A24 and 1232F1 from Arthrobacter sp. and their role in the metabolic pathway of papaverine. *Journal of Biochemistry, 166*, 51—66.
Klenk, J. M., Nebel, B. A., Porter, J. L., Kulig, J. K., Hussain, S. A., Richter, S. M., et al. (2017). The self-sufficient P450 RhF expressed in a whole cell system selectively catalyses the 5-hydroxylation of diclofenac. *Biotechnology Journal, 12*.
Klingenberg, M. (1958). Pigments of rat liver microsomes. *Archives of Biochemistry and Biophysics, 75*, 376—386.
Kumar, S. (2010). Engineering cytochrome P450 biocatalysts for biotechnology, medicine and bioremediation. *Expert Opinion on Drug Metabolism and Toxicology, 6*.

Lamb, D. C., Follmer, A. H., Goldstone, J. V., Nelson, D. R., Warrilow, A. G., Price, C. L., et al. (2019). On the occurrence of cytochrome P450 in viruses. *Proceedings of the National Academy of Sciences, 116*, 12343–12352.

Lamb, D. C., Ikeda, H., Nelson, D. R., Ishikawa, J., Skaug, T., Jackson, C., et al. (2003). Cytochrome p450 complement (CYPome) of the avermectin-producer Streptomyces avermitilis and comparison to that of Streptomyces coelicolor A3(2). *Biochemical and Biophysical Research Communications, 307*, 610–619.

Lamb, D. C., Lei, L., Warrilow, A. G., Lepesheva, G. I., Mullins, J. G., Waterman, M. R., et al. (2009). The first virally encoded cytochrome p450. *Journal of Virology, 83*, 8266–8269.

Lamb, D. C., Lei, L., Zhao, B., Yuan, H., Jackson, C. J., Warrilow, A. G. S., et al. (2010). Streptomyces coelicolor A3(2) CYP102 protein, a novel fatty acid hydroxylase encoded as a heme domain without an N-terminal redox partner. *Applied and Environmental Microbiology, 76*, 1975–1980.

Lamb, D. C., & Waterman, M. R. (2013). Unusual properties of the cytochrome P450 superfamily. *Philosophical Transactions of the Royal Society B: Biological Sciences, 368*, 20120434.

Lentz, O., Urlacher, V., & Schmid, R. D. (2004). Substrate specificity of native and mutated cytochrome P450 (CYP102A3) from Bacillus subtilis. *Journal of Biotechnology, 108*, 41–49.

Lewis, D. F. (2004). 57 varieties: The human cytochromes P450. *Pharmacogenomics, 5*, 305–318.

Lewis, D. F. V., & Wiseman, A. (2005). A selective review of bacterial forms of cytochrome P450 enzymes. *Enzyme and Microbial Technology, 36*, 377–384.

Li, Z., Ding, H., Yang, Y., Chen, X., Zhao, Y., & Zhou, Q. (2012). Activities of cytochrome P450 enzyme CYP102A16 from Bacillus sp. C3 (In Chinese). *Journal of the Zhejiang University - Agriculture and Life Science, 38*, 662–668.

Liu, L., Schmid, R. D., & Urlacher, V. B. (2006). Cloning, expression, and characterization of a self-sufficient cytochrome P450 monooxygenase from Rhodococcus ruber DSM 44319. *Applied Microbiology and Biotechnology, 72*, 876–882.

Li, R.-J., Xu, J.-H., Yin, Y.-C., Wirth, N., Ren, J.-M., Zeng, B.-B., et al. (2016). Rapid probing of the reactivity of P450 monooxygenases from the CYP116B subfamily using a substrate-based method. *New Journal of Chemistry, 40*, 8928–8934.

Ma, M., Bell, S. G., Yang, W., Hao, Y., Rees, N. H., Bartlam, M., et al. (2011). Structural analysis of CYP101C1 from novosphingobium aromaticivorans DSM12444. *ChemBioChem, 12*, 88–99.

Maddigan, N. K., & Bell, S. G. (2017). The self-sufficient CYP102 family enzyme, Krac9955, from Ktedonobacter racemifer DSM44963 acts as an alkyl- and alkyloxy-benzoic acid hydroxylase. *Archives of Biochemistry and Biophysics, 615*, 15–21.

Mclean, K. J., Leys, D., & Munro, A. W. (2015a). Microbial cytochromes P450. In P. R. Ortiz de Montellano (Ed.), *Cytochrome P450: Structure, mechanism, and biochemistry*. Cham: Springer International Publishing.

Mclean, K. J., Luciakova, D., Belcher, J., Tee, K. L., & Munro, A. W. (2015b). Biological diversity of cytochrome P450 redox partner systems. *Advances in Experimental Medicine and Biology, 851*, 299–317.

Meinhold, P., Peters, M. W., Hartwick, A., Hernandez, A. R., & Arnold, F. H. (2006). Engineering cytochrome P450 BM3 for terminal alkane hydroxylation. *Advanced Synthesis and Catalysis, 348*, 763–772.

Mertens, A. (2015). *Untersuchung verschiedener Ein- und Dreikomponenten-P450-Systeme für die Anwendung in der Biokatalyse*. Dissertation/PhD Thesis. RWTH Aachen.

Minerdi, D., Sadeghi, S. J., DI Nardo, G., Rua, F., Castrignano, S., Allegra, P., et al. (2015). CYP116B5: A new class VII catalytically self-sufficient cytochrome P450 from acinetobacter radioresistens that enables growth on alkanes. *Molecular Microbiology, 95*, 539–554.
Miura, Y., & Fulco, A. J. (1974). (Omega -2) hydroxylation of fatty acids by a soluble system from bacillus megaterium. *Journal of Biological Chemistry, 249*, 1880–1888.
Mizutani, M. (2012). Impacts of diversification of cytochrome P450 on plant metabolism. *Biological and Pharmaceutical Bulletin, 35*, 824–832.
Momoi, K., Hofmann, U., Schmid, R. D., & Urlacher, V. B. (2006). Reconstitution of beta-carotene hydroxylase activity of thermostable CYP175A1 monooxygenase. *Biochemical and Biophysical Research Communications, 339*, 331–336.
Moody, S. C., & Loveridge, E. J. (2014). CYP105—diverse structures, functions and roles in an intriguing family of enzymes in Streptomyces. *Journal of Applied Microbiology, 117*, 1549–1563.
Moody, T. S., & Taylor, S. J. C. (2016). *Cytochrome P450 (CYP) progress in biocatalysis for synthetic organic chemistry*.
Munday, S. D., Maddigan, N. K., Young, R. J., & Bell, S. G. (2016). Characterisation of two self-sufficient CYP102 family monooxygenases from Ktedonobacter racemifer DSM44963 which have new fatty acid alcohol product profiles. *Biochimica et Biophysica Acta, 1860*, 1149–1162.
Münzer, D. F., Meinhold, P., Peters, M. W., Feichtenhofer, S., Griengl, H., Arnold, F. H., et al. (2005). Stereoselective hydroxylation of an achiral cyclopentanecarboxylic acid derivative using engineered P450s BM-3. *Chemical Communications*, 2597–2599.
Nagy, I., Schoofs, G., Compernolle, F., Proost, P., Vanderleyden, J., & de Mot, R. (1995). Degradation of the thiocarbamate herbicide EPTC (S-ethyl dipropylcarbamothioate) and biosafening by Rhodococcus sp. strain NI86/21 involve an inducible cytochrome P-450 system and aldehyde dehydrogenase. *Journal of Bacteriology, 177*, 676–687.
Nakayama, N., Takemae, A., & Shoun, H. (1996). Cytochrome P450foxy, a catalytically self-sufficient fatty acid hydroxylase of the fungus Fusarium oxysporum. *Journal of Biochemistry, 119*(3), 435–440.
Narhi, L. O., & Fulco, A. J. (1986). Characterization of a catalytically self-sufficient 119,000-dalton cytochrome P-450 monooxygenase induced by barbiturates in Bacillus megaterium. *Journal of Biological Chemistry, 261*, 7160–7169.
Nebert, D. W., & Gonzalez, F. J. (1987). P450 genes: Structure, evolution, and regulation. *Annual Review of Biochemistry, 56*, 945–993.
Nebert, D. W., Nelson, D. R., Adesnik, M., Coon, M. J., Estabrook, R. W., Gonzalez, F. J., et al. (1989). The P450 superfamily: Updated listing of all genes and recommended nomenclature for the chromosomal loci. *DNA, 8*, 1–13.
Nebert, D. W., Nelson, D. R., Coon, M. J., Estabrook, R. W., Feyereisen, R., Fujii-Kuriyama, Y., et al. (1991). The P450 superfamily: Update on new sequences, gene mapping, and recommended nomenclature. *DNA and Cell Biology, 10*, 1–14.
Nelson, D. R. (2006). Cytochrome P450 nomenclature, 2004. *Methods in Molecular Biology, 320*, 1–10.
Nelson, D. R. (2009). The cytochrome P450 homepage. *Human Genomics, 4*, 59–65.
Nelson, D. R. (2017). Cytochrome P450 diversity in the tree of life. *Biochimica et Biophysica Acta, 1866*, 141–154.
Nelson, D. R., & Nebert, D. W. (2018). *Cytochrome P450 (CYP) gene superfamily. eLS*. John Wiley & Sons, Ltd.
Nelson, D., & Werck-Reichhart, D. (2011). A P450-centric view of plant evolution. *The Plant Journal, 66*.
Nie, Y., Chi, C. Q., Fang, H., Liang, J. L., Lu, S. L., Lai, G. L., et al. (2014). Diverse alkane hydroxylase genes in microorganisms and environments. *Scientific Reports, 4*, 4968.

O'reilly, E., Corbett, M., Hussain, S., Kelly, P. P., Richardson, D., Flitsch, S. L., et al. (2013). Substrate promiscuity of cytochrome P450 RhF. *Catalysis Science and Technology, 3*, 1490–1492.

O'reilly, E., Kohler, V., Flitsch, S. L., & Turner, N. J. (2011). Cytochromes P450 as useful biocatalysts: Addressing the limitations. *Chemical Communications, 47*, 2490–2501.

Omura, T., & Sato, R. (1964). The carbon monoxide-binding pigment of liver microsomes: I. Evidence for its hemoprotein nature. *Journal of Biological Chemistry, 239*, 2370–2378.

Ouellet, H., Guan, S., Johnston, J. B., Chow, E. D., Kells, P. M., Burlingame, A. L., et al. (2010). *Mycobacterium tuberculosis* CYP125A1, a steroid C27 monooxygenase that detoxifies intracellularly generated cholest-4-en-3-one. *Molecular Microbiology, 77*, 730–742.

Pandey, B. P., Roh, C., Choi, K. Y., Lee, N., Kim, E. J., Ko, S., et al. (2010). Regioselective hydroxylation of daidzein using P450 (CYP105D7) from Streptomyces avermitilis MA4680. *Biotechnology and Bioengineering, 105*, 697–704.

Pendyala, B., Chaganti, S. R., Thadikamala, S., & Reddy Shetty, P. (2013). Heterologous expression of CYP102A5 variant from Bacillus cereus CYPPB-1: Validation of model for predicting drug metabolism of human P450 probe substrates. *Applied Microbiology and Biotechnology, 97*, 8107–8119.

Peters, M. W., Meinhold, P., Glieder, A., & Arnold, F. H. (2003). Regio- and enantioselective alkane hydroxylation with engineered cytochromes P450 BM-3. *Journal of the American Chemical Society, 125*, 13442–13450.

Porter, T. D., & Kasper, C. B. (1986). NADPH-cytochrome P-450 oxidoreductase: Flavin mononucleotide and flavin adenine dinucleotide domains evolved from different flavoproteins. *Biochemistry, 25*, 1682–1687.

Porter, J. L., Manning, J., Sabatini, S., Tavanti, M., Turner, N. J., & Flitsch, S. L. (2017). Characterisation of CYP102A25 from B. Marmarensis and CYP102A26 from P. Halophilus: P450 homologues of BM3 with preference towards hydroxylation of medium chain fatty acids. *ChemBioChem, 19*, 513–520.

Porter, J. L., Sabatini, S., Manning, J., Tavanti, M., Galman, J. L., Turner, N. J., et al. (2018). Cloning, expression and characterisation of P450-hal1 (CYP116B62) from Halomonas sp. NCIMB 172: A self-sufficient P450 with high expression and diverse substrate scope. *Enzyme and Microbial Technology, 113*, 1–8.

Poulos, T. L., Finzel, B. C., Gunsalus, I. C., Wagner, G. C., & Kraut, J. (1985). The 2.6-A crystal structure of Pseudomonas putida cytochrome P-450. *Journal of Biological Chemistry, 260*, 16122–16130.

Poulos, T. L., Finzel, B. C., & Howard, A. J. (1987). High-resolution crystal structure of cytochrome P450cam. *Journal of Molecular Biology, 195*, 687–700.

Puchkaev, A. V., & Ortiz de Montellano, P. R. (2005). The *Sulfolobus solfataricus* electron donor partners of thermophilic CYP119: An unusual non-NAD(P)H-dependent cytochrome P450 system. *Archives of Biochemistry and Biophysics, 434*, 169–177.

Quaderer, R., Omura, S., Ikeda, H., & Cane, D. E. (2006). Pentalenolactone biosynthesis. Molecular cloning and assignment of biochemical function to PtlI, a cytochrome P450 of Streptomyces avermitilis. *Journal of the American Chemical Society, 128*, 13036–13037.

Quesenberry, M. S., & Lee, Y. C. (1996). A rapid formaldehyde assay using purpald reagent: Application under periodation conditions. *Analytical Biochemistry, 234*, 50–55.

Roberts, G. A., Celik, A., Hunter, D. J., Ost, T. W., White, J. H., Chapman, S. K., et al. (2003). A self-sufficient cytochrome p450 with a primary structural organization that includes a flavin domain and a [2Fe-2S] redox center. *Journal of Biological Chemistry, 278*, 48914–48920.

Roberts, G. A., Grogan, G., Greter, A., Flitsch, S. L., & Turner, N. J. (2002). Identification of a new class of cytochrome P450 from a Rhodococcus sp. *Journal of Bacteriology, 184*, 3898–3908.

Robin, A., Kohler, V., Jones, A., Ali, A., Kelly, P. P., O'reilly, E., et al. (2011). Chimeric self-sufficient P450cam-RhFRed biocatalysts with broad substrate scope. *Beilstein Journal of Organic Chemistry, 7*, 1494–1498.

Rylott, E. L., Budarina, M. V., Barker, A., Lorenz, A., Strand, S. E., & Bruce, N. C. (2011a). Engineering plants for the phytoremediation of RDX in the presence of the co-contaminating explosive TNT. *New Phytologist, 192*, 405–413.

Rylott, E. L., Jackson, R. G., Sabbadin, F., Seth-Smith, H. M., Edwards, J., Chong, C. S., et al. (2011b). The explosive-degrading cytochrome P450 XplA: Biochemistry, structural features and prospects for bioremediation. *Biochimica et Biophysica Acta, 1814*, 230–236.

Sabbadin, F., Hyde, R., Robin, A., Hilgarth, E. M., Delenne, M., Flitsch, S., et al. (2010). Licred: A versatile drop-in vector for rapid generation of redox-self-sufficient cytochrome P450s. *ChemBioChem, 11*, 987–994.

Savino, C., Sciara, G., Miele, A. E., Kendrew, S. G., & Vallone, B. (2008). Cloning, expression, purification, crystallization and preliminary X-ray crystallographic analysis of C-12 hydroxylase EryK from Saccharopolyspora erythraea. *Protein and Peptide Letters, 15*, 1138–1141.

Schallmey, A., den Besten, G., Teune, I. G., Kembaren, R. F., & Janssen, D. B. (2011). Characterization of cytochrome P450 monooxygenase CYP154H1 from the thermophilic soil bacterium Thermobifida fusca. *Applied Microbiology and Biotechnology, 89*, 1475–1485.

Seth-Smith, H. M. B., Edwards, J., Rosser, S. J., Rathbone, D. A., & Bruce, N. C. (2008). The explosive-degrading cytochrome P450 system is highly conserved among strains of Rhodococcus spp. *Applied and Environmental Microbiology, 74*, 4550–4552.

Sezutsu, H., Goff, G., & Feyereisen, R. (2013). Origins of P450 diversity. *Philosophical Transactions of the Royal Society of London B Biological Sciences, 368*.

Shafiee, A., & Hutchinson, C. R. (1987). Macrolide antibiotic biosynthesis: Isolation and properties of two forms of 6-deoxyerythronolide B hydroxylase from saccharopolyspora erythraea (Streptomyces erythreus). *Biochemistry, 26*, 6204–6210.

Shen, C., Zhao, W., Liu, X., & Liu, J. (2019). Enzyme-catalyzed regio-selective demethylation of papaverine by CYP105D1. *Biotechnology Letters, 41*, 171–180.

Shrestha, P., Oh, T. J., Liou, K., & Sohng, J. K. (2008). Cytochrome P450 (CYP105F2) from Streptomyces peucetius and its activity with oleandomycin. *Applied Microbiology and Biotechnology, 79*, 555–562.

Sim, S. C., & Ingelman-Sundberg, M. (2010). The human cytochrome P450 (CYP) allele nomenclature website: A peer-reviewed database of CYP variants and their associated effects. *Human Genomics, 4*, 278–281.

Stok, J. E., & de Voss, J. (2000). Expression, purification, and characterization of BioI: A carbon-carbon bond cleaving cytochrome P450 involved in biotin biosynthesis in Bacillus subtilis. *Archives of Biochemistry and Biophysics, 384*, 351–360.

Syed, K., & Yadav, J. S. (2012). P450 monooxygenases (P450ome) of the model white rot fungus Phanerochaete chrysosporium. *Critical Reviews in Microbiology, 38*, 339–363.

Takaya, N., Suzuki, S., Kuwazaki, S., Shoun, H., Maruo, F., Yamaguchi, M., et al. (1999). Cytochrome p450nor, a novel class of mitochondrial cytochrome P450 involved in nitrate respiration in the fungus Fusarium oxysporum. *Archives of Biochemistry and Biophysics, 372*, 340–346.

Tavanti, M., Porter, J. L., Levy, C. W., Gomez Castellanos, J. R., Flitsch, S. L., & Turner, N. J. (2018). The crystal structure of P450-TT heme-domain provides the first structural insights into the versatile class VII P450s. *Biochemical and Biophysical Research Communications, 501*, 846–850.

Tavanti, M., Porter, J. L., Sabatini, S., Turner, N. J., & Flitsch, S. (2018). A panel of new thermostable CYP116B self-sufficient cytochrome P450 monooxygenases catalysing C-H activation with diverse substrate scope. *ChemCatChem, 10*, 1042–1051 (n/a-n/a).

Tomura, D., Obika, E., Fukamizu, K., & Shoun, I. (1994). Nitric oxide reductase cytochrome P-450 gene, CYP 55, of the fungus Fusarium oxysporum containing a potential binding-site for FNR, the transcription factor involved in the regulation of anaerobic growth of *Escherichia coli*. *Journal of Biochemistry, 116*, 88–94.

Ueno, M., Yamashita, M., Takase, S., Hino, M., Kobayashi, M., & Fujie, A. (2014). Enhanced oxidative activities of cytochrome P450Rhf from Rhodococcus sp. expressed using the hyper-inducible expression system. *Journal of Bioscience and Bioengineering, 117*, 19–24.

Unterweger, B., Bulach, D. M., Scoble, J., Midgley, D. J., Greenfield, P., Lyras, D., et al. (2016). CYP101J2, CYP101J3, and CYP101J4, 1,8-cineole-hydroxylating cytochrome P450 monooxygenases from sphingobium yanoikuyae strain B2. *Applied and Environmental Microbiology, 82*, 6507–6517.

Urlacher, V. B., & Girhard, M. (2012). Cytochrome P450 monooxygenases: An update on perspectives for synthetic application. *Trends in Biotechnology, 30*, 26–36.

Venkataraman, H., Verkade-Vreeker, M. C. A., Capoferri, L., Geerke, D. P., Vermeulen, N. P. E., & Commandeur, J. N. M. (2014). Application of engineered cytochrome P450 mutants as biocatalysts for the synthesis of benzylic and aromatic metabolites of fenamic acid NSAIDs. *Bioorganic & Medicinal Chemistry, 22*(20), 5613–5620. https://doi.org/10.1016/j.bmc.2014.06.022.

Volokhan, O., Sletta, H., Ellingsen, T. E., & Zotchev, S. B. (2006). Characterization of the P450 monooxygenase nysl, responsible for C-10 hydroxylation during biosynthesis of the polyene macrolide antibiotic nystatin in Streptomyces noursei. *Applied and Environmental Microbiology, 72*, 2514–2519.

Warman, A. J., Roitel, O., Neeli, R., Girvan, H. M., Seward, H. E., Murray, S. A., et al. (2005). Flavocytochro me P450BM3: an update on structure and mechanism of a biotechnologically important enzyme. *Biochemical Society Transactions, 33*, 747–753. https://doi.org/10.1042/BST0330747.

Warman, A. J., Robinson, J. W., Luciakova, D., Lawrence, A. D., Marshall, K. R., Warren, M. J., et al. (2012). Characterization of Cupriavidus metallidurans CYP116B1–a thiocarbamate herbicide oxygenating P450-phthalate dioxygenase reductase fusion protein. *FEBS Journal, 279*, 1675–1693.

Watanabe, I., Nara, F., & Serizawa, N. (1995). Cloning, characterization and expression of the gene encoding cytochrome P-450sca-2 from Streptomyces carbophilus involved in production of pravastatin, a specific HMG-CoA reductase inhibitor. *Gene, 163*, 81–85.

Werck-Reichhart, D., & Feyereisen, R. (2000). Cytochromes P450: A success story. *Genome Biology, 1*. REVIEWS3003.

Whitehouse, C. J., Bell, S. G., & Wong, L. L. (2012). P450(BM3) (CYP102A1): Connecting the dots. *Chemical Society Reviews, 41*, 1218–1260.

Whitman, W. B., Coleman, D. C., & Wiebe, W. J. (1998). Prokaryotes: The unseen majority. *Proceedings of the National Academy of Sciences of the United States of America, 95*, 6578–6583.

Williams, P. A., Cosme, J., Sridhar, V., Johnson, E. F., & Mcree, D. E. (2000). Mammalian microsomal cytochrome P450 monooxygenase: Structural adaptations for membrane binding and functional diversity. *Molecules and Cells, 5*, 121–131.

Wong, N. R., Liu, X., Lloyd, H., Colthart, A. M., Ferrazzoli, A. E., Cooper, D. L., et al. (2018). A new approach to understanding structure-function relationships in cytochromes P450 by targeting terpene metabolism in the wild. *Journal of Inorganic Biochemistry, 188*, 96–101.

Wright, R. L., Harris, K., Solow, B., White, R. H., & Kennelly, P. J. (1996). Cloning of a potential cytochrome P450 from the archaeon *Sulfolobus solfataricus*. *FEBS Letters, 384*, 235–239.

Xu, L. H., Ikeda, H., Liu, L., Arakawa, T., Wakagi, T., Shoun, H., et al. (2015). Structural basis for the 4'-hydroxylation of diclofenac by a microbial cytochrome P450 monooxygenase. *Applied Microbiology and Biotechnology, 99*, 3081–3091.

Yin, Y. C., Yu, H. L., Luan, Z. J., Li, R. J., Ouyang, P. F., Liu, J., et al. (2014). Unusually broad substrate profile of self-sufficient cytochrome P450 monooxygenase CYP116B4 from Labrenzia aggregata. *ChemBioChem, 15*, 2443–2449.

Zanger, U. M., & Schwab, M. (2013). Cytochrome P450 enzymes in drug metabolism: Regulation of gene expression, enzyme activities, and impact of genetic variation. *Pharmacology and Therapeutics, 138*, 103–141.

Zawodny, W., Marshall, J. R., Finnigan, J. D., Turner, N. J., Clayden, J., & Montgomery, S. L. (2018). Chemoenzymatic synthesis of substituted azepanes by sequential biocatalytic reduction and organolithium-mediated rearrangement. *Journal of the American Chemical Society, 140*, 17872–17877.

Zhang, J. D., Li, A. T., Yang, Y., & Xu, J. H. (2010). Sequence analysis and heterologous expression of a new cytochrome P450 monooxygenase from Rhodococcus sp. for asymmetric sulfoxidation. *Applied Microbiology and Biotechnology, 85*, 615–624.

Zhao, B., Lin, X., Lei, L., Lamb, D. C., Kelly, S. L., Waterman, M. R., et al. (2008). Biosynthesis of the sesquiterpene antibiotic albaflavenone in. *Journal of Biological Chemistry, 283*, 8183–8189.

FSC
www.fsc.org
MIX
Papier | Fördert
gute Waldnutzung
FSC® C083411

Kolibri 360 GmbH
Ferdinand-Jühlke-Straße 7
99095 Erfurt, Deutschland
produktsicherheit@kolibri360.de